MECHANICS OF
NON-NEWTONIAN FLUIDS

RELATED PERGAMON JOURNALS

COMPUTERS & FLUIDS
An International Journal
EDITOR: *M. H. BLOOM*,
*Polytechnic Institute of
Brooklyn, New York, U.S.A.*

COMPUTERS & FLUIDS is designed to be multi-disciplinary. It interprets the term "fluid" in a broad sense. Hydrodynamics, high-speed gas dynamics, turbulence, multiphase flow, rheology, kinetic theory, and flows coupled to chemical reactions, radiation and electromagnetics, are all phenomena of interest here—provided that computer technique plays a significant role in the associated studies or design methodology. Applications will be found in all branches of engineering and science: mechanical, civil, chemical, aeronautical, medical, geophysical, vehicular, and oceanographic. It will involve problems of air, sea and land vehicle motion, energy conversion and power, chemical reactors and transport processes, atmospheric effects and pollution, biomedicine, noise and acoustics, and magnetohydrodynamics among others.
1978 Subscription Rate: $95.00
Two Year Rate (1978/79): $180.50
Published Quarterly. Also available in Microform
Prices include postage and insurance.

MULTIPHASE FLOW
An International Journal
EDITORS: *G. HETSRONI*,
*Technion-Israel Institute of
Technology, Haifa, Israel*
H. C. SIMPSON,
*University of Strathclyde,
Glasgow, Scotland*

MULTIPHASE FLOW publishes original, theoretical and experimental, investigations of multiphase flow (solid–fluid, fluid–fluid) which are of relevance and permanent interest. Regular features also include brief communications such as findings of current investigations, discussion of previous publications and occasional authoritative review articles.
1978 Subscription Rate: $90.00
Two Year Rate (1978/79): $171.00
Published Bi-Monthly. Also available in Microform
Prices include postage and insurance.

RHEOLOGY ABSTRACTS
A Survey of World Literature
EDITOR: *D. E. MARSHALL*,
*University of Technology,
Loughborough, England*

RHEOLOGY ABSTRACTS, published for the British Society of Rheology, aims at including a reference to any paper which describes work likely to be of interest to a practising rheologist. The abstracts are grouped under the following headings: Theoretical studies; Instruments and techniques; Metals and other solids; Polymers and other viscoelastic materials; Solutions, pastes and suspensions; Pure fluids; Miscellaneous; most of which are subdivided for greater precision.
1978 Subscription Rate: $54.00
Two Year Rate (1978/79): $102.60
Published Quarterly. Also available in Microform
Prices include postage and insurance.

Free Specimen Copies of all Pergamon Research Journals are available on request from the Publisher.
All subscription enquiries and specimen copy requests should be addressed to the Subscription Fulfillment Manager at your nearest Pergamon Press Office.

Frontispiece: Photograph courtesy of Dr. Andrew Kraynik.

MECHANICS OF
NON-NEWTONIAN FLUIDS

BY

WILLIAM R. SCHOWALTER
Professor of Chemical Engineering, Princeton University , U.S.A.

PERGAMON PRESS
OXFORD · NEW YORK · TORONTO · SYDNEY
PARIS · FRANKFURT

U.K.	Pergamon Press Ltd., Headington Hill Hall, Oxford OX3 0BW, England
U.S.A.	Pergamon Press Inc., Maxwell House, Fairview Park, Elmsford, New York 10523, U.S.A.
CANADA	Pergamon of Canada Ltd., 75 The East Mall, Toronto, Ontario, Canada
AUSTRALIA	Pergamon Press (Aust.) Pty. Ltd., 19a Boundary Street, Rushcutters Bay, N.S.W. 2011, Australia
FRANCE	Pergamon Press SARL, 24 rue des Ecoles, 75240 Paris, Cedex 05, France
FEDERAL REPUBLIC OF GERMANY	Pergamon Press GmbH, 6242 Kronberg-Taunus, Pferdstrasse 1, Federal Republic of Germany

First edition 1978

Library of Congress Cataloging in Publication Data

Schowalter, William Raymond, 1929–
Mechanics of non-Newtonian fluids.

Bibliography: p.
1. Non-Newtonian fluids. I. Title.
QC189.5.S36   1977   531'.11   76–51440
ISBN 0-08-021778-8 Hard Cover

Printed in Great Britain by William Clowes & Sons Limited London, Beccles and Colchester

CONTENTS

PREFACE   xi

1. INTRODUCTION   1
   1.1   What the Words Mean   1
   1.2   The Pervasiveness of Non-Newtonian Fluids   2
   1.3   The Perverseness of Non-Newtonian Fluids   2
   1.4   Non-Newtonian Fluid Mechanics and the Polymer Industry   3
   1.5   Some Other Applications   4
   1.6   Relevance   4
   References   5

2. LINEAR TRANSFORMATIONS, VECTOR SPACES, VECTORS   6
   2.1   Introduction   6
   2.2   An Elementary Example   7
   2.3   Vectors and Vector Spaces   10
   2.4   Linear Independence and Dimension of a Vector Space   11
   2.5   Basis   11
   2.6   Scalar Multiplication of Vectors   13
   2.7   Orthogonality and Orthonormality   14
   2.8   Contravariant and Covariant Components: The Reciprocal of a Basis   14
   2.9   The Natural Basis   15
   References   16
   Problems   16

3. TENSORS   18
   3.1   Introduction   18
   3.2   Definition of a Tensor   18
   3.3   Covariant and Contravariant Components of Tensors   19
   3.4   Transformation of Tensor Components   19
   3.5   Tensor Algebra   21
   3.6   The Metric Tensor   22

v

3.7   Raising and Lowering of Indices   24
3.8   Christoffel Symbols, Covariant Derivative   25
3.9   ε-Systems: The Generalized Kronecker Deltas   27
3.10   Tensor Expressions for Elementary Operations of Vector Analysis   27
3.11   Physical Components of Vectors and Tensors   28
3.12   Linear Transformations as Tensors   29
3.13   Identity Tensor, Inverse, Transpose   30
3.14   Transformations of Tensors   31
3.15   Orthogonal Transformations   31
3.16   Principal Invariants   33
3.17   Principal Values (Eigenvalues) and Principal Directions
       (Eigenvectors) [8]   34
3.18   Principal Coordinates: The Importance of Principal Invariants   36
References   37
Problems   37

4. THE CONSERVATION EQUATIONS OF CONTINUUM MECHANICS   38
4.1   Introduction   38
4.2   The Continuum   38
4.3   Kinematics   39
4.4   Conservation of Mass   40
4.5   Conservation of Linear Momentum   41
4.6   Physical Interpretation of the Stress Tensor   44
4.7   Symmetry of the Stress Tensor   44
4.8   Balance of Energy   45
References   45
Problems   46

5. DEFORMATION   47
5.1   Introduction   47
5.2   Deformation Gradient   47
5.3   Polar Decomposition Theorem   49
5.4   The Stretch Tensor   50
5.5   The Cauchy–Green Strain Tensor   51
5.6   The Relative Deformation Gradient   52
5.7   Time Differentiation   53
5.8   Time Derivatives of the Deformation Gradient   54
5.9   Rate of Stretching and Rate of Spin   56
5.10   Physical Interpretation   56
5.11   Higher Time Derivatives   57
References   57
Problems   57

6. PRINCIPLES GOVERNING FORMULATION OF CONSTITUTIVE
    EQUATIONS   60
6.1   Introduction   60

6.2   Coordinate Invariance   60
6.3   Determinism of the Stress   61
6.4   Principle of Material Objectivity   61
6.5   Application   62
6.6   Other Principles   63
      Appendix   63
      References   63
      Problems   64

7.   SIMPLE FLUIDS AND SIMPLE FLOWS   65
7.1   Introduction   65
7.2   The Simple Fluid   65
7.3   The Rest History   68
7.4   Some Steady Laminar Shear Flows   68
7.5   A Few Examples   71
7.6   Viscometric Flows   76
7.7   Other Approaches to Simple Flows   80
      Appendix   80
      References   82
      Problems   83

8.   MEASUREMENT OF THE VISCOMETRIC FUNCTIONS   84
8.1   Introduction   84
8.2   Plane Couette Flow   85
8.3   Laminar Flow Through Tubes   86
8.4   The Parallel-Plate Viscometer   87
8.5   Cone-and-Plate Viscometer   91
8.6   Flow in Annular Spaces   94
8.7   Jet Devices   102
8.8   Experimental Difficulties   105
8.9   Normal Stresses by the Method of Jackson and Kaye   109
      References   112
      Problems   113

9.   NONVISCOMETRIC FLOWS OF SIMPLE FLUIDS   117
9.1   Introduction   117
9.2   Motions with Constant Stretch History   117
9.3   Extensional Motions of Simple Fluids   120
9.4   Steady Extensions   122
9.5   Nearly Viscometric Flows   126
9.6   Concluding Remarks   130
      Appendix   130
      9A1 The Cayley–Hamilton Theorem   130
      9A2 Proof of Equation (9.13)   131
      References   133
      Problems   133

10. CONSTITUTIVE EQUATIONS: ELEMENTARY MODELS   134
    10.1   Introduction   134
    10.2   The Constitutive Equation for a Class of Purely Viscous Fluids 134
    10.3   Some Specific Examples for Reiner–Rivlin Fluids   137
    10.4   Forms of the Rivlin–Ericksen Fluid   141
    10.5   Linear Viscoelasticity   142
    10.6   Some Linear Viscoelastic Models   144
    10.7   A Spectrum of Relaxation Times   147
    10.8   Higher Time Derivatives   148
    10.9   Integral Models: Their Relation to Differential Models   149
    10.10 Connections Between Experiments   151
    10.11 Molecular Theories of Viscoelasticity: The Dumbbell Model 152
    10.12 The Bead–Spring Models of Rouse and Zimm   152
    10.13 Network Theories: The Rubberlike Liquid   157
    10.14 Concluding Remarks   159
    Appendix   159
       10A1   The Gaussian Spring   159
       10A2   The Brownian Force   160
       10A3   Hydrodynamic Interaction   163
       10A4   Contribution of the Bead–Spring System to the Stress   166
       10A5   Constitutive Behavior of the Zimm Model   169
    References   176
    Problems   177

11. CONSTITUTIVE EQUATIONS: GENERALIZATIONS OF
    ELEMENTARY MODELS   178
    11.1   Introduction   178
    11.2   An Integral Expansion for Simple Fluids   179
    11.3   Reduction to Infinitesimal and Finite Linear Viscoelasticity 180
    11.4   A Differential Expansion: The Relation to an Integral
           Expansion   181
    11.5   Relation to Earlier Work   182
    11.6   The Oldroyd Approach to Constitutive Equations.   183
    11.7   Time Differentiation of Time Dependent Base Vectors   184
    11.8   Relevance of Time Dependent Bases to Constitutive Equations 187
    11.9   Oldroyd's Formulation of Constitutive Equations   188
    11.10 A Choice of Time Derivatives   190
    11.11 A Comparison of Approaches   194
    11.12 The BKZ Model   195
    11.13 A Return to Molecular Models   196
    11.14 Concluding Remarks   201
    References   202
    Problems   203

12. NONVISCOMETRIC FLOWS REVISITED: APPLICATIONS   204
    12.1   Introduction   204
    12.2   Boundary-Layer Theory for Inelastic Fluids   205

12.3   Dimensionless Groups for Non-Newtonian Fluids   211
12.4   Boundary-Layer Theory for Viscoelastic Fluids   213
12.5   Flows in which Inertial Effects are Small   217
12.6   Extensional Flows   220
12.7   Flow through Porous Media   225
12.8   Unsteady Flows   227
12.9   More on Secondary Flows   231
12.10   Hydrodynamic Stability   235
12.11   Drag Reduction   244
12.12   Heat Transfer   249
12.13   Remarks on Polymer Processing   250
12.14   Other Measuring Devices: The Orthogonal Rheometer and the
        Inclined Open Channel   252
References   258
Problems   261

13. SUSPENSION RHEOLOGY   264
13.1   Introduction   264
13.2   Bulk Properties in the Absence of Brownian Motion:
        The Method of Averaging   265
13.3   An Example: The Viscosity of a Dilute Suspension of Rigid Spheres   268
13.4   Deformable Particles: The Slightly Deformed Drop   271
13.5   Other Analyses for Deformed Particles   275
13.6   A Rigid Ellipsoid in a Homogeneous Shearing Field   276
13.7   Rotary Brownian Motion   277
13.8   Elongational Flow: The Importance of Particle Interactions   283
13.9   Nondilute Systems   286
13.10   Conclusion   288
References   289

AUTHOR INDEX   291

SUBJECT INDEX   295

PREFACE

This is a book about fluid mechanics. In contrast to the usual books on that subject, the present volume is an exposition of the mechanics of non-Newtonian fluids, such as polymer melts, polymer solutions, and suspensions. More than 10 years ago, when the first notes were compiled, I was convinced that a need existed for a book which could be used to prepare beginning graduate students for reading current literature associated with the flow of rheologically complex materials. An intervening decade of teaching and research in the subject of non-Newtonian fluid mechanics has sharpened my own interest and knowledge, and it is hoped that the fruits of this experience will, in some measure, be passed on to the reader.

Although needs of beginning graduate students have been a primary motivation for this volume, its contents are intended for all research engineers who wish to learn the fundamentals governing flow of polymer melts, polymer solutions, and suspensions. The book is for people in research because one will not find an extensive treatment of the process technology of non-Newtonian fluids. That is not to say, however, that the subject has been approached as a mathematical exercise, free of real fluids and real fluid mechanical phenomena. Hence the book is also addressed to those with an engineering outlook— students and practitioners, academicians and industrial research people—who wish to learn what it is, at this writing, that we know about non-Newtonian fluids.

I have said that process technology is not covered in detail. Nevertheless, it is hoped that most readers will have an interest in the technology of rheologically complex fluids. It is my belief that a fundamental appreciation of the problems of technology requires firm grounding in the mathematical and physical language that describes mechanics of non-Newtonian fluids, i.e., in rheology. A comprehensive exposition of rheology requires a somewhat higher level of mathematical notation than is necessary for classical fluid mechanics. This fact has been accepted at the outset, and two full chapters are devoted to development of vectors and tensors. Their inclusion means that the volume should be essentially self-contained for one who is familiar with undergraduate-level fluid mechanics or transport phenomena.

Following an introduction and the mathematical developments of Chapters 2 and 3, Chapter 4 is a concise statement of the conservation equations of continuum mechanics. The next two chapters contain principles relating to description of deformation and of constitutive equations. These chapters draw heavily from the work of Truesdell and Noll.

In Chapter 7 one finally reaches the stage where a fluid model can be discussed. Noll's concept of the simple fluid is developed.

Chapter 8 is a discussion of means for measurement of the viscometric functions noted in Chapter 7.

Simple-fluid behavior in some prototype nonviscometric flows is treated in Chapter 9.

To this point the book has concentrated on continuum behavior. In Chapter 10 more specialized constitutive equations, including those which have been developed from elementary discrete models and lead to linear viscoelasticity, are discussed.

Principles enunciated in earlier chapters are used to generalize some of these linear models in Chapter 11. Several choices are available for development of constitutive equations which account for nonlinear behavior. It is possible to explain the subject with a single formalism. On the other hand, one can emphasize the multiplicity of approaches that have been used in the important research literature. The culmination of the latter course would be an encyclopedic listing of all constitutive equations which have enjoyed some measure of success. A middle ground has been followed in Chapter 11 in hopes that the reader will clearly see the fundamental concepts which permeate all of the "systems", and will also become familiar with several approaches to the subject, so that papers developed from an Oldroyd, Noll–Coleman, or Rivlin formalism will all be accessible.

Chapter 12 is a wide-ranging description of fluid mechanical phenomena exhibited by non-Newtonian materials. It is especially true in this chapter that selection from an almost limitless supply of examples and points of view represents a combination of the author's biased interests and his feeling of responsibility for some degree of balance.

The book concludes with an introduction to the subject of suspension rheology. Suspension models are particularly useful as a conceptual aid to the fluid mechanist because one can show, through application of classical fluid mechanics on a microscale, a rigorous physical basis for bulk phenomena. Many of the bulk phenomena predicted from suspension theory also occur with polymer melts and solutions. Although the fundamental physics of the latter is surely different in detail from that governing a suspension, suspensions do have useful modeling properties and are therefore included.

Non-Newtonian fluid mechanics is a quantitative subject and, as is also true of classical fluid mechanics, cannot be learned passively and without practice at theorem proving and problem solving. To this end, a few problems have been suggested at the end of most chapters. In early portions of the book these tend to be amplification or verification of derivations in the text. In later chapters, numerical problems dealing with viscometry and design have been included. It is anticipated that these problems will be supplemented as needed.

It is a challenge to write about a subject that is active. However, the day must come when an author decides to put his material into print without yet another revision to include recent work. I believe that the main ideas of this book have an importance sufficient to transcend the inevitable new theories and experiments which will appear in the literature between the time of writing and the time of publication.

It would be impossible to acknowledge all of those who have contributed to the creation of this volume. Thoughtful suggestions have consistently come from numerous Princeton graduate students. I would be remiss, however, to neglect this opportunity to record the critical reading of an early manuscript by Professor Martin Feinberg, and the great help in matters pedagogical and editorial provided by Dr. Andrew Kraynik.

Colleagues here and elsewhere have been gracious in responding to my requests for critiques of various chapters, and many of them will, I hope, recognize improvements prompted by their suggestions. Professor Roger Tanner's thorough review of all but the last chapter has resulted in a great many changes. He provided a timely voice of con-science against the author's urge to call the manuscript "finished".

Mrs. Loretta Leach has labored with patience, effectiveness, and even temperament far beyond that to be expected from a typist faced with the bewildering array of symbols appearing here.

My family and friends have suffered the indignities and neglect known all too well to families and friends of authors.

Finally, there is the quiet room at 1836 and all that it connotes. Without it the book could not have been born.

Princeton, New Jersey   W. R. SCHOWALTER
  March 1976

CHAPTER 1

INTRODUCTION

1.1. WHAT THE WORDS MEAN

Those who are familiar with the subject know that the words *Fluid Mechanics* in a book title generally mean something less than that. How much less varies from volume to volume, but, in many instances, the term is understood to mean the mechanics of *Newtonian* fluids. A Newtonian fluid is one for which a linear relation exists between stress and the spatial variation of velocity. If changes in fluid density are not important, the constant of proportionality is the viscosity, a characteristic constant of the material at a given temperature and pressure. *Non-Newtonian fluid mechanics* is the mechanics of fluids for which the stress at a given temperature and pressure is *not* a linear function of the spatial variation of velocity.

Newtonian fluid mechanics underwent a transformation during the first half of this century. Primary causes for the transformation were the concept of a boundary layer, put forward by Prandtl in 1904 [1], and the development of the aircraft industry. The latter made it necessary for engineers to achieve an understanding of exterior flows of air past objects and to develop design procedures to deal with these flows. In contrast, non-Newtonian fluid mechanics has evolved more recently. To a large extent its origins are found in tests of polymeric materials by physical chemists who wished to relate the bulk-flow behavior of polymers to molecular structure. A driving force, not unlike that supplied earlier for Newtonian fluids by the infant aircraft industry, was provided by the commercial development of polymeric materials and the resulting need for rational design procedures and correlations. At this juncture, chemical engineers became substantially involved. They were largely responsible for integrating the discipline of classical fluid mechanics with the chemists' studies of stress response of polymeric systems under strain. Over the past 20 years this has led to a mechanics of non-Newtonian fluids, a subject which has developed largely outside the main avenues of activity in the field of Newtonian fluid mechanics.

Rheology, the study of flow of materials, includes classification of various types of non-Newtonian flow behavior. The classification is necessary because of the negative sense in which a non-Newtonian fluid is defined; that is to say, one must know what a fluid *is*, not what it is *not*, before useful equations describing the motion can be written. Rheologists seek to classify flow behavior in terms sufficiently specific to permit prediction of the flow

1

behavior of real systems, but also sufficiently general to avoid useless subdivision and redundancy.

1.2. THE PERVASIVENESS OF NON-NEWTONIAN FLUIDS

Although non-Newtonian fluids were largely bypassed by those responsible for development of the applied science of fluid mechanics, fluids which exhibit rheological response vastly different from such Newtonian fluids as air or water are encountered daily. Non-Newtonian response is typically observed in concentrated suspensions and in high molecular-weight materials. One of the best opportunities to observe non-Newtonian behavior is found in the kitchen. Examples of non-Newtonian fluids include salad dressings, butter, whipped cream, and doughs. Anyone who has separated eggs (see *frontispiece*) is aware of the "strange" elastic and tensile properties of egg white. The resistance of egg white to stretching is characteristic of polymer solutions and melts and is a phenomenon important in, for example, the drawing of molten nylon filaments during the production of synthetic fiber.

Most biologically important fluids contain high molecular-weight components and are, therefore, non-Newtonian. The rheology of blood has received much study. Blood is rheologically complex on two counts: it is a suspension because erythrocytes with characteristic dimensions of several micrometers are present in excess of 40 vol%, and the suspending fluid itself exhibits non-Newtonian behavior because of the presence of high molecular-weight protein. The importance of rheological properties of other body fluids is now recognized. In particular, the rheological response of mucous in respiratory systems of both infants and adults is an important factor for proper respiratory behavior. The lubricating action of synovial fluid in joints is, likewise, strongly dependent on rheological properties.

Anyone who has played with that archetype of rheologically complex material, "silly putty", is aware that classical distinctions between solids and fluids are not always helpful. Indeed, non-Newtonian materials are often classified by the term "visco-elastic", indicating that they display both the properties of viscous fluids and elastic solids. Given this, the term non-Newtonian *fluid* mechanics is also open to interpretation, and, in Chapter 7, we shall state precisely what we mean by a fluid. The fact that certain materials, commonly thought of as solids, nevertheless exhibit flow properties usually associated with liquids, is also evident in geological phenomena: the motion of everything from sediment beds to mountain ranges is governed by rheological characteristics.

1.3. THE PERVERSENESS OF NON-NEWTONIAN FLUIDS

For the academician, non-Newtonian fluids offer a new challenge for description and understanding. Although we have seen that examples are commonplace, most of our intuition concerning the behavior of fluids is centered on Newtonian fluids. Those who have completed formal study of (Newtonian) fluid mechanics must sometimes ignore their intuition if they hope to predict non-Newtonian behavior. The subject is sufficiently new so that many "classical" flows of viscous fluids are still to be studied systematically. Those which have been examined have often revealed surprising new phenomena. We present a few examples below.

(i) Jet Stability

Stability of free and enclosed jets has engaged the attention of many great fluid mechanists, the analysis of Rayleigh being an excellent example. He showed that interfacial forces cause a free inviscid jet to be unstable when the characteristic wavelength of a disturbance on the jet surface exceeds the circumference of the jet. Subsequent work, experimental and theoretical, has led to an understanding of the growth of surface disturbances of certain wavelengths, which leads to the breakup of viscous Newtonian jets [2]. In jets with appreciable elasticity, however, breakup does not occur by clearly defined waves. An example is presented in Fig. 1.1.

(ii) Jet Expansion

Classical analysis of a free jet, roughly confirmed experimentally for viscous Newtonian jets at sufficiently high flow rates, indicates that the jet diameter will decrease upon exit from a tube [4]. Beyond effects due to gravity, the contraction is a consequence of momentum conservation during adjustment of the velocity distribution in the jet to a flat profile (Fig. 1.2a). Viscoelastic jets, however, typically swell upon exit from a tube as a result of relaxation of elastic forces (Fig. 1.2b).

(iii) Drag Reduction

For reasons which are far from understood, it is an experimental fact that small amounts of polymer dissolved in a liquid can drastically reduce the skin friction of a fluid in turbulent flow. A graphic practical example is shown in Fig. 1.3. One notes that it is possible for perversity to be turned to advantage.

1.4. NON-NEWTONIAN FLUID MECHANICS AND THE POLYMER INDUSTRY

The unusual flow properties of polymer melts and solutions, together with the desirable attributes of many polymeric solids, have resulted in development of the huge worldwide industry of polymer processing. We have already referred to the manufacture of synthetic fiber from polymer melts. In a typical installation, fiber is made by forcing a molten polymer, such as nylon, through a die containing perhaps a few hundred holes, each with a diameter of approximately 0.01 in. As individual filaments of molten polymer are drawn away from the die; they are cooled by the surrounding air and are simultaneously stretched to a smaller diameter. Following cooling and solidification the filaments are wound together to form a composite filament on a bobbin or take-up reel. Filament speeds in excess of 5000 ft/min are not uncommon (Fig. 1.4).

Large parts of automobiles and domestic appliances are often formed by injection molding. This is a highly unsteady process in which a molten polymer is forced into a mold and then allowed to solidify (Fig. 1.5). Often the whole process is repeated by a machine at intervals of only a few seconds.

Another important polymer processing operation is film blowing (Fig. 1.6) in which the deformation approaches biaxial straining. This is to be contrasted to injection molding operations in which the predominant motion experienced by the molten polymer is often laminar shearing, or with fiber spinning, where the flow is primarily uniaxial stretching. If one is to perform laboratory flow experiments that will be helpful in predicting behavior in polymer-processing operations, it is clear that the kinematic and dynamic distinctions between these three processes should be understood.

One cannot fail to note the importance of non-Newtonian fluid mechanics in the polymerization process itself. During the polymerization, which is generally carried out under batch conditions with transient transfer of heat and mass, the batch viscosity changes from that of water to perhaps 10^6 poise.

1.5. SOME OTHER APPLICATIONS

In many cases the marketability of a polymer is due to the rheologically complex behavior of the material. An example can be found in the compounding of materials for coating of surfaces. The non-Newtonian behavior of paints is an important factor in determining the "brushability" of a paint. Interesting tests have been devised to measure this quality in the laboratory. One wishes to have a paint that will not show brush marks after drying; on the other hand, if the paint is too thin it will not adequately cover a surface. Similarly, the art of paper coating is highly dependent on the rheology of the coating material. In conventional coating applications the coating "color", as it is called, is subjected to extreme variations of high and low shearing.

The petroleum industry uses large quantities of "drilling muds" to lubricate the drill bit and to carry rock chips out of the hole during drilling of oil wells. It is important to have muds which exhibit low viscosity under shearing but which are very thick at rest, thus preventing rapid settling of chips when the drilling unit is not in operation.

Further examples can be cited almost without limit. We note in closing this section that most foodstuffs are non-Newtonian. This is important in respect both to food processing and to the preparation of acceptable natural food substitutes.

1.6. RELEVANCE

The foregoing illustrations of non-Newtonian fluid mechanics in the polymer and related industries are not included as a prelude to the unfolding of design equations suitable for each practical engineering need. It has already been stated in the Preface that this is not a book on polymer processing. Nevertheless, it is often useful to see the possible scope for application of the fundamentals of a subject. At present, the fundamentals of non-Newtonian fluid mechanics are known to a group of academic and industrial engineers and scientists which, given the pervasiveness of industrially important operations involving non-Newtonian flow, is numerically small. It is generally recognized by educators and practitioners alike that a command of the fundamentals of classical fluid mechanics is essential for those who must deal, albeit in an approximate way, with such complex problems as ocean dynamics, aircraft design, and flow through porous media. This book is motivated by the corresponding belief that those who are engaged with non-Newtonian fluids should be conversant with the fundamentals of that subject.

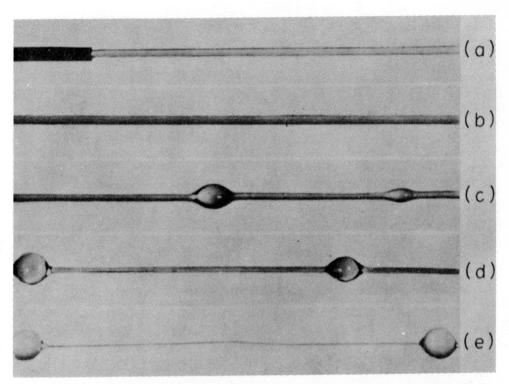

Fig. 1.1. Breakup of a jet containing 0.25% Separan (polyacrylamide) dissolved in water. Magnification, 4.2 × ; nozzle diameter, 0.0414 cm ; jet velocity, 792 cm/s. Distance from nozzle tip to mid-point of photograph: (a) 0.6 cm, (b) 60.5 cm, (c) 80.9 cm, (d) 115.7 cm, (e) 145 cm [3].

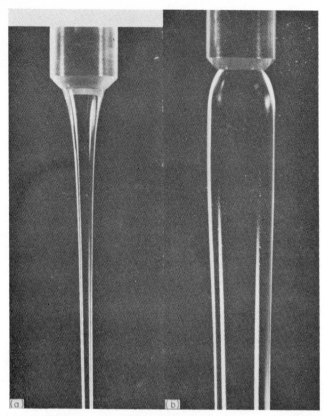

Fig. 1.2. (a) Contraction of a glycerine jet. (b) Expansion of a jet formed from a dilute solution of polyvinyl alcohol and sodium borate in water. (Photograph, courtesy of Dr. Andrew Kraynik.)

Fig. 1.3. Enhancement of fire-hose range by addition of small amounts of polyethylene oxide to water. (Photograph, courtesy of Union Carbide Corporation.)

Fig. 1.4. Spinning of a synthetic fiber. (Photograph, courtesy of Franz Fourné K.G., Impekoven bei Bonn, West Germany.)

Fig. 1.5. Injection molding of an agitator for a clothes washer. (Photograph, courtesy of Whirlpool Corporation.)

FIG. 1.6. Manufacture of polymer film by film blowing. (Photograph, courtesy of Egan Machinery Company, Somerville, New Jersey.)

REFERENCES

1. Schlichting, H., *Boundary-layer Theory*, 6th edn., p. 117 *et seq.* (New York: McGraw-Hill, 1968).
2. Meister, B. J. and Scheele, G. F., *AIChE Jl* **13**, 682 (1967).
3. Goldin, M., Yerushalmi, J., Pfeffer, R. and Shinnar, R., *J. Fluid Mech.* **38**, 689 (1969).
4. Gavis, J. and Modan, M., *Phys. Fluids* **10**, 487 (1967).

CHAPTER 2

LINEAR TRANSFORMATIONS, VECTOR SPACES, VECTORS

2.1. INTRODUCTION

One might argue that there has been sufficient proliferation of guides to vectors and tensors, either in book form [1, 2] or as adjuncts to books of interest to rheologists [3–6], to make another attempt unnecessary. Least of all, the argument might continue, should a monograph which is claimed to be an exposition of the physical behavior of fluids begin with two chapters devoted largely to an explanation of mathematical symbols.

A decision to the contrary has been prompted by several factors. It is believed that some mathematical tools unlikely to be in the firm grasp of most engineers and chemists can nevertheless be very useful in explaining problems in non-Newtonian fluid mechanics. Also, one obtains the impression from both engineering students and practitioners that an understanding of the tensor concept and its application to problems of rheological interest are difficult to obtain from any single source. It would be presumptuous to hope that appearance of one more volume will remedy the matter. Nevertheless, there does seem to be ample justification for another presentation of the subject. It is hoped that the mathematical background presented in the first two chapters will help to provide a book which is reasonably self-contained for most readers.

This preamble has been written with the hope that it will help to motivate the reader to spend time on the early chapters even though primary interests lie in the physical aspects of rheology. Those who do not find the mathematical notation or operations of this book unfamiliar, are encouraged to proceed to Chapter 4 or to some of the references cited at the end of Chapter 3, which provide more completeness and rigor than is offered here.

The purpose of this chapter is to introduce the concept of a *vector* in a manner somewhat more precise than may have been previously experienced by the reader. The reason for this is to lay a serviceable foundation for the *tensor* concept of Chapter 3. Of particular importance in subsequent paragraphs are the notions of and distinctions between vectors, base vectors, vector components, covariant components, and contravariant components. However, before launching into a host of definitions, we establish a point of departure with a familiar example.

2.2. AN ELEMENTARY EXAMPLE

A simple example is first offered which demonstrates the utility as well as the physical meaning of some of the notation which will be used in this book.

Consider a vector **v** defined in the elementary sense as an entity exhibiting both direction and magnitude. For simplicity we suppose that the vector is embedded in a two-dimensional space and that points in the space are referred to a rectangular Cartesian coordinate system Y with axes y_1 and y_2 as shown in Fig. 2.1a. We note that the vector **v** has components v_{y_1} and v_{y_2} with respect to the Y-coordinate system. Next suppose that we wish to describe components of the same vector **v** with respect to a second rectangular Cartesian system X which is formed by a counterclockwise rotation of Y through an angle θ, as shown in Fig. 2.1b. In particular, we are interested in the relation between the components v_{y_1}, v_{y_2}, and v_{x_1}, v_{x_2}. It is apparent that

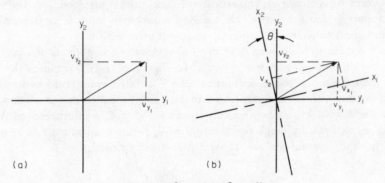

(a)   (b)

Fig. 2.1. Transformation of coordinates.

$$v_{x_1} = v_{y_1} \cos \theta + v_{y_2} \sin \theta \tag{2.1}$$

$$v_{x_2} = v_{y_1} (-\sin \theta) + v_{y_2} \cos \theta \tag{2.2}$$

and

$$|\mathbf{v}|^2 = v_{y_1}^2 + v_{y_2}^2 = v_{x_1}^2 + v_{x_2}^2 \tag{2.3}$$

The system of (2.1) and (2.2) can be expressed in a concise way by associating the trigonometric functions with the components a_{ij} of a matrix which we shall call **A**. For our present purposes we define the matrix **A** as a representation of the ordered array.

$$\begin{bmatrix} a_{11} & a_{12} \\ a_{21} & a_{22} \end{bmatrix} = \begin{bmatrix} \cos \theta & \sin \theta \\ -\sin \theta & \cos \theta \end{bmatrix}$$

Then if we consider $\mathbf{v}_y$ to represent the column matrix

$$\begin{bmatrix} v_{y_1} \\ v_{y_2} \end{bmatrix}$$

the system of (2.1) and (2.2) can be compactly represented symbolically by

$$\mathbf{v}_x = \mathbf{A}\mathbf{v}_y \tag{2.4}$$

The juxtaposition of symbols $\mathbf{A}\mathbf{v}_y$ is a matrix with elements

$$v_{x_i} = \sum_{\alpha=1}^{\alpha=2} a_{i\alpha}v_{y\alpha} \tag{2.5}$$

which will be recognized as the rule for multiplication of matrix elements. Equation (2.5) can be written more simply by introducing a convenience known as the *summation convention*. The summation convention is an agreement that whenever an index belonging to factors in a product appears twice in the product, that index is to be summed over an unstated but obvious number of values. Thus, using the summation convention, we can write

$$v_{x_i} = \sum_{\alpha=1}^{\alpha=2} a_{i\alpha}v_{y\alpha} = a_{i\alpha}v_{y\alpha} = a_{ij}v_{yj} \tag{2.6}$$

From the context it is clear that the repeated index is to be summed over the values 1, 2. Also, it is apparent that the symbol chosen for the repeated index is immaterial. We shall use the summation convention frequently in subsequent sections.

The above equations have been developed to relate the components of a given constant vector in two different coordinate systems. In this case (2.4) represents the change in vector components when the coordinate frame is rotated counterclockwise through an angle θ. Note that the same formalism can be used to interpret the change in vector components with respect to a *fixed* coordinate system if (2.4) is interpreted as $\mathbf{A}$ operating on a vector $\mathbf{v}_y$ to form, by clockwise rotation of $\mathbf{v}_y$ through an angle θ, a new vector $\mathbf{v}_x$ (Fig. 2.2). Both interpretations are important and will be used.

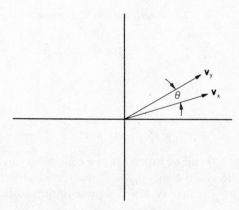

Fig. 2.2. Rotation of a vector.

Note that the determinant of $\mathbf{A}$, (det $\mathbf{A}$), has a value of $+1$. We shall see later that this is a characteristic of operators the sole effect of which is a rotation of coordinates or, alternatively, a rotation of a vector. Note also that the above example is easily extended to a vector with three components.

If we interpret $\mathbf{A}$ as an operator which, when operating on $\mathbf{v}_y$ transforms $\mathbf{v}_y$ into $\mathbf{v}_x$, we should expect that the operator $\mathbf{B}$, which transforms $\mathbf{v}_x$ into $\mathbf{v}_y$ and is defined by

$$\mathbf{v}_y = \mathbf{B}\mathbf{v}_x \tag{2.7}$$

would bear a definite relation to $\mathbf{A}$. One can in fact show simply from the geometry of Fig. 2.1 or its three-dimensional equivalent that the elements b_{ij} of $\mathbf{B}$, are given by

$$b_{ij} = \frac{A_{ij}}{\det \mathbf{A}} \qquad (2.8)$$

where A_{ij} is the cofactor of the element a_{ji} of $\mathbf{A}$. Recall that the cofactor of a_{ji} is the product of $(-1)^{i+j}$ and the determinant formed from the matrix with elements a_{kl} when the jth row and ith column have been removed. We can define the inverse of $\mathbf{A}$, $\mathbf{A}^{-1}$, as the matrix with elements

$$(a^{-1})_{ij} = b_{ij} = \frac{A_{ij}}{\det \mathbf{A}} \qquad (2.9)$$

Then

$$\mathbf{v}_y = \mathbf{A}^{-1}\mathbf{v}_x \qquad (2.10)$$

The inverse of a matrix $\mathbf{A}$ is defined only for cases where $\det \mathbf{A} \neq 0$.

If we define the product of two matrices $\mathbf{AB} = \mathbf{C}$ as a matrix with elements $c_{ij} = a_{ik}\, b_{kj}$, one can easily show that

$$\mathbf{AA}^{-1} = \mathbf{A}^{-1}\mathbf{A} = \mathbf{I} \qquad (2.11)$$

where $\mathbf{I}$ is the unit matrix with elements

$$\begin{bmatrix} 1 & 0 & 0 \\ 0 & 1 & 0 \\ 0 & 0 & 1 \end{bmatrix}$$

We are now in a position formally to obtain the inverse transformation (2.10) from (2.4). Premultiply (2.4) by $\mathbf{A}^{-1}$

$$\mathbf{A}^{-1}\mathbf{v}_x = \mathbf{A}^{-1}\mathbf{A}\mathbf{v}_y$$

But since $\mathbf{A}^{-1}\mathbf{A} = \mathbf{I}$, we have

$$\mathbf{v}_y = \mathbf{A}^{-1}\mathbf{v}_x \qquad (2.12)$$

It is also clear that if $\mathbf{v}_y$ is the sum of two vectors $\mathbf{u}_y + \mathbf{w}_y$ as shown in Fig. 2.3,

$$\mathbf{v}_x = \mathbf{A}\mathbf{v}_y = \mathbf{A}(\mathbf{u}_y + \mathbf{w}_y) = \mathbf{A}\mathbf{u}_y + \mathbf{A}\mathbf{w}_y \qquad (2.13)$$

If we define some scalar multiple of $\mathbf{v}_y$, say $s\mathbf{v}_y$, one can immediately see that

$$\mathbf{A}(s\mathbf{v}_y) = s\mathbf{A}\mathbf{v}_y \qquad (2.14)$$

The properties of $\mathbf{A}$ shown by (2.13) and (2.14) are those used to define a *linear transformation*. Thus transformation of $\mathbf{v}_y$ into $\mathbf{v}_x$ by operation of $\mathbf{A}$ on $\mathbf{v}_y$ is a *linear* transformation.

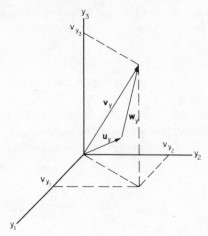

Fig. 2.3. Vector addition.

2.3. VECTORS AND VECTOR SPACES[†]

From the concrete example considered in the preceding section we now proceed to an abstract discussion of vectors and vector spaces. We say that E constitutes a vector space over the field of real numbers[‡] if E is a set of elements in arbitrary order $\mathbf{x}$, $\mathbf{y}$, . . . called vectors such that a sum of two elements is an element of E. Also, the operation of addition is defined by the following properties:

(1) $\mathbf{x} + \mathbf{y}$ is commutative; i.e., $\mathbf{x} + \mathbf{y} = \mathbf{y} + \mathbf{x}$.
(2) $\mathbf{x} + (\mathbf{y}+\mathbf{z})$ is associative; i.e., $\mathbf{x} + (\mathbf{y}+\mathbf{z}) = (\mathbf{x}+\mathbf{y}) + \mathbf{z}$.
(3) The set E contains a zero vector; i.e., $\mathbf{x} + \mathbf{0} = \mathbf{x}$.
(4) For every vector $\mathbf{x}$ in E there corresponds a negative vector $(-\mathbf{x})$; i.e.
$\mathbf{x} + (-\mathbf{x}) = \mathbf{0}$.

Furthermore, we require that multiplication of a vector $\mathbf{x}$ by a real scalar number a yield a new vector element $a\mathbf{x}$ which is contained in E. The operation of multiplication by a scalar has the following characteristics:

(5) Multiplication by unity leaves the vector unchanged; i.e., $1\mathbf{x} = \mathbf{x}$.
(6) Multiplication is associative; i.e. $a(b\mathbf{x}) = (ab)\mathbf{x}$.
(7) Multiplication is distributive over scalar addition; i.e., $(a+b)\mathbf{x} = a\mathbf{x} + b\mathbf{x}$.
(8) Multiplication is distributive over vector addition; i.e., $a(\mathbf{x}+\mathbf{y}) = a\mathbf{x} + a\mathbf{y}$.

By means of these eight properties we have defined a vector space and a vector.

If we suppose that the vectors $\mathbf{x}$, $\mathbf{y}$, . . . are ordered arrays of real numbers $\mathbf{x} = (x_1, x_2, x_3)$, $\mathbf{y} = (y_1, y_2, y_3)$, . . ., and that

$$\mathbf{x} + \mathbf{y} = (x_1 + y_1, x_2 + y_2, x_3 + y_3)$$

[†] This and subsequent sections of Chapter 2 follow closely the exposition of Lichnerowicz [2].
[‡] That is to say, we assume validity of the familiar operations of addition and multiplication with real numbers [7].

along with

$$a\mathbf{x} = (ax_1,\ ax_2,\ ax_3)$$

we than have vectors of the form familiar from elementary applications in three-dimensional space.

2.4. LINEAR INDEPENDENCE AND DIMENSION OF A VECTOR SPACE

Consider n vectors $\mathbf{x}_1, \mathbf{x}_2, \ldots, \mathbf{x}_n$ in a vector space E. If it is possible to satisfy the condition

$$a_i\mathbf{x}_i = \mathbf{0} \quad \text{(using summation convention over } n) \tag{2.15}$$

without requiring $a_i = 0$ for $i = 1, 2, \ldots, n$, then the vectors $\mathbf{x}_1, \mathbf{x}_2, \ldots, \mathbf{x}_n$ are *linearly dependent*. On the other hand, if it is *not* possible to satisfy (2.15) unless all of the a_i are zero, the set of vectors is *linearly independent*.

As a simple example consider the three unit vectors $\mathbf{i}, \mathbf{j}, \mathbf{k}$ customarily used in a rectangular Cartesian coordinate system embedded in physical space. Clearly

$$a_1\mathbf{i} + a_2\mathbf{j} + a_3\mathbf{k} \neq \mathbf{0}$$

unless $a_1 = a_2 = a_3 = 0$. However, the idea of representing a vector in terms of its components is grounded in the requirement that for a vector $\mathbf{v}$ and basis $(\mathbf{i}, \mathbf{j}, \mathbf{k})$ there is a unique set of coefficients a_i such that

$$a_1\mathbf{i} + a_2\mathbf{j} + a_3\mathbf{k} + (-1)\mathbf{v} = \mathbf{0}$$

Thus the vectors $(\mathbf{i}, \mathbf{j}, \mathbf{k}, \mathbf{v})$ are linearly dependent.

We can generalize these notions by defining the dimension n of a vector space as an integer such that one can form a linearly independent system of n vectors, but not of $(n+1)$ vectors. The vector space can then be characterized by the symbol E_n.

2.5. BASIS

From the definition of dimension we know that one can write for some arbitrary vector $\mathbf{v}$ in E_n

$$\mathbf{v} = \sum_{i=1}^{n} a_i\mathbf{e}_i \tag{2.16}$$

where $\mathbf{e}_1, \mathbf{e}_2, \ldots, \mathbf{e}_n$ is a linearly independent set of vectors. Equation (2.16) follows from the fact that a proper choice of b_i other than $b_i = 0$ for all i yields

$$\sum_{i=1}^{n} b_i\mathbf{e}_i + b_{n+1}\mathbf{v} = \mathbf{0} \tag{2.17}$$

The combination of a_i in (2.16) is unique since if we had

$$\mathbf{v} = \sum_{i=1}^{n} a_i' \mathbf{e}_i \tag{2.18}$$

then

$$\sum_{i=1}^{n} (a_i - a'_i)\mathbf{e}_i = \mathbf{0} \tag{2.19}$$

which violates the linear independence of the $\mathbf{e}_i$ if $a_i \neq a_i'$.

Any set of linearly independent vectors $(\mathbf{e}_1, \mathbf{e}_2, \ldots, \mathbf{e}_n)$ constitutes a *basis* for a vector space E_n. It follows that any vector in E_n can be expressed as a unique linear combination of the basis vectors, as in (2.16). There is, of course, an infinity of bases for any vector space. The example of the preceding section showed how $\mathbf{i}, \mathbf{j}, \mathbf{k}$, form a basis for E_3. It is important to remember that the components of a vector $\mathbf{v}$, given by a_i in (2.16) have meaning *only* when the basis $\mathbf{e}_i$ to which the a_i refer has been given.

A basis is composed of vectors, and this fact can be used to show the relations between different bases of E_n. Given two bases $(\mathbf{e}_1, \mathbf{e}_2, \ldots, \mathbf{e}_n)$ and $(\mathbf{e}_1', \mathbf{e}_2', \ldots, \mathbf{e}_n')$, we can write

$$\mathbf{e}_I' = b_I{}^j \mathbf{e}_j \quad \text{(summation convention over } n\text{)} \tag{2.20}$$

$$\mathbf{e}_i = c_i{}^J \mathbf{e}_J' \tag{2.21}$$

where upper-case indices have been used to show that they are associated with the primed basis. By using the superscript and subscript notation on the coefficients we are anticipating a convention which will be explained later. Note that the displacement of superscript from subscript indicates the order of indices. Thus, employing the matrix notation introduced in Section 2.2, the element $b_I{}^j$ is in the Ith row and the jth column of a matrix $\mathbf{B}$.

From (2.20) and (2.21) it is easy to relate components of different bases. Thus if

$$\mathbf{v} = v^i \mathbf{e}_i = v^I \mathbf{e}_I' \tag{2.22}$$

and

$$\mathbf{e}_I' = b_I{}^j \mathbf{e}_j$$

we have

$$(v^j - v^I b_I{}^j)\mathbf{e}_j = \mathbf{0}$$

which requires, since the $\mathbf{e}_j$ are linearly independent, that

$$v^j = v^I b_I{}^j \tag{2.23}$$

This equation can be rewritten in terms of the elements formed by interchanging rows and columns of all the elements $b_I{}^j$ of $\mathbf{B}$. The resulting matrix will be recognized as the transpose of the matrix $\mathbf{B}$ and will have elements $b^j{}_I$. Then (2.23) becomes

$$v^j = b^j{}_I v^I$$

or, in matrix form,

$$[\mathbf{v}] = [\mathbf{B}]^T [\mathbf{v}]'$$

where the square brackets have been used to emphasize that we are dealing with a matrix equation. The symbols $[\mathbf{v}]$ and $[\mathbf{v}]'$ represent column matrices, the elements of which are components of $\mathbf{v}$ with respect to $(\mathbf{e}_1, \mathbf{e}_2, \ldots, \mathbf{e}_n)$ and $(\mathbf{e}_1', \mathbf{e}_2', \ldots, \mathbf{e}_n')$, respectively. If we define $[\mathbf{A}] \equiv [\mathbf{B}]^T$ we have

$$[\mathbf{v}] = [\mathbf{A}] \, [\mathbf{v}]' \tag{2.24}$$

$$[\mathbf{v}]' = [\mathbf{A}]^{-1}[\mathbf{v}] \tag{2.25}$$

Equations (2.24) and (2.25) correspond to changes in components of a given vector due to a change of basis. However, as was noted in Section 2.2, we can alternatively view $\mathbf{A}$ as a linear transformation which, when operating on a vector $\mathbf{v}'$, generates a *new* vector $\mathbf{v}$. Thus

$$\mathbf{v} = \mathbf{A} \, \mathbf{v}' \tag{2.26}$$

$$\mathbf{v}' = \mathbf{A}^{-1} \mathbf{v} \tag{2.27}$$

In Section 2.2 we associated $\mathbf{A}$ with a transformation of components of one vector to components of another, each set of components being referred to the same basis. We now broaden the interpretation of (2.26) and (2.27) and view (2.26), for example, as a transformation of a vector $\mathbf{v}'$ into a new vector $\mathbf{v}$ through operation on $\mathbf{v}'$ by the linear transformation $\mathbf{A}$. Note that no reference to a basis is necessary. Equation (2.25) then provides relationships between components of $\mathbf{v}$ and $\mathbf{v}'$ when the components of both vectors are referred to the basis $\mathbf{e}$. Equation (2.24) has a similar meaning with respect to $\mathbf{e}'$.

2.6. SCALAR MULTIPLICATION OF VECTORS

We next define scalar or dot multiplication of two vectors as an operation which yields a number and which has the following properties:

(1) $\mathbf{u} \cdot \mathbf{v} = \mathbf{v} \cdot \mathbf{u}$.
(2) $(a\mathbf{u}) \cdot (\mathbf{v}+\mathbf{w}) = \mathbf{u} \cdot (a\mathbf{v}) + a(\mathbf{u} \cdot \mathbf{w})$.
(3) If $\mathbf{u} \cdot \mathbf{v} = 0$ for arbitrary $\mathbf{u}$, then $\mathbf{v} = \mathbf{0}$.
(4) $\mathbf{u} \cdot \mathbf{u} = u^2 > 0$ for $\mathbf{u} \neq \mathbf{0}$.

From these rules we have

$$\mathbf{u} \cdot \mathbf{v} = u^i v^j \mathbf{e}_i \cdot \mathbf{e}_j \tag{2.28}$$

As a matter of convenience we define

$$g_{ij} = g_{ji} = \mathbf{e}_i \cdot \mathbf{e}_j \tag{2.29}$$

Thus the scalar product can be written

$$\mathbf{u} \cdot \mathbf{v} = g_{ij} u^i v^j \tag{2.30}$$

Using the scalar product as a primary definition, the angle θ between two vectors $\mathbf{u}$ and $\mathbf{v}$ is defined by

$$\cos \theta = \frac{\mathbf{u} \cdot \mathbf{v}}{|\mathbf{u}| \, |\mathbf{v}|} \tag{2.31}$$

where $|\mathbf{u}|$ is the magnitude of $\mathbf{u}$, given by

$$|\mathbf{u}| = (\mathbf{u} \cdot \mathbf{u})^{1/2}$$

2.7. ORTHOGONALITY AND ORTHONORMALITY

Two vectors $\mathbf{u}$ and $\mathbf{v}$ are orthogonal if $\mathbf{u} \cdot \mathbf{v} = 0$. If the individual vectors are nonzero, (2.31) shows that orthogonality implies mutual perpendicularity. We also define $\mathbf{u}$ and $\mathbf{v}$ as *orthonormal* if

$$\mathbf{u} \cdot \mathbf{v} = 0 \quad \text{for} \quad \mathbf{u} \neq \mathbf{v} \quad \text{and} \quad \mathbf{u} \cdot \mathbf{u} = \mathbf{v} \cdot \mathbf{v} = 1$$

Generalizing, we can define an orthonormal basis $(\mathbf{e}_1, \mathbf{e}_2, \ldots, \mathbf{e}_n)$ in E_n as one for which

$$\mathbf{e}_i \cdot \mathbf{e}_j = \delta_{ij}$$

where δ_{ij} is the Kronecker delta and has the value

$$\delta_{ij} = 1 \quad \text{for} \quad i = j$$
$$\delta_{ij} = 0 \quad \text{for} \quad i \neq j$$

The rectangular Cartesian basis $(\mathbf{i}, \mathbf{j}, \mathbf{k})$ of Section 2.2 is an example of an orthonormal basis.

2.8. CONTRAVARIANT AND COVARIANT COMPONENTS: THE RECIPROCAL OF A BASIS

Given an arbitrary basis in $E_n(\mathbf{e}_1, \mathbf{e}_2, \ldots, \mathbf{e}_n)$, the contravariant components of a vector $\mathbf{v}$ in E_n are those numbers a^i for which

$$\mathbf{v} = a^i \mathbf{e}_i \tag{2.32}$$

Let us consider the set of scalar quantities a_i defined by

$$a_i = \mathbf{v} \cdot \mathbf{e}_i = a^k g_{ki} \tag{2.33}$$

We now look for the conditions necessary to ensure that a basis $(\mathbf{e}^1, \mathbf{e}^2, \ldots, \mathbf{e}^n)$, which we denote by $\{\mathbf{e}^i\}$, exists such that

$$\mathbf{v} = a_i \mathbf{e}^i \tag{2.34}$$

where the a_i are designated as the covariant components of $\mathbf{v}$.

Combining (2.32), (2.33), and (2.34) we obtain

$$a^k g_{kj} = a^k g_{ki}(\mathbf{e}^i \cdot \mathbf{e}_j)$$

Since this must be true for an arbitrary vector $\mathbf{v}$ and arbitrary basis $(\mathbf{e}_1, \mathbf{e}_2, \ldots, \mathbf{e}^n)$, we conclude that

$$\mathbf{e}^i \cdot \mathbf{e}_j = \delta_j{}^i \tag{2.35}$$

where $\delta_j{}^i \equiv \delta_{ij}$ for present purposes.

Thus the set of vectors $\{\mathbf{e}^i\}$ is uniquely defined (to within a factor of ± 1) in terms of $(\mathbf{e}_1, \mathbf{e}_2, \ldots, \mathbf{e}_n)$, denoted by $\{\mathbf{e}_i\}$. Since the $\{\mathbf{e}_i\}$ are linearly independent, we conclude that the set $\{\mathbf{e}^i\}$ is also linearly independent and forms a basis. We call the basis $\{\mathbf{e}^i\}$ which satisfies (2.35) a reciprocal basis to $\{\mathbf{e}_i\}$.[†]

2.9. THE NATURAL BASIS

In this book we shall associate a basis, designated by $(\mathbf{e}_1, \mathbf{e}_2, \ldots, \mathbf{e}_n)$, with the *natural basis* of a coordinate system. To explain the meaning of a natural basis we again draw upon the familiar concepts of points and vectors in the three-dimensional physical space discussed in Section 2.2. Points in this space can be located relative to some origin O by specifying a basis and components with respect to the basis. For example, we can define a point M in terms of a set of three constant orthonormal base vectors and three numbers (y^1, y^2, y^3) such that the position vector drawn from the origin to M is given by

$$\overrightarrow{OM} = \mathbf{M} = y^i \mathbf{k}_i$$

The y^i are, of course, the rectangular Cartesian coordinates of vector $\mathbf{M}$.

Let us now suppose that the y^i are uniquely defined by an invertible function of a set of variables (x^1, x^2, x^3)

$$y^i = y^i(x^1, x^2, x^3)$$
$$x^i = x^i(y^1, y^2, y^3)$$

If the point M moves in such a way that all of the x^i except one, say x^j, remain at some constant value, then movement of M traces out an x^j coordinate line. A natural basis $(\mathbf{e}_1, \mathbf{e}_2, \mathbf{e}_3)$ for the coordinate lines (x^1, x^2, x^3) is defined by

$$\mathbf{e}_i = \frac{\partial \mathbf{M}}{\partial x^i}$$

where the $\mathbf{e}_i$ can be found, relative to the $\mathbf{k}_i$, once $y^i = y^i(x^j)$ and its inverse have been given. Since the transformation between y^i and x^j was assumed invertible, the determinant $\left| \dfrac{\partial y^i}{\partial x^j} \right|$ is nonzero, and from this fact one can readily show that $(\mathbf{e}_1, \mathbf{e}_2, \mathbf{e}_3)$ constitutes a linearly independent system. Note that in contrast to the basis $\mathbf{k}_i$ of the familiar rectangular Cartesian coordinate system, the base vectors $\mathbf{e}_i$ are in general functions of position; i.e., $\mathbf{e}_i = \mathbf{e}_i(x^j)$.

The geometrical meaning of a natural basis is shown in Fig. 2.4 for a space E_3. It is apparent that $\mathbf{e}_i$ is tangent to the coordinate curve x^i at the point of definition. Also, since $\mathbf{M} = \mathbf{M}(x^i)$, we have

$$d\mathbf{M} = \mathbf{e}_i dx^i$$

which indicates that the dx^i are contravariant components of $d\mathbf{M}$. Note the consistency of the superscript notation for the coordinates x^i.

We have developed the concept of a natural basis in E^3 by drawing upon ideas familiar from elementary vector analysis. If one formally extends the concept of a point to spaces

---

[†] A more complete treatment of the meaning of a reciprocal basis is available through discussion of the concept of dual vector spaces. The interested reader is referred to Halmos [7]. Our needs are met by the restricted context given above.

of higher dimension, a natural basis may be similarly defined and interpreted for E_n [2]. Thus

$$\mathbf{e}_i = \frac{\partial \mathbf{M}}{\partial x^i} \quad (i = 1, 2, 3, \ldots, n) \tag{2.36}$$

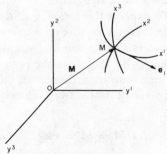

Fig. 2.4. Natural basis.

REFERENCES

1. Block, H. D. *Introduction to Tensor Analysis* (Columbus, Ohio: Charles E. Merrill, 1962).
2. Lichnerowicz, A. *Elements of Tensor Calculus* (London: Methuen, 1962).
3. Leigh, D. C. *Nonlinear Continuum Mechanics* (New York: McGraw-Hill, 1968).
4. Fredrickson, A. G. *Principles and Applications of Rheology* (Englewood Cliffs, New Jersey: Prentice-Hall 1964).
5. Eringen, A. C. *Nonlinear Theory of Continuous Media* (New York: McGraw-Hill, 1962).
6. Middleman, S., *The Flow of High Polymers* (New York: Interscience, 1968).
7. Halmos, P. R., *Finite-Dimensional Vector Spaces*, 2nd edn. (Princeton: D. Van Nostrand, 1958).

PROBLEMS

2.1. Consider the natural basis $(\mathbf{e}_1, \mathbf{e}_2, \mathbf{e}_3)$. Prove that the reciprocal basis

$$\mathbf{e}^1 = \frac{\mathbf{e}_2 \times \mathbf{e}_3}{\mathbf{e}_1 \cdot (\mathbf{e}_2 \times \mathbf{e}_3)}; \quad \mathbf{e}^2 = \frac{\mathbf{e}_3 \times \mathbf{e}_1}{\mathbf{e}_2 \cdot (\mathbf{e}_3 \times \mathbf{e}_1)}; \quad \mathbf{e}^3 = \frac{\mathbf{e}_1 \times \mathbf{e}_2}{\mathbf{e}_3 \cdot (\mathbf{e}_1 \times \mathbf{e}_2)}$$

is consistent with the definition given in (2.35). Properties of the cross product ($\times$) of elementary vector analysis may be assumed to be known. (The above expressions are the usual definitions of a reciprocal basis in three dimensions.)

2.2. Show in detail how (2.35) follows from (2.32) to (2.34), and how (2.35) uniquely (to a factor ± 1) defines a basis reciprocal to $\{\mathbf{e}_i\}$.

2.3. Consider the nonorthogonal basis $(\mathbf{e}_1, \mathbf{e}_2)$ and a vector $\mathbf{A}$ in a two-dimensional space as shown below. Construct the reciprocal basis $(\mathbf{e}^1, \mathbf{e}^2)$, draw the components $A_i \mathbf{e}^i$, and show geometrically that $A_i = \mathbf{e}_i \cdot \mathbf{A}$

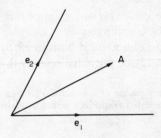

2.4. An n-dimensional vector space is spanned by a set of constant orthonormal base vectors $(\mathbf{k}_1, \mathbf{k}_2, \ldots \mathbf{k}_n)$. A corresponding set of coordinate lines $(y^1, y^2, \ldots, y^n)$ is defined so that a vector $\mathbf{M}$ from the origin to point M is given by

$$\mathbf{M} = \mathbf{k}_i\, y^i.$$

Given a new set of coordinate lines $(x^1, x^2, \ldots, x^n)$ such that a unique invertible transformation

$$x^i = x^i(y^1, y^2, \ldots, y^n)$$

exists, show that the n vectors

$$\mathbf{e}_i = \frac{\partial \mathbf{M}}{\partial x^i}$$

constitute a basis.

2.5. In (2.20) and (2.21) it was shown how a transformation of bases can be defined. Note the contrast between (2.21)

$$\mathbf{e}_i = c_i{}^J \mathbf{e}_J' \tag{2.21}$$

and the corresponding equation following from (2.23) for components of a vector $\mathbf{v}$

$$v^i = b^i{}_J v^J.$$

Hence the name *contravariant* for v^i. Show that

$$v_i = c_i{}^J v_J$$

and thus can be considered *covariant* with respect to (2.21).

CHAPTER 3

TENSORS

3.1. INTRODUCTION

With the apparatus of Chapter 2 at our disposal we are able to develop in a meaningful way the properties of that elusive yet ubiquitous entity known as a tensor. There are many possible primary definitions of a tensor [1–5]. A definition is chosen here which makes use of the concept of a vector space discussed in Chapter 2.

3.2. DEFINITION OF A TENSOR

Consider a vector space with dimension $n_1 n_2$ which is related to the vector spaces E_{n1} and E_{n2} and is denoted by the symbol $E_{n_1} \otimes E_{n_2}$. Elements of $E_{n_1} \otimes E_{n_2}$ are formed from arbitrary vectors $\mathbf{u}_i$ and $\mathbf{v}_j$ of E_{n_1} and E_{n_2}, respectively, by defining a *tensor product* $\mathbf{u}_i \otimes \mathbf{v}_j$ as an operation with the following properties when applied to vectors $\mathbf{u}_i$ and $\mathbf{v}_j$:

(1) $\otimes$ is distributive with respect to addition

$$\mathbf{u}_i \otimes (\mathbf{v}_j + \mathbf{v}_k) = \mathbf{u}_i \otimes \mathbf{v}_j + \mathbf{u}_i \otimes \mathbf{v}_k$$
$$(\mathbf{u}_i + \mathbf{u}_j) \otimes \mathbf{v}_k = \mathbf{u}_i \otimes \mathbf{v}_k + \mathbf{u}_j \otimes \mathbf{v}_k$$

(2) $\otimes$ is associative with respect to multiplication by a scalar

$$a(\mathbf{u}_i \otimes \mathbf{v}_j) = a\mathbf{u}_i \otimes \mathbf{v}_j = \mathbf{u}_i \otimes a\mathbf{v}_j$$

(3) If $(\mathbf{u}_1, \mathbf{u}_2, \ldots, \mathbf{u}_{n_1})$ and $(\mathbf{v}_1, \mathbf{v}_2, \ldots, \mathbf{v}_{n_2})$ are two arbitrary bases for E_{n_1} and E_{n_2}, respectively, the $n_1 n_2$ elements $\mathbf{u}_i \otimes \mathbf{v}_j$ are a basis for $E_{n_1} \otimes E_{n_2}$. A *tensor* is an element of the vector space $E_{n_1} \otimes E_{n_2}$ formed by a linear combination of the basis $\mathbf{u}_i \otimes \mathbf{v}_j$. Thus a tensor $\mathbf{T}$ may be written

$$\mathbf{T} = t^{ij} \mathbf{e}_{ij} \tag{3.1}$$

where $\mathbf{e}_{ij} = (\mathbf{u}_i \otimes \mathbf{v}_j)$. Since the $\mathbf{e}_{ij}$ constitute a basis, we know from Chapter 2 that $\mathbf{T} = \mathbf{0}$ if and only if all t^{ij}, which we designate the *components* of $\mathbf{T}$, are zero.

The tensor concept can be generalized to multiple products of vector spaces to give a more general definition of a tensor as any element of the vector space $E_{n_1} \otimes E_{n_2} \otimes \ldots$

constructed from the individual vector spaces E_{n_1}, E_{n_2}, ... We also make the stipulation that

$$(E_{n_1} \otimes E_{n_2}) \otimes E_{n_3} = E_{n_1} \otimes (E_{n_2} \otimes E_{n_3})$$

3.3. COVARIANT AND CONTRAVARIANT COMPONENTS OF TENSORS

The exposition of this chapter is limited to vector spaces formed from products of the space E_n with itself. Furthermore, the basis used to span the space E_n will be restricted to a basis $\{\mathbf{e}_i\}$ and its reciprocal basis $\{\mathbf{e}^i\}$, the two bases being related by (2.35). Consider as an example the vector space

$$E_n \otimes E_n \otimes E_n{}^*$$

where we use the notation $E_n{}^*$ to indicate that the reciprocal basis $\{\mathbf{e}^i\}$ is being used to span the space. From the previous paragraph we see that the basis for this space is

$$\mathbf{e}_{ij}{}^k = \mathbf{e}_i \otimes \mathbf{e}_j \otimes \mathbf{e}^k$$

and the tensor **T** formed by a linear combination of the basis $\mathbf{e}_{ij}{}^k$ is

$$\mathbf{T} = t^{ij}{}_k \, \mathbf{e}_{ij}{}^k \tag{3.2}$$

We designate $t^{ij}{}_k$ as a component of **T**. Furthermore, we state that the component is *contravariant* of order two (referring to i and j) and covariant of order 1 (referring to k), so that the total order of the tensor component is three. This convention is easily extended to arbitrary order. Tensor components containing both covariant and contravariant indices are known as *mixed* tensor components. Just as the components of a vector in (2.32) and (2.34) have meaning only when referred to a stated basis, (3.2) shows that the components of a tensor must likewise be referred to a basis in order to permit expression of the relation of tensor components to the tensor itself. In the literature the term "tensor" is frequently used interchangeably to describe both a tensor, as we have defined it here, and to describe one or more tensor components. For brevity we shall continue this rather loose procedure when it is felt that the meaning intended is clear from the context.

3.4. TRANSFORMATION OF TENSOR COMPONENTS

Suppose that we wish to write the components of **T** in (3.2) with respect to a basis other than $\mathbf{e}_{ij}{}^k$. Specifically, suppose, following (2.26), we write

$$\mathbf{e}'_I = a^j{}_I \, \mathbf{e}_j \tag{3.3}$$

where capital letters have been used as subscripts to refer to the primed basis. Also, we define the coefficients $c^I{}_j$ by the transformation

$$\mathbf{e}'^I = c^I{}_j \, \mathbf{e}^j \tag{3.4}$$

Then one can write

$$\mathbf{T} = t^{IJ}{}_K \, \mathbf{e}'_{IJ}{}^K \tag{3.5}$$

But

$$\mathbf{e'}_{IJ}{}^{K} = a^{l}{}_{I} a^{m}{}_{J} c^{K}{}_{n} \left[\mathbf{e}_{l} \otimes \mathbf{e}_{m} \otimes \mathbf{e}^{n} \right] \tag{3.6}$$

Now suppose that the base vectors $\mathbf{e}_i$ and $\mathbf{e'}_I$ are taken to be the natural bases for two different coordinate systems $(x^1, x^2, \ldots, x^n)$ and $(y^1, y^2, \ldots, y^n)$, respectively, in the n-dimensional vector space under consideration. Then, following Section 2.9,

$$\mathbf{e}_i = \frac{\partial \mathbf{M}}{\partial x^i}; \qquad \mathbf{e'}_I = \frac{\partial \mathbf{M}}{\partial y^I} \tag{3.7}$$

We assume here and elsewhere in the book, unless contrary statements are made, that an invertible transformation exists between the two coordinate systems. Then

$$x^i = x^i(y^1, y^2, \ldots, y^n); \, y^I = y^I(x^1, x^2, \ldots, x^n)$$

or, for brevity

$$x^i = x^i(y^J); \, y^I = y^I(x^j) \tag{3.8}$$

From (3.7) and (3.8) it follows that

$$d\mathbf{M} = \frac{\partial \mathbf{M}}{\partial x^j} dx^j = \mathbf{e}_j dx^j = \frac{\partial \mathbf{M}}{\partial y^I} dy^I = \mathbf{e'}_I dy^I$$

$$dx^j = \frac{\partial x^j}{\partial y^I} dy^I$$

Thus

$$\mathbf{e'}_I = \mathbf{e}_j \frac{\partial x^j}{\partial y^I}$$

and from (3.3)

$$a^j{}_I = \frac{\partial x^j}{\partial y^I} \tag{3.9}$$

From (3.3), (3.4), and (2.35) we form

$$\mathbf{e'}^I \cdot \mathbf{e'}_J = \delta^I{}_J = c^I{}_k \mathbf{e}^k \cdot \mathbf{e'}_J = c^I{}_k a^l{}_J \mathbf{e}^k \cdot \mathbf{e}_l$$

or

$$\delta^I{}_J = c^I{}_k a^l{}_J \delta^k{}_l = c^I{}_k a^k{}_J$$

Since

$$\frac{\partial y^I}{\partial x^k} \frac{\partial x^k}{\partial y^J} = \delta^I{}_J$$

we conclude, recalling (3.9), that

$$a^k{}_J \left(\frac{\partial y^I}{\partial x^k} - c^I{}_k \right) = 0 \qquad (3.10)$$

and that for this to be true in general we must have

$$c^I{}_k = \frac{\partial y^I}{\partial x^k} \qquad (3.11)$$

Hence, we can rewrite (3.6) as

$$\mathbf{e}'_{IJ}{}^K = \frac{\partial x^l}{\partial y^I} \frac{\partial x^m}{\partial y^J} \frac{\partial y^K}{\partial x^n} \, \mathbf{e}_{lm}{}^n \qquad (3.12)$$

Then from (3.5) we have the transformation rule for tensor components

$$t^{lm}{}_n = t^{IJ}{}_K \frac{\partial x^l}{\partial y^I} \frac{\partial x^m}{\partial y^J} \frac{\partial y^K}{\partial x^n} \qquad (3.13)$$

The spacing of indices of tensor components is cumbersome and we shall dispense with it in cases where the order of appearance of base vectors in the vector product which forms the tensor basis is not important to the specific problem. Furthermore, the use of upper- and lower-case indices is an awkward notation and we can avoid it by noting that $t(y)^{ij}_k$ refers to a tensor component associated with the natural basis for coordinates (y^1, $y^2, \ldots, y^n$). Then (3.13) can be rewritten

$$t^l{}_m(x) = \frac{\partial x^l}{\partial y^i} \frac{\partial x^m}{\partial y^j} \frac{\partial y^k}{\partial x^n} \, t^{ij}_k(y) \qquad (3.14)$$

Equation (3.14) is readily extended to tensor components of arbitrary order. Thus, given $x^i = x^i(y^j)$, we have

$$t^{i_1 i_2 \cdots i_m}_{j_1 j_2 \cdots j_n}(x) = \frac{\partial x^{i_1}}{\partial y^{k_1}} \cdots \frac{\partial x^{i_m}}{\partial y^{k_m}} \frac{\partial y^{l_1}}{\partial x^{j_1}} \cdots \frac{\partial y^{l_n}}{\partial x^{j_n}} \, t^{k_1 \cdots k_m}_{l_1 \cdots l_n}(y) \qquad (3.15)$$

Equation (3.15) is the transformation law for an absolute[†] tensor or arbitrary order.

3.5. TENSOR ALGEBRA

1. *Addition of tensors.* It is clear from the law for transformation of tensor components as well as from the definition of a tensor as a linear combination of a set of base vectors

---

[†] Detailed expositions of tensor calculus [3] include the general transformation law for a relative tensor of weight M, which is

$$t^{i_1 i_2 \cdots i_m}_{j_1 j_2 \cdots j_n}(x) = \left| \frac{\partial y^r}{\partial x^s} \right|^M \frac{\partial x^{i_1}}{\partial y^{k_1}} \cdots \frac{\partial x^{i_m}}{\partial y^{k_m}} \frac{\partial y^{l_1}}{\partial x^{j_1}} \cdots \frac{\partial y^{l_n}}{\partial x^{j_n}} \, t^{k_1 \cdots k_m}_{l_1 \cdots l_n}(y) \qquad (3.15A)$$

where $\left| \frac{\partial y^r}{\partial x^s} \right|$ refers to the Jacobian of the coordinate transformation. The Jacobian is a scalar weighting function which is advantageously associated with the tensor *component* rather than with the corresponding basis which when multiplied by the component yields a true tensor quantity (see, for example, (3.2)).

(see, for example, (3.2)), that the sum of two tensors of equal order is a tensor of the same order. Thus, for example

$$s^i_{jk} + t^i_{jk} = u^i_{jk} \tag{3.16}$$

2. *Two tensors of any order can be multiplied to produce a tensor the order of which is the sum of the orders of the two tensors.* This fact also follows directly from (3.15). For example,

$$t^i_{jk} u^{lm}_n = v^{ilm}_{jkn} \tag{3.17}$$

This is sometimes called *outer multiplication* and corresponds, of course, to the tensor component which results from taking the tensor product of $t^i_{jk}\mathbf{e}^{jk}_i$ and $u^l{}_m \mathbf{e}^n_{lm}$.

3. *Inner multiplication* is performed by equating one or more pairs of covariant and contravariant indices. The result is a tensor the total order of which is two less than the corresponding outer product for every pair of indices equated. For example,

$$t^i_{jk} u^{km}_n = v^{im}_{jn} \tag{3.18}$$

This result is also compatible with (3.15). It is readily seen that inner multiplication of tensor components corresponds to an extension, from vectors to tensors, of the notion of a scalar product considered in Section 2.6.

Note that from the right-hand side of (3.18) it is not clear which indices of the tensors t^i_{jk} and $u^l{}_n{}^m$ have been contracted. One could also write

$$t^i_{jk} u^{lk}_n = v^{il}_{jn} \tag{3.19}$$

Various schemes have been devised to remove this ambiguity in cases where it is important [3].

The operation of contraction can be very useful in determining the distribution of covariant and contravariant order of a tensor component. Thus if one has

$$t(ijk)u^k_l = v^{ij}_l$$

where $t(ijk)$ is a component of a third-order tensor, it is apparent that

$$t(ijk) = t^{ij}_k$$

This is often called the "quotient rule" and is proved formally in several standard texts [3].

3.6. THE METRIC TENSOR

In Section 2.7 we associated the natural basis $\mathbf{e}_i$ of a coordinate system x_i with a vector **M**. We see from (2.36) that the square of an element of length between two points connected by $d\mathbf{M}$ can be written

$$ds^2 = d\mathbf{M} \cdot d\mathbf{M} = \mathbf{e}_i \cdot \mathbf{e}_j dx^i dx^j = g_{ij} dx^i dx^j \tag{3.20}$$

Equation (3.20) is an excellent example of the utility of index notation. The double contraction on the right-hand side indicates that the result is a scalar, independent of reference to any coordinate system. Such is indeed the case for the element of length ds. In fact (3.20) is a useful equation for relating components of the element of length in any two coordinate systems. If we represent a rectangular Cartesian coordinate system by y^i, then, by the Pythagorean theorem,

$$ds^2 = dy^k \, dy^k = g_{ij} \, dx^i \, dx^j \tag{3.21}$$

But we have from the chain rule of differentiation

$$dy^k = \frac{\partial y^k}{\partial x^l} \, dx^l \tag{3.22}$$

which leads to

$$g_{ij}(x^l) = \frac{\partial y^k}{\partial x^i} \frac{\partial y^k}{\partial x^j} \tag{3.23}$$

One can readily show that the g_{ij} transform in accordance with the rule for a covariant component of a second-order tensor. Then the tensor **G** is the linear combination

$$\mathbf{G} = g_{ij}(x^l)\mathbf{e}^i \otimes \mathbf{e}^j$$

where $\{\mathbf{e}^k\}$ is the reciprocal to $\{\mathbf{e}_k\}$, and $\mathbf{e}_i = \partial\mathbf{M}/\partial x^i$. We see from (3.20) that components of **G** fix the connection between elements of length in space and differential changes along coordinate lines. For this reason the g_{ij} are called components of the *metric tensor*.

If a coordinate system is orthogonal we have, by definition of an orthogonal coordinate system,

$$g_{ij} = \mathbf{e}_i \cdot \mathbf{e}_j = 0 \quad \text{for} \quad i \neq j$$

Also, it is clear that for any coordinate system g_{ij} is *symmetric*; i.e.,

$$g_{ij} = g_{ji} \tag{3.24}$$

As a simple (and standard!) example, we show how one may determine the components of the metric tensor and the element of length in a cylindrical polar coordinate system x^i, related to a rectangular Cartesian system by

$$x^1 = [(y^1)^2 + (y^2)^2]^{1/2}$$
$$x^2 = \tan^{-1}\left[\frac{y^2}{y^1}\right] \tag{3.25}$$
$$x^3 = y^3$$

From (3.23) we find

$$g_{11} = 1; \quad g_{22} = (x^1)^2; \quad g_{33} = 1 \tag{3.26}$$

and

$$ds^2 = (dx^1)^2 + (x^1)^2(dx^2)^2 + (dx^3)^2 \tag{3.27}$$

Note that the contravariant components x^i do not all have the same units. Using the customary notation for cylindrical coordinates we make the associations

$$x^1 \equiv r; \quad x^2 \equiv \theta; \quad x^3 \equiv z$$

At this point it is also convenient to define

$$g^{ij} = \mathbf{e}^i \cdot \mathbf{e}^j \tag{3.28}$$

Then we may deduce an important result for contraction of the metric tensor. For any vector **u**, write

$$\mathbf{u} = u^i \, \mathbf{e}_i = u_k \mathbf{e}^k$$

Taking the dot product of both sides with $\mathbf{e}_j$

$$\mathbf{u} \cdot \mathbf{e}_j = u^i g_{ij} = u_k \mathbf{e}^k \cdot \mathbf{e}_j = u_j$$

Thus

$$u_j = g_{ij} u^i \tag{3.29}$$

Then

$$\mathbf{u} = g_{ij} u^i \, \mathbf{e}^j = u^k \, \mathbf{e}_k$$

Dotting both sides with $\mathbf{e}^l$,

$$g_{ij} u^i g^{jl} = u^k \, \delta_k^l = u^l$$

Since the components of $\mathbf{u}$ are independent, we conclude that

$$g_{ij} \, g^{jl} = \delta_i^l \tag{3.30}$$

For orthogonal systems one can show that, for $i=j$,

$$g^{ij} = \frac{1}{g_{ij}} \tag{3.31}$$

3.7. RAISING AND LOWERING OF INDICES

We can always write one base vector as a linear combination of another set of base vectors over the same space. Thus

$$\mathbf{e}^i = a^{ij} \mathbf{e}_j; \quad \mathbf{e}^i \cdot \mathbf{e}^k = a^{ij} \mathbf{e}_j \cdot \mathbf{e}^k = a^{ij} \delta_j^k = a^{ik} \tag{3.32}$$

or

$$g^{ik} \equiv a^{ik}$$

and (3.32) becomes

$$\mathbf{e}^i = g^{ij} \mathbf{e}_j \tag{3.33}$$

Similarly, one can show

$$\mathbf{e}_i = g_{ij} \mathbf{e}^j \tag{3.34}$$

By contraction of an arbitrary tensor with the metric tensor, one can form *associated tensors* which are of the same total order as the tensors from which they were derived but which are of different covariant and contravariant order. Thus

$$t^{ij} g_{jk} = t_k^i; \quad t_j^i g^{jk} = t^{ik}$$

The components t_k^i and t^{ik} refer of course to the same tensor $\mathbf{T}$. However, the components are associated with different bases formed from the tensor product described in Section 3.2.

Thus we have from (3.34)

$$\mathbf{T} = t^{ij} \mathbf{e}_i \otimes \mathbf{e}_j = t^{ij} g_{jk} \mathbf{e}_i \otimes \mathbf{e}^k \tag{3.35}$$

3.8. CHRISTOFFEL SYMBOLS, COVARIANT DERIVATIVE

Since $\mathbf{e}_i$ was associated with the natural basis (2.36) it follows that the $\mathbf{e}_i$ will in general depend upon position. Thus $\partial \mathbf{e}_i/\partial x^j$ is a vector and can be expressed in terms of a set of base vectors and a set of components which we designated by $\left\{ {k \atop ij} \right\}$. Then

$$\frac{\partial \mathbf{e}_i}{\partial x^j} = \left\{ {k \atop ij} \right\} \mathbf{e}_k \quad \text{(sum over } k) \tag{3.36}$$

The quantities $\left\{ {k \atop ij} \right\}$ are called Christoffel symbols of the second kind. From (3.36) and (2.35),

$$\left\{ {k \atop ij} \right\} = \frac{\partial \mathbf{e}_i}{\partial x^j} \cdot \mathbf{e}^k \tag{3.37}$$

Now consider some position dependent vector $\mathbf{u}(x^j)$. In order to form the partial derivative $\partial \mathbf{u}/\partial x^j$ of this vector we must remember that base vectors as well as components are position dependent. Hence

$$\frac{\partial \mathbf{u}}{\partial x^j} = \left[\frac{\partial u^k}{\partial x^j} + u^i \left\{ {k \atop ij} \right\} \right] \mathbf{e}_k \tag{3.38}$$

The quantity in brackets

$$\left[\frac{\partial u^k}{\partial x^j} + u^i \left\{ {k \atop ij} \right\} \right]$$

is seen to be the kth component of the vector $\partial \mathbf{u}/\partial x^j$ and is known as the covariant derivative of u^k. One frequently writes

$$\frac{\partial u^k}{\partial x^j} + u^i \left\{ {k \atop ij} \right\} = u^k{}_{,j} \tag{3.39}$$

Similarly, one can show that

$$\frac{\partial \mathbf{u}}{\partial x^j} = u_{k,j} \mathbf{e}^k \tag{3.40}$$

where

$$u_{k,j} = \frac{\partial u_k}{\partial x^j} - \left\{ {i \atop kj} \right\} u_i \tag{3.41}$$

Equation (3.40) is readily proved. From (2.35) and (3.37) we have

$$\frac{\partial \mathbf{e}^i}{\partial x^j} \cdot \mathbf{e}_k = -\mathbf{e}^i \cdot \frac{\partial \mathbf{e}_k}{\partial x^j} = -\left\{ {i \atop kj} \right\} = -\left\{ {i \atop lj} \right\} \delta^l_k = -\left\{ {i \atop lj} \right\} \mathbf{e} \cdot \mathbf{e}_k$$

from which we obtain

$$\frac{\partial \mathbf{e}^i}{\partial x^j} = - \left\{ {i \atop lj} \right\} \mathbf{e}^l \qquad\qquad 3.42$$

We now apply (3.42) to $\partial \mathbf{u}/\partial x^j$:

$$\frac{\partial \mathbf{u}}{\partial x^j} = \frac{\partial u_i}{\partial x^j} \mathbf{e}^i + u_i \frac{\partial \mathbf{e}^i}{\partial x^j} = \left[\frac{\partial u_k}{\partial x^j} - u_i \left\{ {i \atop kj} \right\} \right] \mathbf{e}^k$$

thus demonstrating the validity of (3.40). These definitions are easily extended to absolute tensors of arbitrary order. For example,

$$u^{ij}_{k,l} = \frac{\partial u^{ij}_k}{\partial x^l} + \left\{ {i \atop ml} \right\} u^{mj}_k + \left\{ {j \atop ml} \right\} u^{im}_k - \left\{ {m \atop kl} \right\} u^{ij}_m \qquad (3.43)$$

The covariant derivative of a scalar is understood to be $S_{,k} = \dfrac{\partial S}{\partial x^k}$. One can show from (3.15) that the notation for the covariant derivative is consistent with the tensor character of the result. Thus $S_{,i}$ is a covariant vector, $u_{i,j}$ is a covariant tensor of order two, etc.

We now wish to find explicit expressions for computation of the Christoffel symbols. From (2.29) and (3.34) we have

$$\frac{\partial g_{ij}}{\partial x^k} = \frac{\partial}{\partial x^k} (\mathbf{e}_i \cdot \mathbf{e}_j) = \mathbf{e}_i \cdot \frac{\partial \mathbf{e}_j}{\partial x^k} + \frac{\partial \mathbf{e}_i}{\partial x^k} \cdot \mathbf{e}_j$$

$$= \frac{\partial \mathbf{e}_j}{\partial x^k} \cdot \mathbf{e}^l g_{il} + \frac{\partial \mathbf{e}_i}{\partial x^k} \cdot \mathbf{e}^l g_{jl}$$

$$\frac{\partial g_{ij}}{\partial x^k} = \left\{ {l \atop jk} \right\} g_{il} + \left\{ {l \atop ik} \right\} g_{jl} \qquad (3.44)$$

By permuting indices

$$\frac{\partial g_{ki}}{\partial x^j} = \left\{ {l \atop ij} \right\} g_{kl} + \left\{ {l \atop kj} \right\} g_{il} \qquad (3.45)$$

$$\frac{\partial g_{jk}}{\partial x^i} = \left\{ {l \atop ki} \right\} g_{jl} + \left\{ {l \atop ji} \right\} g_{kl} \qquad (3.46)$$

However, since

$$\left\{ {k \atop ij} \right\} = \frac{\partial \mathbf{e}_i}{\partial x^j} \cdot \mathbf{e}^k = \frac{\partial^2 \mathbf{M}}{\partial x^j \partial x^i} \cdot \mathbf{e}^k = \frac{\partial^2 \mathbf{M}}{\partial x^i \partial x^j} \cdot \mathbf{e}^k$$

and

$$\left\{ {k \atop ji} \right\} = \frac{\partial \mathbf{e}_j}{\partial x^i} \cdot \mathbf{e}^k = \frac{\partial^2 \mathbf{M}}{\partial x^i \partial x^j} \cdot \mathbf{e}^k$$

we see that the Christoffel symbols are symmetric with respect to the two lower indices and

$$\begin{Bmatrix} k \\ ij \end{Bmatrix} = \begin{Bmatrix} k \\ ji \end{Bmatrix}$$

Then from (3.44), (3.45), and (3.46)

$$\frac{\partial g_{ij}}{\partial x^k} - \frac{\partial g_{ki}}{\partial x^j} = \begin{Bmatrix} l \\ ik \end{Bmatrix} g_{jl} - \begin{Bmatrix} l \\ ij \end{Bmatrix} g_{kl}$$

$$\frac{\partial g_{ij}}{\partial x^k} - \frac{\partial g_{ki}}{\partial x^j} - \frac{\partial g_{jk}}{\partial x^i} = -2 \begin{Bmatrix} l \\ ij \end{Bmatrix} g_{kl}$$

Multiplying by g^{km}, we finally obtain

$$\begin{Bmatrix} m \\ ij \end{Bmatrix} = \frac{1}{2} g^{km} \left[\frac{\partial g_{ik}}{\partial x^j} + \frac{\partial g_{jk}}{\partial x^i} - \frac{\partial g_{ij}}{\partial x^k} \right] \tag{3.47}$$

from which the Christoffel symbols may be computed as soon as the metric tensor is known.

An important property of the metric tensor components is their invariance to covariant differentiation. The reader is urged to demonstrate that

$$g_{ij,k} = g^{ij}_{,k} = 0 \tag{3.48}$$

3.9. ϵ-SYSTEMS: THE GENERALIZED KRONECKER DELTAS

We define the useful components e^{ijk} and e_{ijk}, which have the values:

(a) $+1$ if the indices can be ordered 123 by an *even* number of permutations;
(b) -1 if the indices can be ordered 123 by an *odd* number of permutations;
(c) zero under all other conditions.

It can be shown that e^{ijk} and e_{ijk} are relative tensors of weight $+1$ and -1, respectively [2, 3]. From them one can form the generalized Kronecker delta

$$\delta^{ijk}_{lmn} = e^{ijk} e_{lmn} \tag{3.49}$$

which is seen to be plus or minus one depending upon whether an even or odd number of permutations is required to match the order of superscripts to subscripts when each contains a permutation of 123. Under all other conditions it is zero.

The determinant of the metric tensor is a relative scalar of weight two. Thus one can form the absolute tensors

$$\varepsilon_{ijk} = e_{ijk} \ \sqrt{g} \tag{3.50}$$

$$\varepsilon^{ijk} = e^{ijk} \frac{1}{\sqrt{g}} \tag{3.51}$$

where $g = |g_{ij}|$.

3.10 TENSOR EXPRESSIONS FOR ELEMENTARY OPERATIONS OF VECTOR ANALYSIS

We have already seen in (2.30) how the scalar or dot product may be computed. We now consider several other operations from elementary vector analysis. One way to formulate these results is to recognize from (3.15) and (3.16) that if an equation involving tensor components is correct in one coordinate system, say y^i, it is correct in any other coordinate system, x^i, where $x^i = x^i(y^j)$. This means that if we write a tensorially correct statement in a familiar coordinate system, such as a rectangular Cartesian system, the statement is correct for any system related to the Cartesian system by the transformation given above. We use this fact to find:

(a) The vector or cross-product of two vectors.

Recall from elementary vector analysis that

$$\mathbf{u} \times \mathbf{v} = \begin{vmatrix} \mathbf{i} & \mathbf{j} & \mathbf{k} \\ u_x & u_y & u_z \\ v_x & v_y & v_z \end{vmatrix}$$

It is readily seen that the *i*th covariant component of this cross-product can be written

$$(\mathbf{u} \times \mathbf{v})_i = \varepsilon_{ijk} u^j v^k \tag{3.52}$$

In rectangular Cartesian coordinates there is no distinction between covariant, contravariant, and physical components (see Section 3.11) since $g_{ij} = \delta_{ij}$. Equation (3.52) is a proper tensor expression and hence must generate the *i*th covariant component of the cross-product in any coordinate system.

(b) Gradient of a scalar ∇S.

We immediately see, applying the scheme outlined above, that the covariant *i*th component is merely $S_{,i}$.

(c) Divergence of a vector.

$$\nabla \cdot \mathbf{u} = u^i_{,i} = g^{ij} u_{i,j} \tag{3.53}$$

since the components of the metric tensor behave as constants under covariant differentiation.

3.11. PHYSICAL COMPONENTS OF VECTORS AND TENSORS

Previous paragraphs have dealt at length with covariant and contravariant components of vectors and tensors. However, neither of these is necessarily equivalent to the components familiar from elementary applications of vectors and tensors—the *physical components*. In fact we saw in Section 3.6 that different contravariant components of a curvilinear coordinate system did not even necessarily have the same physical dimension. The link between covariant or contravariant components and physical components is readily established in any orthogonal coordinate system.

A physical component $v(i)$ of a vector $\mathbf{v}$ is defined with respect to a basis $\mathbf{k}_i$ by

$$\mathbf{v} = v(i)\mathbf{k}_i = v^i \mathbf{e}_i \tag{3.54}$$

where $\mathbf{k}_i$ is a unit vector parallel to $\mathbf{e}_i$ at the point of interest. Since the $\mathbf{k}_i$ must have values of unity, we set

$$\mathbf{k}_i = \frac{\mathbf{e}_i}{\sqrt{\mathbf{e}_i \cdot \mathbf{e}_i}} = \frac{\mathbf{e}_i}{\sqrt{g_{ii}}} \quad \text{(no sum)} \tag{3.55}$$

If the basis $\{\mathbf{e}_i\}$ is orthogonal, we readily obtain

$$v(i) = v^i \sqrt{g_{ii}} = v_i/\sqrt{g_{ii}} \quad \text{(no sum)} \tag{3.56}$$

This definition is consistent with that which relates physical components to the parent vector through the parallelogram law of vector addition.

By analogy the physical components of tensors in an orthogonal coordinate system are defined by, for example,

$$t(ijk) = t^i_{jk} \frac{\sqrt{g_{ii}}}{\sqrt{g_{jj}}\sqrt{g_{kk}}} \quad \text{(no sum)} \tag{3.57}$$

An excellent account of physical components has been given by Truesdell [6], who also discusses the complications which attend definition of physical components of tensors in nonorthogonal coordinate systems.

3.12. LINEAR TRANSFORMATIONS AS TENSORS

In previous sections considerable emphasis has been placed on the importance of tensor components. Indeed, for computational purposes they are indispensable. However, we have seen in Section 3.3 that tensor components are unambiguous only when referred to a basis. An obvious example can be drawn from vector components. The reader is familiar with the convenience frequently afforded by using the invariant vector $\mathbf{u}$ in equations rather than the components u_i. Similarly, it is useful, and often enhances clarity of a presentation, to describe physical concepts in terms of some tensor (e.g., $\mathbf{T}$), avoiding any explicit reference to a basis. We shall be particularly concerned with second-order tensors.

In Sections 2.2 and 2.5 we saw how a vector $\mathbf{u}$ could be transformed into a vector $\mathbf{v}$ through operation on $\mathbf{u}$ with $\mathbf{A}$:

$$\mathbf{v} = \mathbf{Au} \tag{3.58}$$

where $\mathbf{A}$ has the properties of a linear transformation. We now show that the components of $\mathbf{A}$ obey the law of tensor transformation (3.15).

Equation (3.58) is a vector equation. In component form it may be written

$$v^i = a^i_j u^j \tag{3.59}$$

This is consistent with the notions of Chapter 2 and the notation of (2.23). Suppose that the components v^i and u^j are with respect to the same basis $\mathbf{e}_i(y) = \dfrac{\partial \mathbf{M}}{\partial y^i}$ where y^i refers to some curvilinear coordinate system. The elements a^i_j are in general position dependent. We denote this by writing $a^i_j(y)$. If we now introduce a new natural basis defined with respect to a different curvilinear system $x^i = x^i(y^j)$, $\mathbf{e}_i(x) = \dfrac{\partial \mathbf{M}}{\partial x^i}$ so that

$$\mathbf{v} = v^i(y)\mathbf{e}_i(y) = v^i(x)\mathbf{e}_i(x)$$

it is clear that (3.59) becomes

$$v^i(y) = a^i_j(y)u^j(y) = \frac{\partial y^i}{\partial x^k} v^k(x) = a^i_j(y)\frac{\partial y^j}{\partial x^l} u^l(x) \tag{3.60}$$

from which we find

$$v^k(x) = a_j(y) \frac{\partial y^j}{\partial x^l} \frac{\partial x^k}{\partial y^i} u^l(x) \tag{3.61}$$

Setting

$$b_l^k(x) = \frac{\partial x^k}{\partial y^i} \frac{\partial y^j}{\partial x^l} a_j^i(y) \tag{3.62}$$

we have

$$v^k(x) = b_l^k(x)u^l(x) \tag{3.63}$$

Equation (3.62) shows that the a_j^i are indeed the mixed components of a second-order tensor, and our anticipated placement of superscript and subscript indices was correct. For the example chosen, namely the operator which transforms one vector to a new vector, we have established the equivalence of a second-order tensor and a linear transformation.

Since the operation of (3.58) can be viewed as a contraction between a second-order tensor and a vector to form another vector, one could write that equation as

$$\mathbf{v} = \mathbf{A} \cdot \mathbf{u} \tag{3.64}$$

We emphasize once more that (3.58) and (3.64) are "true" irrespective of the base vectors with which components may be associated. To simplify matters and also to be consistent with much of the literature on the subject, we shall normally use the notation of (3.58) rather than (3.64). We also reserve the notation $\mathbf{A}$ for a second-order tensor unless a contrary statement is made.

3.13. IDENTITY TENSOR, INVERSE, TRANSPOSE

In Section 2.2 we defined the identity or unit matrix in a limited context. We now define the identity or unit tensor as one which has the following property when applied to an arbitrary vector $\mathbf{u}$:

$$\mathbf{Iu} = \mathbf{u} \tag{3.65}$$

If $\mathbf{v} = \mathbf{Bu}$ and $\mathbf{w} = \mathbf{Av}$, then we define the product transformation $\mathbf{AB}$ by $\mathbf{w} = (\mathbf{AB})\mathbf{u}$ It is clear that the notation $\mathbf{AB}$ signifies a succession of linear transformations, i.e., $\mathbf{w} = \mathbf{A}(\mathbf{Bu}) = (\mathbf{AB})\mathbf{u}$. Components of $\mathbf{AB}$ are formed, as shown in Section 3.5, by a contraction of components of $\mathbf{A}$ and components of $\mathbf{B}$. In dyadic notation we have $\mathbf{AB} \equiv \mathbf{A} \cdot \mathbf{B}$. A special product transformation is that which defines the *inverse* of a tensor $\mathbf{A}$. The inverse of $\mathbf{A}$ is denoted by $\mathbf{A}^{-1}$ and is defined by the properties

$$\mathbf{AA}^{-1} = \mathbf{A}^{-1}\mathbf{A} = \mathbf{I} \tag{3.66}$$

If an inverse does not exist, $\mathbf{A}$ is said to be *singular*.

In elementary applications the *transpose* of a tensor is usually defined in terms of the tensor components. The components of the transpose are formed by interchanging the indices of components of the original tensor. We now define the transpose without reference to components by writing

$$(\mathbf{Au}) \cdot \mathbf{v} = \mathbf{u} \cdot (\mathbf{A}^T\mathbf{v}) \tag{3.67}$$

where $\mathbf{A}^T$ is the transpose of $\mathbf{A}$, and $\mathbf{u}$ and $\mathbf{v}$ are arbitrary vectors. If $\mathbf{A} = \mathbf{A}^T$, $\mathbf{A}$ is a *symmetric* tensor.

3.14. TRANSFORMATIONS OF TENSORS

In Section 3.4 rules for transformation of tensor components from one basis to another were determined. Thus components of a mixed second-order tensor in two different coordinate frames can be related by

$$t^i{}_j(x) = a^i{}_k \, b^l{}_j \, t^k{}_l(y) \tag{3.68}$$

where

$$a^i{}_k = \frac{\partial x^i}{\partial y^k}; \quad b^l{}_j = \frac{\partial y^l}{\partial x^j}$$

It is evident that $a^i{}_k \, b^k{}_l = b^i{}_k \, a^k{}_l = \delta^i{}_l$. Hence we can write

$$b^l{}_j = (a^{-1})^l{}_j \tag{3.69}$$

and (3.68) becomes

$$t^i{}_j(x) = a^i{}_k \, t^k{}_l(y)(a^{-1})^l{}_j \tag{3.70}$$

Recall that in our discussion of vectors we could associate changes in components with changes in the basis for a given vector, or alternatively we could look upon the operation as a change in the vector itself for a given basis. The same approach can be applied to second-order tensors. From (3.70) we can write a basis-free operator equation indicating an operation on $\mathbf{T}$ to give a new tensor $\bar{\mathbf{T}}$ by

$$\bar{\mathbf{T}} = \mathbf{A}\mathbf{T}\mathbf{A}^{-1} \tag{3.71}$$

3.15. ORTHOGONAL TRANSFORMATIONS

We shall be especially interested in transformations of the type

$$\mathbf{u} = \mathbf{Q}\mathbf{v} \tag{3.72}$$

where $\mathbf{Q}^T = \mathbf{Q}^{-1}$. This is known as an orthogonal transformation, and the symbol $\mathbf{Q}$ will be used henceforth to imply a tensor for which $\mathbf{Q}^T = \mathbf{Q}^{-1}$. The transformation $\mathbf{Q}$ can be interpreted as a rotation or reflection of the vector $\mathbf{v}$. We first prove [7] that an orthogonal transformation does not affect the angle between two vectors, and also leaves the magnitude of a vector unchanged. Let

$$\mathbf{w} = \mathbf{Q}\mathbf{u} \tag{3.73}$$

$$\mathbf{x} = \mathbf{Q}\mathbf{v} \tag{3.74}$$

Then

$$\mathbf{w} \cdot \mathbf{x} = \mathbf{Q}\mathbf{u} \cdot \mathbf{Q}\mathbf{v} \tag{3.75}$$

But from (3.67) we can write

$$\mathbf{w} \cdot \mathbf{x} = \mathbf{u} \cdot (\mathbf{Q}^T \mathbf{Q} \mathbf{v}) = \mathbf{u} \cdot \mathbf{v} \tag{3.76}$$

In the special case where $\mathbf{u} = \mathbf{v}$ we see from (3.76) that the magnitude of a vector is not changed by an orthogonal transformation, i.e.,

$$|\mathbf{Q}\mathbf{u}| = |\mathbf{u}| \tag{3.77}$$

We further note that since

$$\mathbf{Q}\mathbf{Q}^T = \mathbf{I} \tag{3.78}$$

$$\det(\mathbf{Q}) = \pm 1 \tag{3.79}^\dagger$$

One can show that if $\det(\mathbf{Q}) = +1$, $\mathbf{Q}$ corresponds to a rotation, while if $\det(\mathbf{Q}) = -1$, $\mathbf{Q}$ corresponds to a reflection (Section 3.17). Unless otherwise stated, we shall assume $\det(\mathbf{Q}) = +1$. Thus the results (3.73) and (3.74) can be interpreted as a rotation in space of the two vectors $\mathbf{u}$ and $\mathbf{v}$, each being rotated through the same angle. Hence an orthogonal set of base vectors remains orthogonal under the transformation $\mathbf{Q}$.

Next consider the orthogonal transformation of tensors. As an example, let

$$\mathbf{T} = t^{ij} \mathbf{e}_i \otimes \mathbf{e}_j \tag{3.80}$$

Suppose for the moment we consider the change in tensor components from t^{ij} to $\bar{t}^{ij}$ as we refer $\mathbf{T}$ to a new set of base vectors $\bar{\mathbf{e}}_i$ defined by

$$\bar{\mathbf{e}}_j = q_j{}^i \mathbf{e}_i \tag{3.81}$$

$$\mathbf{e}_i = p_i{}^j \bar{\mathbf{e}}_j \tag{3.82}$$

$q_j{}^i$ being the elements of the transformation $\mathbf{Q}$ and the components of $\bar{\mathbf{e}}_j$ being referred to the basis $\mathbf{e}_i$. Since (3.82) is the inverse transformation of (3.81), if the $p_i{}^j$ are elements of $\mathbf{P}$, we must have

$$\mathbf{P} = \mathbf{Q}^{-1} = \mathbf{Q}^T \tag{3.83}$$

In terms of the new basis (3.80) becomes

$$\mathbf{T} = t^{ij} p_i{}^k p_j{}^l \bar{\mathbf{e}}_k \otimes \bar{\mathbf{e}}_l \tag{3.84}$$

Thus

$$\bar{t}^{kl} = t^{ij} p_i{}^k p_j{}^l \tag{3.85}$$

Equation (3.85) represents the change in components of $\mathbf{T}$ when the base vectors are rotated by $\mathbf{Q}$. As in the case of vectors we could look upon (3.85) as a component representation of a new tensor $\bar{\mathbf{T}}$ with components referred to the original basis:

$$\bar{\mathbf{T}} = \bar{t}^{ij} (\mathbf{e}_i \otimes \mathbf{e}_j) \tag{3.86}$$

and from (3.85) we see that $\bar{\mathbf{T}}$ can be represented by

$$\bar{\mathbf{T}} = \mathbf{P}^T \mathbf{T} \mathbf{P} \tag{3.87}$$

where the components of $\mathbf{P}^T$ are formed by replacing $p_i{}^j$ in $\mathbf{P}$ by $p^j{}_i$. We rewrite (3.87) as

$$\bar{\mathbf{T}} = \mathbf{Q} \mathbf{T} \mathbf{Q}^T \tag{3.88}$$

---

† Evaluation of the determinant of a tensor $\mathbf{A}$ is accomplished with the *mixed* components a^i_j of $\mathbf{A}$ (cf. Section 3.16). To prove (3.79) one uses two properties of determinants, viz., $\det(\mathbf{AB}) = \det(\mathbf{A})\det(\mathbf{B})$, and $\det(\mathbf{A}) = \det(\mathbf{A}^T)$.

or, premultiplying by $\mathbf{Q}^T$ and postmultiplying by $\mathbf{Q}$, we obtain

$$\mathbf{T} = \mathbf{Q}^T \,\overline{\mathbf{T}}\, \mathbf{Q} \tag{3.89}$$

We shall have frequent occasion to employ transformations of the type (3.88) and (3.89) in the case where $\det(\mathbf{Q}) = +1$. In such cases it is convenient to speak of (3.88) as a *rotation* of $\mathbf{T}$ to $\overline{\mathbf{T}}$.

An isotropic orthogonal transformation is one which leaves the tensor unchanged:

$$\mathbf{T} = \mathbf{Q}\,\mathbf{T}\,\mathbf{Q}^T \tag{3.90}$$

A tensor $\mathbf{T}$ which obeys (3.90) for any $\mathbf{Q}$ is an *isotropic tensor*.

3.16. PRINCIPAL INVARIANTS

In succeeding chapters we shall find that a central role is played by certain functions which are invariant to transformation of coordinates. These are scalar-valued tensor functions and are known as "invariants" of a given tensor. One way to introduce these invariants is through a discussion of the characteristic (or eigenvalue) equation for a second-order tensor $\mathbf{T}$. We consider the equation

$$\mathbf{Tu} = \lambda\mathbf{u} \tag{3.91}$$

and inquire about those values of λ and $\mathbf{u}$ for which (3.91) can be satisfied. If a nontrivial solution of (3.91) exists it must, of course, be true that the determinant

$$\det(\mathbf{T} - \mathbf{I}\lambda) = 0 \tag{3.92}$$

where $\mathbf{I}$ is the identity tensor.

The determinant can be expanded to give

$$-\lambda^3 + I_1\lambda^2 - I_2\lambda + I_3 = 0 \tag{3.93}$$

where I_1, I_2, and I_3 are functions of $\mathbf{T}$. By writing out the elements of the determinant, it is easily verified that

$$\begin{aligned} I_1 &= \operatorname{tr}\mathbf{T} \\ I_2 &= \tfrac{1}{2}\left[(\operatorname{tr}\mathbf{T})^2 - \operatorname{tr}(\mathbf{T}^2)\right] \\ I_3 &= \det(\mathbf{T}) \end{aligned} \tag{3.94}$$

The notation $\operatorname{tr}\mathbf{T}$ refers to the trace of $\mathbf{T}$, which is the sum of the diagonal elements of the components of $\mathbf{T}$. (Recall that we should specify the elements t^i_j of $\mathbf{T}$ with respect to a *mixed* basis.) The notation used implies that both the determinant and the trace are independent of the mixed basis to which the components are referred. That this is indeed so is readily proved. With respect to a basis $\mathbf{e}^j_i$ we have

$$\operatorname{tr}\mathbf{T} = t^i_i \tag{3.95}$$

Transforming to a new basis $\bar{\mathbf{e}}^i_j$ and denoting $\bar{t}^i_i$ by $\operatorname{tr}\overline{\mathbf{T}}$, we have, from (3.70)

$$\bar{t}^i_j = a^i_k\, t^k_l\, (a^{-1})^l_j \tag{3.96}$$

$$\operatorname{tr}\overline{\mathbf{T}} = \bar{t}^i_i = a^i_k\, (a^{-1})^l_i\, t^k_l = \delta^l_k\, t^k_l = t^l_l \tag{3.97}$$

Equation (3.96) can also be used to establish the invariance of the determinant.

$$\det(\overline{\mathbf{T}}) = \det[a_k^i\, t_i^k\, (a^{-1})_j^l] = \det(\mathbf{A}\mathbf{A}^{-1})\det(\mathbf{T}) = \det(\mathbf{T}) \tag{3.98}$$

where we have used the fact that $\det(\mathbf{AB}) = (\det\mathbf{A})(\det\mathbf{B})$. From (3.94), (3.97), and (3.98) we see that I_1, I_2, and I_3 are indeed invariant to the coordinate system in which they are expressed. These three scalar quantities are essential for formulation of appropriate equations describing response of materials to stresses. Referred to any tensor, $\mathbf{T}$, I_1, I_2, and I_3 are known as the *principal* invariants of $\mathbf{T}$.

3.17. PRINCIPAL VALUES (EIGENVALUES) AND PRINCIPAL DIRECTIONS (EIGENVECTORS) [8]

Equation (3.93) can be solved for the three roots λ_1, λ_2, λ_3. These are known as the principal values of the tensor $\mathbf{T}$. Alternately, one can consider I_1, I_2, and I_3 as unknowns in a system of three simultaneous equations of the form (3.93), this equation being written for each of the three principal values. The result is:

$$I_1 = \lambda_1 + \lambda_2 + \lambda_3 \tag{3.99}$$

$$I_2 = \lambda_1\lambda_2 + \lambda_1\lambda_3 + \lambda_2\lambda_3 \tag{3.100}$$

$$I_3 = \lambda_1\,\lambda_2\,\lambda_3 \tag{3.101}$$

We next inquire about the nature of the roots λ_i of (3.93) and the corresponding principal directions given by $\mathbf{u}$. The answers to these questions will be seen to depend greatly on the properties of $\mathbf{T}$. First of all, we take special note of the fact that tensors which are related by transformations of the form (3.88) have the same principal values and related principal directions. Thus let

$$\mathbf{T}\mathbf{u} = \lambda\mathbf{u}$$

and define a vector $\bar{\mathbf{u}}$ by $\mathbf{u} = \mathbf{Q}^T\bar{\mathbf{u}}$. Then

$$\mathbf{T}\mathbf{Q}^T\bar{\mathbf{u}} = \lambda\mathbf{Q}^T\bar{\mathbf{u}}$$

Operating on both sides with $\mathbf{Q}$ and defining

$$\overline{\mathbf{T}} = \mathbf{Q}\,\mathbf{T}\,\mathbf{Q}^T$$

we see that

$$\overline{\mathbf{T}}\bar{\mathbf{u}} = \lambda\bar{\mathbf{u}} \tag{3.102}$$

Equation (3.102) indicates that any principal value λ of $\mathbf{T}$ is also a principal value of $\overline{\mathbf{T}}$. Furthermore, the corresponding principal directions $\mathbf{u}$ and $\bar{\mathbf{u}}$ of $\mathbf{T}$ and $\overline{\mathbf{T}}$, respectively, are related by

$$\bar{\mathbf{u}} = \mathbf{Q}\mathbf{u} \tag{3.103}$$

Returning to (3.91) consider several special cases:

(1) $\mathbf{T}$ is orthogonal, i.e., we set $\mathbf{T} \equiv \mathbf{Q}$.

We know from (3.79) that $\det(\mathbf{Q}) = \pm 1$. Three subcases are:

(a) $\mathbf{Q}$ is a rotation (also called a *proper* transformation, meaning that $\det(\mathbf{Q}) = +1$). From (3.94) and (3.101) we know that $\lambda_1\,\lambda_2\,\lambda_3 = 1$. The very special case where $\lambda_1 = \lambda_2 = \lambda_3 = 1$ is treated in paragraph (c) below. For the moment we note the physical meaning of the case where $\lambda_1 = 1$ and $\lambda_2\,\lambda_3 = 1$ but λ_2 and λ_3 are not both equal to $+1$.

$$Qu_1 = \lambda_1 u_1 = u_1 \tag{3.104}$$

Thus vectors in the direction u_1 or $-u_1$ undergo no change when operated on by Q. Since we also know from (3.77) that the magnitude of a vector is unchanged by orthogonal transformation, we conclude that Q corresponds to rotation of a vector about an axis parallel to u_1. It is of interest to note in passing that if λ_2 and λ_3 are to be real, the requirements $\lambda_2 \lambda_3 = 1$ and $|Qu| = |u|$ indicate that $\lambda_2 = \lambda_3 = -1$ and the rotation is through an angle π in a plane perpendicular to u_1.

(b) Now suppose the transformation is an *improper* one, i.e., $\det(Q) = -1$. Excluding the case where all of the λ's are equal, we consider the result when $\lambda_1 = -1$, $\lambda_2 \lambda_3 = +1$. Then

$$Qu_1 = -u_1 \tag{3.105}$$

$$Q(-u_1) = u_1 \tag{3.106}$$

illustrating the fact that, for λ_i real, Q corresponds to a reflection in a plane perpendicular to u_1.

(c) $\lambda_1 = \lambda_2 = \lambda_3 = \pm 1$. In this case the characteristic equation is satisfied by $Q = \pm I$, and $\pm Iu = u$ for any choice of u.

(2) T is symmetric.

We shall restrict our discussion to *real* tensors. A real tensor T is defined by

$$(Tu)^* = Tu^* \tag{3.107}$$

where * refers to the complex conjugate. Equation (3.107) can be used to prove that the principal values of a real symmetric tensor T are real.[†] Let

$$Tu = \lambda u$$

Then

$$Tu^* = \lambda^* u^*$$

which shows that λ^* is a principal value corresponding to the principal direction u^*.

Consider the scalar product

$$\lambda \, (u^* \cdot u)$$

Using (3.67) and the symmetry of T,

$$\lambda(u^* \cdot u) = u^* \cdot \lambda u = u^* \cdot (Tu) = Tu^* \cdot u = \lambda^*(u^* \cdot u)$$

or

$$\lambda(u^* \cdot u) = \lambda^*(u^* \cdot u) \tag{3.108}$$

Excluding the trivial case $u = 0$, it is seen from (3.108) that

$$\lambda = \lambda^*$$

and one concludes that λ is real. Then it follows immediately from (3.99) through (3.101) that the principal invariants of T must be real.

[†] This is actually a special case of the more general theorem: principal values of a Hermitian tensor are real. A tensor A is Hermitian if $A = (A^T)^*$.

Some special cases of interest for symmetric $\mathbf{T}$ are:

(a) $\lambda_1 \neq \lambda_2 \neq \lambda_3$. Each principal value λ_i will have associated with it a different principal direction according to

$$\mathbf{Tu}_i = \lambda_i \mathbf{u}_i \quad \text{(no sum)} \tag{3.109}$$

Also, the principal directions are mutually orthogonal. We prove this by recalling from (3.67) that we can write for two principal directions $\mathbf{u}_i$ and $\mathbf{u}_j$

$$(\mathbf{Tu}_i) \cdot \mathbf{u}_j = \lambda_i \mathbf{u}_i \cdot \mathbf{u}_j = \mathbf{u}_i \cdot (\mathbf{Tu}_j) = \lambda_j \mathbf{u}_i \cdot \mathbf{u}_j \tag{3.110}$$

Since $\lambda_i \neq \lambda_j$, $\mathbf{u}_i \cdot \mathbf{u}_j = 0$. One readily sees that if we consider the $\mathbf{u}_i$ as an orthonormal basis, then the components of $\mathbf{T}$ are $t^i_j = 0$ for $i \neq j$, and $t^i_i = \lambda_i$ (no sum).

(b) Two of the principal values are equal. Suppose $\lambda_1 = \lambda_2 \neq \lambda_3$. Then, as in the previous case, we have a unique principal direction $\mathbf{u}_3$ associated with λ_3. However, we see from (3.110) that $\mathbf{u}_1$ and $\mathbf{u}_2$ may lie anywhere in a plane orthogonal to $\mathbf{u}_3$. This means that the components of $\mathbf{T}$ may be diagonalized and set equal to the principal values in any orthonormal basis one vector of which is along $\mathbf{u}_3$.

(c) All three principal values are equal: $\lambda_1 = \lambda_2 = \lambda_3$. In this case (3.110) places no restriction on the orientation of the principal directions. We note that any direction is a principal direction and

$$\mathbf{T} = \lambda \mathbf{I} \tag{3.111}$$

This $\mathbf{T}$ is an example of an isotropic tensor since $\mathbf{QTQ}^T = \mathbf{T}$.

3.18. PRINCIPAL COORDINATES: THE IMPORTANCE OF PRINCIPAL INVARIANTS

In conclusion we note two very useful properties of symmetric tensors $\mathbf{T}$ and their real principal invariants.

First, the examples above have shown that components of $\mathbf{T}$ are particularly simple when referred to a basis formed from tensor products of unit base vectors aligned in the principal directions. Then $\mathbf{T}$ can be written in diagonal form:

$$\begin{bmatrix} \lambda_1 & 0 & 0 \\ 0 & \lambda_2 & 0 \\ 0 & 0 & \lambda_3 \end{bmatrix} \tag{3.112}$$

Such a system of base vectors defines a *principal coordinate system* for $\mathbf{T}$.

Second, we show that any real single-valued scalar function of $\mathbf{T}$ which is invariant under changes in the coordinate system (i.e., a scalar invariant of $\mathbf{T}$) can be expressed as a function of the three *principal* invariants of $\mathbf{T}$. To demonstrate this we note from (3.99) through (3.101) that any single-valued function of the (real) principal values must be a single-valued function of the principal invariants. Thus if the scalar function of $\mathbf{T}$ is written using a principal coordinate system so that the components of $\mathbf{T}$ are expressed by (3.112), the scalar function $\mathcal{N}$ becomes $\mathcal{N} = \mathcal{N}(\lambda_i)$. However, since the λ_i are determined by the principal invariants, we conclude

$$\mathcal{N} = f(I_1, I_2, I_3) \tag{3.113}$$

REFERENCES

1. Aris, R., *Vectors, Tensors, and the Basic Equations of Fluid Mechanics* (Englewood Cliffs, New Jersey: Prentice-Hall, 1962).
2. Sokolnikoff, I. S., *Tensor Analysis: Theory and Applications* 2nd edn. (New York: Wiley, 1964).
3. McConnell, A. J., *Applications of Tensor Analysis* (New York: Dover Publications, 1957).
4. Ericksen, J. L., Tensor fields, in *Handbuch der Physik*, Band III/1, pp. 794–858. (S. Flügge, ed.), (Berlin: Springer Verlag, 1960).
5. Lichnerowicz, A., *Elements of Tensor Calculus* (London: Methu en, 1962).
6. Truesdell, C., *Z. ang. Math. Mech.* **33**, 345–356 (1953).
7. Leigh, D. C., *Nonlinear Continuum Mechanics*, p. 38. (New York: McGraw-Hill, 1968).
8. *Ibid.*, pp. 59 *et seq.*

PROBLEMS

3.1. Prove that the components of any tensor can be expressed as the sum of two tensor components, one of which is symmetric and the other antisymmetric. In this context *symmetry* implies that interchange of two stated covariant (or contravariant) indices does not alter the value of the component. The same operation applied to components of an antisymmetric tensor results in a change in sign of the component. The definitions of symmetry and antisymmetry also include the requirement that the properties described above be invariant to cooordinate transformation.

3.2. The square of an element of length is
$$ds^2 = g_{ij}(x^l)dx^i dx^j = h_{ij}(z^l)dz^i dz^j$$
where $g_{ij}(x^l)$ and $h_{ij}(z^l)$ are defined by (3.23), and $x^i = x^i(z^j)$ is an invertible transformation. Show that:
(a) the transformation between $g_{ij}(x^l)$ and $h_{ij}(z^l)$ is that of covariant components of a second-order tensor (the metric tensor); and that
(b) the determinant $|g_{ij}|$ is a relative scalar of weight two.

3.3. Prove that the metric tensor is a constant under covariant differentiation (see (3.48)).

3.4. Show that for orthogonal coordinate systems
$$\begin{Bmatrix} k \\ i\ j \end{Bmatrix} = 0 \quad \text{for } i, j, k \text{ distinct}$$
and show that for $i \neq j$
$$\begin{Bmatrix} i \\ i\ i \end{Bmatrix} = \frac{1}{2}\frac{\partial}{\partial x^i}(\ln g_{ii}); \quad \begin{Bmatrix} i \\ i\ j \end{Bmatrix} = \frac{1}{2}\frac{\partial}{\partial x^j}(\ln g_{ii}); \quad \begin{Bmatrix} i \\ j\ j \end{Bmatrix} = -\frac{1}{2g_{ii}}\frac{\partial g_{jj}}{\partial x^i} \quad \text{(no sum)}$$

3.5. Sketch a nonorthogonal curvilinear coordinate system (x^1, x^2) in two dimensions. Recall that the natural basis for this system is defined by
$$\mathbf{e}_i(x^j) = \frac{\partial \mathbf{M}(x^j)}{\partial x^i} \quad \text{(see Fig. 2.4)}$$
On the sketch indicate the distinction between *components* M^i of $\mathbf{M}$, where
$$\mathbf{M} = M^i \mathbf{e}_i$$
and *coordinates* x^1, x^2 of $\mathbf{M}(x^1, x^2)$.

3.6. The Laplacian of a scalar ϕ is defined in elementary vector analysis by
$$\left(\frac{\partial^2}{\partial x^2} + \frac{\partial^2}{\partial y^2} + \frac{\partial^2}{\partial z^2}\right)\phi$$
where (x, y, z) constitutes a rectangular Cartesian coordinate system. Show that the correct general tensor expression is $g^{ij}\phi_{,ij}$.

3.7. In (3.90) an isotropic tensor was defined as a tensor which, for any orthogonal transformation $\mathbf{Q}$, is unchanged, i.e.,
$$\mathbf{T} = \mathbf{Q}\,\mathbf{T}\,\mathbf{Q}^T$$
Clearly $\mathbf{T} = k\mathbf{I}$, where k is a constant, is a sufficient condition for $\mathbf{T}$ to be isotropic. Show, by choosing $\mathbf{Q}$'s to effect, successively, reflections of a rectangular Cartesian basis, that $\mathbf{T} = k\mathbf{I}$ is also a necessary condition for isotropy of $\mathbf{T}$.

CHAPTER 4

THE CONSERVATION EQUATIONS OF CONTINUUM MECHANICS

4.1. INTRODUCTION

The subject of concern in this chapter is formulation of basic differential equations which are statements of conservation of mass, momentum, and energy for a system. These are the equations which an engineer or applied scientist wishes to solve over a macroscopic volume for given initial and boundary conditions in order to make predictions about the behavior of a flow system. To begin one needs to make some postulates about the nature of the material being described. For example, by assuming that the material is composed of discrete masses interacting with each other, one can derive the balance equations in terms of a molecular kinetic theory [1]. This approach can be especially instructive if one is interested in relating gross material properties (such as viscosity and thermal conductivity) to parameters describing molecular interactions. However, our ability to solve equations relating molecular behavior to gross properties is severely limited both because of the formidable nature of the equations and because of a lack of basic molecular data. Hence the gross material properties are usually measured experimentally and inserted into the appropriate balance equations. Since we are not often in a position to utilize kinetic theoretical descriptions to predict properties of real materials, it is frequently convenient to bypass the molecular model completely and to use a model in which the material is treated as a continuum. The user of a continuum model is in no way denying existence of molecules. He is merely recognizing that representation of material by a continuum model is frequently a useful way to describe phenomena for which a molecular picture is either unnecessary or unprofitable. The kinetic theorist is also using a continuum model of sorts. His basic unit, the molecule, is composed of atoms, which in turn have nuclei, which in turn contain subatomic particles.

4.2. THE CONTINUUM

In continuum mechanics we use as our description of a material an infinity of particles or *material points* (strictly speaking, a differentiable manifold of material points). We postulate that properties may be ascribed to each point in the continuum, and that these properties are continuous as one moves between adjacent material points. Our task in this chapter is to formulate equations which describe the motion of the material points of a system with respect to one or more reference coordinate frames. Knowing how the material points move, we can then also formulate how properties associated with material points will be transported in a given system.

4.3. KINEMATICS

We shall find it mathematically useful and physically meaningful to follow the symbolic notation used by Truesdell and others [2–4] to signify the change in configuration of the material points of a continuum with time. By χ_t we signify the configuration (i.e., the position with respect to some reference) of all of the material points of a body at some time t. Denote by X a label (for example, a number or a color) which identifies a given material point. If $\mathbf{x}$ is used to represent the coordinates (x^1, x^2, x^3) of a position fixed in space, we can then write for the position $\mathbf{x}$ of any particular material point X of the body in configuration χ_t (Fig. 4.1)

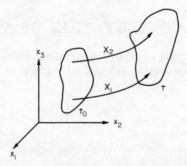

Fig. 4.1. Motion of two material points during the time interval t_0 to t.

$$\mathbf{x} = \chi_t(X) \tag{4.1}$$

In the language of mathematics one can say that (4.1) signifies a *mapping* of X on $\mathbf{x}$. By inverting (4.1) we have a mapping of $\mathbf{x}$ on X. Thus

$$X = \chi_t^{-1}(\mathbf{x}) \tag{4.2}$$

We can compute the velocity of a material point X by finding

$$\dot{\mathbf{x}} = \frac{d\mathbf{x}}{dt} = \frac{d}{dt}\chi_t(X) \tag{4.3}$$

where d/dt implies a time derivative holding X constant, i.e., the time derivative for a given material point. This derivative is also known as the *material* derivative. One can of course take higher time derivatives in similar fashion. For any scalar, vector, or tensor function of a material point and of time, say $\Psi(X, t)$, we can write

$$\Psi(X, t) = \Psi[\chi_t^{-1}(\mathbf{x}), t] = \psi(\mathbf{x}, t) \tag{4.4}$$

Thus ψ is a function of $\mathbf{x}$ and t, and we can write

$$\dot{\psi} = \left(\frac{\partial \psi}{\partial t}\right)_{\mathbf{x}\,=\,\text{const}} + \dot{\mathbf{x}} \cdot \nabla \psi \tag{4.5}$$

using dyadic notation.

For most purposes the labeling of material points X is readily accomplished by associating the points with coordinates $\mathbf{x}$ which they occupy at some reference time t_0. Of course one can also employ a different set of coordinates (X^1, X^2, X^3) to describe the configuration of the body at time t_0. Representing these coordinates by $\mathbf{X}$ we now identify material points by their position with respect to the X^i-coordinates at time t_0. Thus

$$\mathbf{X} = \chi_{t_o}(X) \tag{4.6}$$

and at any time t we can write

$$\mathbf{x} = \chi_t[\chi_{t_o}^{-1}(\mathbf{X})] = \overline{\chi}_{t_o}(\mathbf{X}, t) \tag{4.7}$$

The symbol $\overline{\chi}_{t_o}(\mathbf{X}, t)$ refers to a mapping on $\mathbf{x}$ at time t of the material points which at reference time t_0 occupy positions given by the reference coordinates $\mathbf{X}$.

4.4. CONSERVATION OF MASS

If by χ_t we refer to a mapping of all the material points of some body at time t, then a statement expressing the conservation of mass for the body is

$$\int_{\chi_{t_o}} \rho_0 dV = \int_{\chi_t} \rho dv \tag{4.8}$$

where ρ_0 is the density of the material in the configuration χ_{t_o} at t_0, ρ is the density in the configuration χ_t at any time t, and dV and dv refer to elements of volume in the body at t_0 and t, respectively.[†] Equation (4.8) does not allow for sources or sinks of mass.

We next relate volumes occupied by the material points in the two configurations through the appropriate Jacobian [5]. Referring to coordinates describing the two infinitesimal volumes dV and dv by X^i and x^i, respectively, we can write

$$\mathcal{J} = \det\left[\frac{\partial x^i}{\partial X^j}\right] \tag{4.9}$$

where $dv = \mathcal{J}dV$. Then (4.8) becomes

$$\int_{\chi_{t_o}} [\rho_0 - \rho\mathcal{J}]dV = 0 \tag{4.10}$$

Equation (4.10) must hold for arbitrary configurations χ_{t_o}. Hence we have

$$\rho\mathcal{J} = \rho_0 \tag{4.11}[‡]$$

[†] In accord with the continuum hypothesis we associate a smoothly varying property, the density, with each material point. For some infinitesimal mass dm which occupies volume dv of the body in question, the density may be defined by $dm = \rho dv$. Clearly the density depends upon the configuration.

[‡] This argument, which allows one to proceed from an integral to a differential form of the conservation equation, tacitly assumes that terms in the integrand are continuous.

Taking the material derivative of (4.11)

$$\rho \dot{\mathcal{J}} + \dot{\rho} \mathcal{J} = 0 \tag{4.12}$$

One can show from the definition of a determinant and from some rearrangements [6] that

$$\dot{\mathcal{J}} = \mathcal{J}(\nabla \cdot \dot{\mathbf{x}}) \tag{4.13}$$

Hence

$$\dot{\rho} + \rho \nabla \cdot \dot{\mathbf{x}} = 0 \tag{4.14}$$

We can replace $\dot{\mathbf{x}}$ by $\mathbf{v}$, the usual symbol for velocity of a material point. Rearranging (4.14) one arrives at the familiar continuity equation

$$\frac{\partial \rho}{\partial t} + \nabla \cdot (\rho \mathbf{v}) = 0 \tag{4.15}$$

4.5. CONSERVATION OF LINEAR MOMENTUM

By conservation of linear momentum we mean an expression which relates the change in linear momentum of a body to the forces acting on the body. Our starting point is an equation, analogous to Newton's law of motion for point-mass particles, which states

$$\mathbf{F} = \dot{\mathbf{P}} \tag{4.16}$$

where $\mathbf{F}$ is the net force acting on some body or portion of a body, and $\mathbf{P}$ is the momentum of the body, i.e.,

$$\mathbf{P} = \int_{\chi_t} \rho \dot{\mathbf{x}} dv \tag{4.17}$$

Equation (4.16) is often referred to as *Euler's law of motion*. We find it convenient to subdivide $\mathbf{F}$ into a body force $\mathbf{F}_b$ and a contact force $\mathbf{F}_c$. Working within the continuum hypothesis one supposes that it is possible to define contact forces as those which are exerted on the body at the boundary of the body and the surroundings, and can be described in terms of properties and motion at the boundary. A dilemma is introduced by intermolecular forces which may act over a distance significant with respect to a length scale for the body in question. Nevertheless, one can frequently ignore such complications. In contrast to contact forces, body forces are due to interactions of the whole body with the surroundings. Common examples are the force on the body caused by presence of an external electromagnetic or gravitational field. For our purposes, we usually restrict our formulation to body forces caused by gravitation. If $\mathbf{b}$ is the body force per unit mass exerted on the body at any point, we can write

$$\mathbf{F}_b = \int_{\chi_t} \rho \mathbf{b} dv \tag{4.18}$$

We shall assume that the total force $\mathbf{F}$ is given by

$$\mathbf{F} = \mathbf{F}_b + \mathbf{F}_c \tag{4.19}$$

On the basis of this postulate we can state that at each point on the surface of the body in question there is a vector **s** which represents the force per unit area acting on the body so that

$$\mathbf{F}_c = \int_{\chi_{s_t}} \mathbf{s} \, ds \tag{4.20}$$

where χ_{s_t} refers to the surface of the mapping χ_t. From (4.16) and subsequent equations,

$$\int_{\chi_{s_t}} \mathbf{s} \, ds + \int_{\chi_t} \rho \mathbf{b} \, dv = \frac{d}{dt} \int_{\chi_t} \rho \dot{\mathbf{x}} \, dv \tag{4.21}$$

We wish to form from this initial equation a differential equation that is not dependent on a particular body configuration. To do this one must recognize that **s** is dependent on the orientation of the surface to which it refers, i.e., we postulate $\mathbf{s} = \mathbf{s}(\mathbf{x}, t, \mathbf{n})$, where **n** is the unit normal of surface element ds. (We follow the convention that **n** is positive when it is directed *out* of the body whose surface is given by χ_{s_t}.)

Our first task is to show that **s** is a *linear* vector-valued function of **n**. Though one can demonstrate this fact, known as Cauchy's stress principle, without appeal to rectangular Cartesian coordinates [7], by using them convenience is gained and no generality is lost. Consider the pyramidal volume formed by intersection of a plane with arbitrarily chosen Cartesian coordinate axes (x, y, z), as shown in Fig. 4.2. As we make the tetrahedron

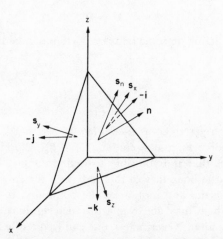

Fig. 4.2. Tetrahedron formed by intersection of plane with rectangular Cartesian axes.

arbitrarily small, the plane with outer unit normal **n** and area δA_n can be identified with a differential element ds of an arbitrary macroscopic surface (Fig. 4.3) which is not necessarily planar but which, at the point in question, has orientation **n**. Note also that, assuming finite values for the integrands, the volume integrals of (4.21) will approach zero as $O(l^3)$, while the surface area will approach zero as $O(l^2)$ when the tetrahedron with characteristic linear dimension l approaches zero size. Then we can write

$$\lim_{l \to 0} \left\{ \frac{1}{O(l^2)} \int_{\chi_{s_t}} \mathbf{s} \, ds \right\} = \mathbf{0} \tag{4.22}$$

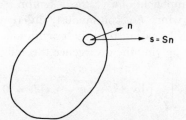

Fig. 4.3. Stress vector at a surface.

Approximating the stress vector on each of the four planes by its mean value,[†] we have

$$\lim_{l \to 0} \left\{ \frac{1}{\delta A_n} [\mathbf{s}_n \delta A_n + \mathbf{s}_{-x} \delta A_x + \mathbf{s}_{-y} \delta A_y + \mathbf{s}_{-z} \delta A_z] \right\} = 0$$

where $\delta A_x = \delta A_n(\mathbf{i} \cdot \mathbf{n})$, etc., so that in the limit we have at the point $\mathbf{x}$ and time t

$$-\mathbf{s}_n = \mathbf{s}_{-x}(\mathbf{i} \cdot \mathbf{n}) + \mathbf{s}_{-y}(\mathbf{j} \cdot \mathbf{n}) + \mathbf{s}_{-z}(\mathbf{k} \cdot \mathbf{n})$$

Taking a similar limit for a body composed of surfaces with normals $\mathbf{n}$ and $-\mathbf{n}$, separated by a small distance l, one readily finds as $l \to 0$ that $\mathbf{s}_n = -\mathbf{s}_{-n}$. Then one can write

$$\mathbf{s}_n = \mathbf{s}_x(\mathbf{i} \cdot \mathbf{n}) + \mathbf{s}_y(\mathbf{j} \cdot \mathbf{n}) + \mathbf{s}_z(\mathbf{k} \cdot \mathbf{n})$$

Since $\mathbf{s}_x$, etc., represent stress vectors on planes with *fixed* unit normal those vectors cannot be explicit functions of the orientation $\mathbf{n}$, so we conclude that $\mathbf{s}_n$ is a *linear* vector function of $\mathbf{n}$. We can express this in a general way at some position $\mathbf{x}$ and time t by

$$\mathbf{s}(\mathbf{x}, t, \mathbf{n}) = \mathbf{S}(\mathbf{x}, t)\mathbf{n} \tag{4.23}$$

$\mathbf{S}$ is a *linear transformation* which, when operating on the outer unit surface normal $\mathbf{n}$ at the point in question, generates a contact force per unit area equal to $\mathbf{s}$ (Fig. 4.3). $\mathbf{S}$ is known as the stress tensor, and is seen to be *independent* of $\mathbf{n}$.

Now let us proceed with the formulation of a differential equation for linear momentum. Returning to (4.21) we note that the right-hand side may be rewritten by referring the quantities inside the integral to a reference configuration at t_0. Then

$$\frac{d}{dt} \int_{\chi_t} \rho \dot{\mathbf{x}} dv = \frac{d}{dt} \int_{\chi_{t_0}} \rho_0 \dot{\mathbf{x}} dV = \int_{\chi_{t_0}} \rho_0 \ddot{\mathbf{x}} dV = \int_{\chi_t} \rho \ddot{\mathbf{x}} dv \tag{4.24}$$

We also make use of the divergence theorem [8, p. 127], which may be written in component form as

$$\int_v F_{,i} dv = \int_s F n_i ds \tag{4.25}[‡]$$

[†] One can formally invoke the mean value theorem of integral calculus to arrive at (4.22) and subsequent equations [7].

[‡] This important theorem is derived in most books on vector calculus. The reader who has not encountered a proof which includes (4.25) is referred to Aris [9].

where F is any scalar, or component of a vector or tensor, v represents a volume bounded by the surface s and within which F is continuously differentiable, and n_i is a component of the outer surface normal. Equating $\mathbf{b}$ to the gravitational force $\mathbf{g}$, substituting (4.23) and (4.24) into (4.21) and applying the divergence theorem, we obtain

$$\int_{\chi_t} [\operatorname{div} \mathbf{S}^T + \rho\mathbf{g} - \rho\ddot{\mathbf{x}}] dv = 0$$

where the contravariant components of div $\mathbf{S}^T$ are $s_{,j}^{ij}$. Following the argument of the previous section we finally obtain the differential form of the momentum equation

$$\operatorname{div} \mathbf{S}^T + \rho\mathbf{g} = \rho\ddot{\mathbf{x}} \tag{4.26}$$

This equation, which is known as *Cauchy's first law of motion*, will frequently be referred to simply as the *equation of motion*.

4.6. PHYSICAL INTERPRETATION OF THE STRESS TENSOR

From the definition of the stress tensor in (4.23) we can assign direct physical significance to the components of $\mathbf{S}$. This can be done in an illuminating way by considering an elemental cube, shown in Fig. 4.4, located at the position of interest. From (4.23) we see that the contact force on the face with outer surface normal in the positive x direction is

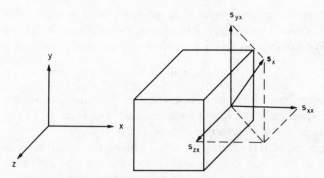

Fig. 4.4. Components of stress tensor on an elemental cube.

$$\mathbf{s}_x \, \Delta y \Delta z = \mathbf{S} \, \mathbf{i} \, \Delta y \Delta z = [\mathbf{i}\, s_{xx} + \mathbf{j}\, s_{yx} + \mathbf{k}\, s_{zx}] \Delta y \Delta z \tag{4.27}$$

where $\mathbf{s}_x$ is the net contact force per unit area on the face under consideration, and $\mathbf{i}, \mathbf{j}, \mathbf{k}$ are the usual unit vectors of elementary vector analysis. Thus s_x can be treated as a vector with components s_{xx}, s_{yx}, s_{zx} in the x-, y-, and z-directions, respectively. A similar interpretation can be placed on the other components of $\mathbf{S}$ to associate them with $\mathbf{s}_y$ and $\mathbf{s}_z$.

4.7. SYMMETRY OF THE STRESS TENSOR

Elementary texts frequently prove symmetry of the stress tensor on the basis of conservation of angular momentum for an elemental volume. Actually, these derivations are misleading since they usually do not state that their validity depends upon absence of extraneous couples such as body couples or couple stresses. One can show [8, p. 136] from the continuity equation and the equation for conservation of linear momentum that

$$\frac{d}{dt}\int_V \rho \mathbf{r} \times \mathbf{v}\,dV = \int_V \rho(\mathbf{r} \times \mathbf{b})\,dV + \int_S \mathbf{r} \times \mathbf{s}\,dS + \int_V \mathbf{a}\,dV \qquad (4.28)^\dagger$$

where $\mathbf{a}$ is an axial vector, the components of which are

$$a^i = e^{ijk}s_{jk} \qquad (4.29)$$

If $\int_V \mathbf{a}\,dV$ is zero for arbitrary volumes (4.29) indicates that $\mathbf{S}$ is symmetric. We shall assume in this book, unless otherwise stated, that the stress tensor is symmetric. Then (4.28) becomes the familiar statement for conservation of angular momentum in a system with no extraneous couples.

4.8. BALANCE OF ENERGY

With the exception, in later chapters, of some applications to heat transport we shall not be concerned explicitly with the equation of energy in this book. That is partly because of the fact that the basic ideas concerning thermodynamics of systems undergoing large deformations are still in development. When our purposes do require an energy equation, we shall be content with the usual expressions for transport of heat through a medium which is subjected to a gradient of temperature and in which thermal energy is being created because of viscous dissipation. For later reference we record here the equation which results from writing a balance for the various forms of energy which can be exchanged across and within a differential volume in a single-component system [10].

$$\rho \frac{dE}{dt} = -\nabla \cdot \mathbf{q} + \mathbf{S}:\nabla\mathbf{v} \qquad (4.30)$$

where E is the internal energy per unit mass, $\mathbf{q}$ is the rate of heat transport by conduction, and $\mathbf{S}:\nabla\mathbf{v}$ represents the double contraction $s^{ij}v_{j,i}$.

REFERENCES

1. Hirschfelder, J. O., Curtiss, C. F. and Bird, R. B., *Molecular Theory of Gases and Liquids*, ch. 7. (New York: Wiley, 1954).
2. Truesdell, C., *The Elements of Continuum Mechanics* (New York: Springer-Verlag New York, 1965).
3. Truesdell, C. and Noll, W., The non-linear field theories of mechanics, in *Handbuch der Physik*, Band III/3. (S. Flügge, ed.) (Berlin: Springer-Verlag, 1965).
4. Leigh, D. C., *Nonlinear Continuum Mechanics* (New York: McGraw-Hill, 1968).
5. Aris, R., *Vectors, Tensors, and the Basic Equations of Fluid Mechanics*, p. 50. (Englewood Cliffs, New Jersey: Prentice-Hall, 1962).
6. *Ibid.*, pp. 83–84.
7. Eringen, A. C., *Nonlinear Theory of Continuous Media*, p. 97 (New York: McGraw-Hill, 1962).

† We no longer reserve V for the volume in a reference configuration. Henceforth V and S refer to a volume and enclosing surface in an arbitrary configuration.

8. Serrin, J., Mathematical principles of classical fluid mechanics, in *Handbuch der Physik*, Band VIII/1, (S. Flügge, ed.), (Berlin: Springer-Verlag, 1959).
9. Aris, R., *op. cit.*, p. 58.
10. Bird, R. B., Stewart, W. E. and Lightfoot, E. N., *Transport Phenomena*, p. 314. (New York: Wiley, 1960).

PROBLEMS

4.1. Using the definitions of Section 4.4, show that

$$dv = \mathcal{J}dV$$

Hint: Apply the fact that

$$dv = d\mathbf{x}_{(1)} \cdot (d\mathbf{x}_{(2)} \times d\mathbf{x}_{(3)})$$

where the $d\mathbf{x}_{(i)}$ are elements of length directed along the natural bases $\mathbf{e}_{(i)} = \partial \mathbf{r}/\partial x^i$ so that

$$d\mathbf{x}_{(i)} = \mathbf{e}_{(i)}dx^i \quad \text{(no sum)}$$

Also note that

$$\mathcal{J} = e_{ijk} \frac{\partial x^i}{\partial X^1} \frac{\partial x^j}{\partial X^2} \frac{\partial x^k}{\partial X^3}$$

4.2. Verify (4.26). Use (4.21), (4.23), (4.24), and the divergence theorem.

4.3. Derive (4.28) from (4.26) and (4.11).

Note: Begin by integrating (4.26) over an arbitrary material volume. Recall from Section (4.4) that the volume integral

$$\frac{d}{dt} \int_V \rho \mathbf{r} \times \mathbf{v}dV$$

can be shifted to a reference configuration

$$\frac{d}{dt} \int_{V_0} \rho \mathbf{r} \times \mathbf{v}\mathcal{J}dV_0$$

Use this, along with (4.11), to show that

$$\frac{d}{dt} \int_V \rho \mathbf{r} \times \mathbf{v}dV = \int_V \rho \mathbf{r} \times \frac{d\mathbf{v}}{dt} dV$$

Next, express

$$\int_V \mathbf{r} \times \nabla \cdot \mathbf{S}^T dV$$

in index notation, and show that, in Cartesian tensors,

$$\frac{\partial}{\partial x_l} [e_{ijk}x_j S_{lk}^T] = e_{ijk}S_{kj} + e_{ijk}x_j \frac{\partial S_{lk}^T}{\partial x_l}$$

From the divergence theorem one can then obtain the last two terms in (4.28).

CHAPTER 5

DEFORMATION

5.1. INTRODUCTION

Elementary expositions of elasticity theory are often based upon the requirement that the change in distance between two neighboring material points in a body undergoing deformation must be infinitesimal with respect to the original distance between the two points. This simplification leads to a unique measure of strain which can be expressed as the change in distance between two neighboring points in a body undergoing deformation divided by either the initial or the final distance between the points. Since the deformation is infinitesimal, the choice of reference state as initial or final position is immaterial. This measure of strain is readily seen to be the limiting case of several different measures of strain, each of which is distinct from the other for *finite* deformations. For a discussion of different measures of strain the reader is referred to Truesdell and Toupin [1].

In this chapter we develop appropriate means for describing deformations that are not necessarily small. It is our goal to develop a convenient description of changes in relative position of neighboring points and the *rate* at which such spatial changes vary with time. The results form a class of physically important quantities which are essential ingredients for the description of material behavior in motion.

5.2. DEFORMATION GRADIENT

In this section we define a measure of the relative motion of two neighboring material points. The fact that we are interested in *relative* motion is important. If the material points of a body undergo a motion composed of uniform translation, the relative position of material points is unchanged during the motion and no deformation (or rotation) occurs.

Consider two neighboring material points X and $X + dX$ in Fig. 5.1. The point X is located with respect to some arbitrary fixed origin by the position vector **P**. The point X has coordinates **X**, where **X** represents (X^1, X^2, X^3) of some arbitrary coordinate system. Hence a neighboring point $X + dX$ has coordinates $\mathbf{X} + d\mathbf{X}$ $(X^1 + dX^1, X^2 + dX^2, X^3 + dX^3)$, and the two points are connected by the vector $d\mathbf{X}$. The body containing X and $X + dX$ is now deformed as shown in Fig. 5.1 so that after the deformation X is located by position vector

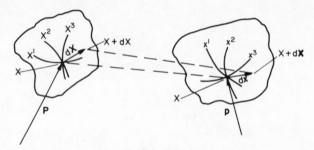

Fig. 5.1. Deformation of two neighboring points.

p with respect to an arbitrary fixed origin. In some **x**-coordinate system the vector joining material points X and $X + dX$ after deformation is $d\mathbf{x}$.

The *deformation gradient* is defined as a linear transformation operator **F** which effects a relative mapping of material points from **X** onto **x**:

$$d\mathbf{x} = \mathbf{F}d\mathbf{X} \tag{5.1}$$

The statement that **F** is a linear transformation is a consequence of limiting ourselves to deformations $\mathbf{x} = \overline{\chi}_{t_o}(\mathbf{X}, t)$ which are differentiable at the point of interest. One can then arrive at (5.1) through a definition of differentiability [2]. Alternatively, components of $\mathbf{x} + d\mathbf{x}$ can be related to $\mathbf{X} + d\mathbf{X}$ through a Taylor series expansion. Then component equations can be combined to form (5.1) [3]. The former procedure, though perhaps less familiar than the latter, is nevertheless very direct. Following Coleman *et al.* [2], we define differentiability of the function $\overline{\chi}_{t_o}$ at position **X** by the requirement that

$$\overline{\chi}_{t_o}(\mathbf{X} + \mathbf{N}, t) = \overline{\chi}_{t_o}(\mathbf{X}, t) + \mathbf{F}\mathbf{N} + \mathbf{o}(\mathbf{N})$$

where **F** is independent of **N** and

$$\lim_{|\mathbf{N}| \to 0} \frac{\mathbf{o}(\mathbf{N})}{|\mathbf{N}|} = \mathbf{0}$$

Thus we are saying that for sufficiently small $|\mathbf{N}|$ the deformation $\overline{\chi}_{t_o}(\mathbf{X}, t)$ approaches a homogeneous deformation. It is the assumption of differentiability or, in the present context, the assumption that in some small region $d\mathbf{X}$ about **X** the deformation is homogeneous, that leads to (5.1).

The component form of (5.1) is written as

$$dx^i = f^i{}_\alpha dX^\alpha \tag{5.2}$$

The components of **F** are of special interest since they are associated with two different base vectors. We have followed the convention of using a Greek index for the **X** coordinate and a Latin index for **x**. Since in writing (5.2) we have assumed a differentiable transformation $x^i = x^i(X^\alpha, t)$, we can immediately write

$$dx^i = \frac{\partial x^i}{\partial X^\alpha} dX^\alpha \tag{5.3}$$

whereupon we have

$$f^i{}_\alpha = \frac{\partial x^i}{\partial X^\alpha} \tag{5.4}$$

From (4.7) one can also write

$$f^i{}_\alpha = \frac{\partial}{\partial X^\alpha} [\overline{\chi}_{t_o}(\mathbf{X}, t)]^i \tag{5.5}$$

The rule of transformation of $f^i{}_\alpha$ is readily found by considering the same deformation as shown in Fig. 5.1 but described by locating the material points with coordinates $\mathbf{Y}$ and $\mathbf{Y} + d\mathbf{Y}$ before deformation and with $\mathbf{y}$ and $\mathbf{y} + d\mathbf{y}$ after deformation. The deformation gradient $\mathbf{K}(\mathbf{Y}, \mathbf{y})$ is then, according to (5.1),

$$d\mathbf{y} = \mathbf{K}d\mathbf{Y} \tag{5.6}$$

and

$$k^i{}_\alpha = \frac{\partial y^i}{\partial Y^\alpha} \tag{5.7}$$

Assuming we have smooth transformations $X \to Y$ and $x \to y$, one can write

$$\frac{\partial y^j}{\partial x^i} dx^i = k^j{}_\beta \frac{\partial Y^\beta}{\partial X^\alpha} dX^\alpha \tag{5.8}$$

Then it is a simple matter to prove that

$$f^i{}_\alpha = \frac{\partial Y^\beta}{\partial X^\alpha} \frac{\partial x^i}{\partial y^j} k^j{}_\beta \tag{5.9}$$

and the components of $\mathbf{F}$ are seen to transform as tensors. Since (5.9) arises from transformation of two *pairs* of coordinates and their corresponding bases $\mathbf{F}$ is known as a *double tensor* [4].

5.3. POLAR DECOMPOSITION THEOREM

We state without proof the *polar decomposition theorem* of linear algebra (see, for example, [5, 6]) as it applies to real vector spaces: any tensor (linear transformation) $\mathbf{F}$ with $\det(\mathbf{F}) \neq 0$ can be uniquely decomposed into the product of two tensors in two different ways; namely,

$$\mathbf{F} = \mathbf{QU} \tag{5.10}$$

$$\mathbf{F} = \mathbf{VQ} \tag{5.11}$$

where $\mathbf{Q}$ is an orthogonal tensor, and $\mathbf{U}$ and $\mathbf{V}$ are positive definite tensors.[†]
If $\mathbf{A}$ is a real positive definite tensor and λ_i are the principal values of the equation

$$(\mathbf{A} - \lambda_i \mathbf{I})\mathbf{u}_i = \mathbf{0} \quad \text{(no sum)} \tag{5.14}$$

we can write

---

† A positive tensor $\mathbf{A}$ is a symmetric tensor which, when operating on an arbitrary nonzero vector $\mathbf{u}$, yields

$$(\mathbf{A}\mathbf{u}) \cdot \mathbf{u} \geqslant 0 \tag{5.12}$$

If $\mathbf{A}$ is positive definite,

$$(\mathbf{A}\mathbf{u}) \cdot \mathbf{u} > 0 \tag{5.13}$$

$$(\mathbf{Au}_i) \cdot \mathbf{u}_i = \lambda_i(\mathbf{u}_i)^2 > \mathbf{0} \quad \text{(no sum)} \tag{5.15}$$

which indicates that the principal values of a positive definite tensor are positive. Then, since we know from (3.101) that

$$I_3 = \lambda_1 \lambda_2 \lambda_3 = \det(\mathbf{A})$$

we have

$$\det(\mathbf{A}) > 0 \tag{5.16}$$

We now apply these facts to the deformation gradient $\mathbf{F}$. For any physically meaningful reference frame $\mathbf{X}$, we expect the transformation $\mathbf{x} = \overline{\boldsymbol{\chi}}_{t_0}(\mathbf{X}, t)$ to be invertible, i.e., there is a one–one relation between $\mathbf{x}$ and $\mathbf{X}$. Then we know that $\mathbf{F}$ in (5.1) is nonsingular and consequently $\det(\mathbf{F}) \neq 0$. If we choose some physically realizable reference configuration, we then must be able to go from $\mathbf{x}$ to $\mathbf{X}$ or vice versa without going through a condition $\det(\mathbf{F}) = 0$. This implies that $\det(\mathbf{F})$ is always positive or always negative. We shall assume that a reference configuration has been chosen such that $\det(\mathbf{F}) > 0$ (cf. Section 5.6). Applying the polar decomposition theorem (5.10) we have

$$\det(\mathbf{F}) = \det(\mathbf{Q}) \det(\mathbf{U}) \tag{5.17}$$

From (5.16) we know $\det(\mathbf{U}) > 0$. Hence $\det(\mathbf{Q}) > 0$. However, because of the orthogonality of $\mathbf{Q}$ we must have

$$\det(\mathbf{Q}) = +1 \tag{5.18}$$

meaning that $\mathbf{Q}$ signifies a rotation. We then rewrite the polar decomposition theorem as it applies to the deformation gradient:

$$\mathbf{F} = \mathbf{RU} = \mathbf{VR} \tag{5.19}$$

where $\mathbf{R}$ is a rotation tensor. We note also that $\mathbf{U}$ and $\mathbf{V}$ are related by an equation of the form (3.71): $\mathbf{V} = \mathbf{RUR}^{-1} = \mathbf{RUR}^T$. Note that the requirement $\det(\mathbf{F}) \neq 0$ is necessary for the polar decomposition theorem, as stated above, to be valid.

5.4. THE STRETCH TENSOR

Application of results from Section 3.17 to eqn. (5.19) provides important physical significance to the deformation gradient. It is apparent that $\mathbf{U}$ and $\mathbf{V}$ have the same principal values, and that their principal directions are related by the rotation

$$\mathbf{v}_i = \mathbf{Ru}_i \tag{5.20}$$

$\mathbf{v}_i$ and $\mathbf{u}_i$ being the ith principal directions of $\mathbf{V}$ and $\mathbf{U}$, respectively.

The principal values of $\mathbf{U}$ have an interesting physical interpretation.

$$(\mathbf{U} - \lambda_i \mathbf{I})\mathbf{u}_i = \mathbf{0} \quad \text{(no sum)} \tag{5.21}$$

If we consider an elemental length $(d\mathbf{X})_i$ along the principal direction $\mathbf{u}_i$ to be operated upon by $\mathbf{U}$, the result is

$$\mathbf{U}(d\mathbf{X})_i = \lambda_i(d\mathbf{X})_i = (d\mathbf{w})_i \quad \text{(no sum)} \tag{5.22}$$

Thus the result is a vector in the same direction as $(d\mathbf{X})_i$ but changed in magnitude by the factor λ_i. Operating on this new vector $(d\mathbf{w})_i$ with $\mathbf{R}$ yields a vector $\mathbf{R}(d\mathbf{w})_i$ which, from (3.77), has the same magnitude as $(d\mathbf{w})_i$ but has been rotated through some angle by $\mathbf{R}$. This pair of operations can be identified with

$$d\mathbf{x} = \mathbf{F}d\mathbf{X} = \mathbf{R}\mathbf{U}d\mathbf{X} \tag{4.1}$$

when $d\mathbf{X}$ is taken along a principal direction of $\mathbf{U}$. Thus $\mathbf{U}(d\mathbf{X})_i$ corresponds to a stretching along the ith principal direction of the element $(d\mathbf{X})_i$ into a new element $(d\mathbf{x})_i$. This stretching is then followed by a rotation of $(d\mathbf{x})_i$ which changes its direction but not its magnitude. The dual process is shown schematically in Fig. 5.2.

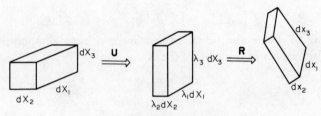

Fig. 5.2. Stretch and rotation.

Alternatively, the same result can be obtained by beginning with a rotation $\mathbf{R}$ followed by a stretching $\mathbf{V}$. Because of this interpretation $\mathbf{U}$ and $\mathbf{V}$ are customarily designated the *right* and *left stretch tensor*, respectively.

5.5. THE CAUCHY–GREEN STRAIN TENSOR

Although the tensor $\mathbf{U}$ is a perfectly legitimate measure of deformation, it is frequently more convenient to work with

$$\mathbf{C} = \mathbf{U}^2 \tag{5.23}$$

which is designated as the *right Cauchy–Green strain tensor*. $\mathbf{C}$ is related to the deformation gradient tensor since

$$\mathbf{F} = \mathbf{R}\mathbf{U} \tag{5.19}$$

$$\mathbf{F}^T = (\mathbf{R}\mathbf{U})^T = \mathbf{U}^T\mathbf{R}^{T\dagger} \tag{5.24}$$

Then

$$\mathbf{F}^T\mathbf{F} = \mathbf{U}^T\mathbf{U} = \mathbf{U}^2 = \mathbf{C} \tag{5.25}$$

which follows from the symmetry of $\mathbf{U}$. Similarly we define the *left Cauchy–Green strain tensor* by

$$\mathbf{B} = \mathbf{V}^2 = \mathbf{F}\mathbf{F}^T \tag{5.26}$$

We shall have frequent need for computation of components of $\mathbf{C}$. These may be expressed from the component representation of $\mathbf{F}^T\mathbf{F}$.

---

† One can easily show, for example by considering components or directly from (3.67), that
$$(\mathbf{AB})^T = \mathbf{B}^T\mathbf{A}^T$$

In (3.3) the discussion of tensor products was limited to those formed from a basis with itself or with its reciprocal. However, the double tensor **F** possesses components which are associated with base vectors for the **X** and **x** coordinate systems. Consequently care must be exercised when forming products of components of **F**. We can only contract indices which refer to the set of base vectors associated with **x** alone or **X** alone. Put in a more utilitarian way, we cannot contract Greek indices with Latin indices. Let us form components of **C** which are associated with base vectors for the **X**-coordinates. Hence we wish to sum over Latin indices. However, since the Latin index appears as a superscript in the component of the deformation gradient shown in (5.4) we must form an associated tensor component to give a subscript Latin index. Thus

$$c_{\alpha\beta} = (f^T)_{\alpha i} f^i{}_\beta = (f^T)_\alpha{}^j g_{ji} f^i{}_\beta = f^j{}_\alpha f^i{}_\beta g_{ji} \tag{5.27}$$

Indices on $c_{\alpha\beta}$ can of course also be raised and lowered using components of the metric tensor $h_{\alpha\beta}(\mathbf{X})$ with respect to the basis of the **X**-coordinate system.

$$c^\alpha{}_\beta = c_{\gamma\beta} h^{\gamma\alpha} \tag{5.28}$$

5.6. THE RELATIVE DEFORMATION GRADIENT

We can avoid some of the complications of multiple sets of base vectors if we choose a particular set of reference coordinates. In (4.1) we wrote an expression for the mapping at time t of material points X onto coordinates **x**:

$$\mathbf{x} = \chi_t(X) \tag{4.1}$$

Suppose we denote the coordinates of material points X in the coordinate system **x** at time τ by ξ.

$$\xi = \chi_\tau(X) \tag{5.29}$$

But from (4.2) we can write

$$\xi = \chi_\tau(\chi_t^{-1}(\mathbf{x})) \tag{5.30}$$

We rewrite the right-hand side of (5.30) as a new mapping $\chi_{(t)}$. Then

$$\xi = \chi_{(t)}(\mathbf{x}, \tau) \tag{5.31}$$

In words (5.31) states that ξ indicates the coordinates of a material point at time τ which at time t was at **x**. Now we define the *relative deformation gradient* $\mathbf{F}_{(t)}$ by

$$d\xi = \mathbf{F}_{(t)} d\mathbf{x} \tag{5.32}$$

which is a special case of (5.1). In (5.32) the reference configuration is denoted by **x**, the position of X at time t. For a given material point $\mathbf{F}_{(t)} = \mathbf{F}_{(t)}(\tau)$.[†] Clearly,

$$\mathbf{F}_{(t)}(t) = \mathbf{I} \tag{5.33}$$

In component form

$$f_{(t)}{}^i{}_j = \frac{\partial \xi^i}{\partial x^j} \tag{5.34}$$

[†] The dependence on **x** is understood.

By analogy to (5.23) and (5.27) we can define the *right relative Cauchy–Green strain tensor*

$$\mathbf{C}_{(t)}(\tau) = (\mathbf{U}_{(t)}(\tau))^2 = \mathbf{F}_{(t)}^T(\tau)\mathbf{F}_{(t)}(\tau) \tag{5.35}$$

with covariant components

$$c_{(t)ij} = (f_{(t)}{}^T)_{ik} f_{(t)}{}^k{}_j = f_{(t)ki} f_{(t)}{}^k{}_j = f_{(t)}{}^l{}_i f_{(t)}{}^k{}_j \gamma_{lk} \tag{5.36}$$

where γ_{lk} are components of the metric tensor defined in terms of the basis with respect to which components of ξ are referred. At $\tau = t$, $\gamma_{lk} = g_{lk}$.

It is easily shown that $\det(\mathbf{F}_{(t)}(\tau)) > 0$. From Section 4.3 we know that $\det(\mathbf{F}) > 0$ or $\det(\mathbf{F}) < 0$. The former is true for $\det(\mathbf{F}_{(t)}(\tau))$ since $\det(\mathbf{F}_{(t)}(t)) = 1$. This result also follows from Section 4.4 if we write, by analogy to (4.9),

$$\mathcal{J}_{(t)}(\tau) = \det\left[\frac{\partial \xi^i}{\partial x^j}\right] = \det(\mathbf{F}_{(t)}(\tau))$$

Since $\mathcal{J}_{(t)}(t) = 1$, we have from (4.11)

$$\rho_{(t)} = \rho_{(\tau)}\mathcal{J}_{(t)}(\tau)$$

Thus $\mathcal{J}_{(t)}(\tau) > 0$.

5.7. TIME DIFFERENTIATION

In subsequent sections time derivatives of the deformation gradient and relative deformation gradient will be required. We now digress briefly to develop some expressions for material time differentiation.

In Section 4.3 the concept of a *material* time derivative was introduced. For some tensor $\mathbf{A}(\mathbf{x}, t)$, of arbitrary order, the material derivative is

$$\dot{\mathbf{A}} = \left(\frac{d\mathbf{A}}{dt}\right)_{X=\text{const}} \tag{5.37}$$

Suppose for the moment that $\mathbf{A}$ is a vector $\mathbf{u}(\mathbf{x}, t)$, where here $\mathbf{x}$ and t refer to independent variables of space and time. We wish to compute the components $\dot{u}^i$ such that they have proper tensor character [7] when referred to base vectors which are time independent. Thus $\dot{u}^i$ is defined by

$$\dot{\mathbf{u}} = \dot{u}^i \mathbf{e}_i$$

By the chain rule

$$\dot{\mathbf{u}} = \frac{\partial \mathbf{u}}{\partial t} + \frac{\partial \mathbf{u}}{\partial x^j}\dot{x}^j \tag{5.38}$$

Since the base vectors may be functions of position but not of time,

$$\frac{\partial \mathbf{u}}{\partial t} = \frac{\partial u^i}{\partial t}\mathbf{e}_i$$

Also, from (3.38)

$$\frac{\partial \mathbf{u}}{\partial x^j} = u^i{}_{,j}\, \mathbf{e}_i$$

Hence

$$\dot{\mathbf{u}} = \left[\frac{\partial u^i}{\partial t} + u^i{}_{,j}\, \dot{x}^j\right] \mathbf{e}_i \tag{5.39}$$

and

$$\dot{u}^i = \frac{\partial u^i}{\partial t} + u^i{}_{,j}\, v^j \tag{5.40}$$

where $v^j = \dot{x}^j$ is the contravariant jth component of the velocity.

In the same way one can determine the tensorially correct form for components of $\dot{\mathbf{A}}$ when $\mathbf{A}$ is a double tensor of arbitrary order and $\mathbf{A} = \mathbf{A}(\mathbf{x}, \mathbf{X}, t)$. One finds [8]

$$\frac{d}{dt} A\,{}^{\alpha_1 \ldots \alpha_k\, a_1 \ldots a_l}_{\ \ \beta_1 \ldots \beta_m\, b_1 \ldots b_n} \equiv \dot{A}\,{}^{\alpha_1 \ldots \alpha_k\, a_1 \ldots a_l}_{\ \ \beta_1 \ldots \beta_m\, b_1 \ldots b_n}$$

$$= \left[\frac{\partial}{\partial t} A\,{}^{\alpha_1 \ldots \alpha_k\, a_1 \ldots a_l}_{\ \ \beta_1 \ldots \beta_m\, b_1 \ldots b_n}\right]_{\mathbf{x}=\text{const}}_{\mathbf{X}=\text{const}} + A\,{}^{\alpha_1 \ldots \alpha_k\, a_1 \ldots a_l}_{\ \ \beta_1 \ldots \beta_m\, b_1 \ldots b_n,\, j}\,\dot{x}^j \tag{5.41}$$

In applications one of course would expect to have a relation between $\mathbf{x}$ and $\mathbf{X}$ such as $\mathbf{x} = \overline{\chi}_{t_0}(\mathbf{X}, t)$. Then we may write $\mathbf{A} = \mathbf{A}(\mathbf{X}, t)$ or $\mathbf{A} = \mathbf{A}(\mathbf{x}, t)$, in which case the time derivative on the right-hand side of (5.41) becomes $(\partial/\partial t)_{\mathbf{x}=\text{const}}$ or $(\partial/\partial t)_{\mathbf{x}=\text{const}}$, respectively.

5.8. TIME DERIVATIVES OF THE DEFORMATION GRADIENT

Since this book is primarily concerned with fluid-like materials, we are especially interested not in the deformation but rather in the *rate* at which deformation changes. In this section we develop some general schemes for time differentiation of quantities of interest.

For example, consider the component form of $\dfrac{d}{dt}\,(\mathbf{F}(\mathbf{X}, t))$. From (5.41)

$$\frac{d}{dt}(f^i{}_\alpha) = \frac{d}{dt}\left(\frac{\partial x^i}{\partial X^\alpha}\right) = \left[\frac{\partial}{\partial t}\left(\frac{\partial x^i}{\partial X^\alpha}\right)\right]_{\mathbf{X}=\text{const}} + \begin{Bmatrix} i \\ ml \end{Bmatrix} \frac{\partial x^l}{\partial X^\alpha}\,\dot{x}^m$$

$$= \frac{\partial \dot{x}^i}{\partial X^\alpha} + \begin{Bmatrix} i \\ lm \end{Bmatrix} \frac{\partial x^l}{\partial X^\alpha}\,\dot{x}^m$$

$$= \left[\frac{\partial \dot{x}^i}{\partial x^l} + \left\{ \begin{matrix} i \\ lm \end{matrix} \right\} \dot{x}^m \right] \frac{\partial x^l}{\partial X^\alpha}$$

$$= \dot{x}^i_{;l} \frac{\partial x^l}{\partial X^\alpha} = v^i_{,l} \frac{\partial x^l}{\partial X^\alpha} \tag{5.42}$$

In particular, it is desired to take the derivative of $\mathbf{F}_{(t)}(\mathbf{x}, \tau)$, the relative deformation gradient, with respect to τ, holding t constant. This is immediately recognized as the material derivative since we are following a given material point by holding t (and therefore $\mathbf{X}$) constant. We wish to evaluate this derivative at $\tau = t$. The tensorially correct component form is readily inferred from (5.42) by replacing x^i by ξ^i and X^α by x^j. Then

for components of $\left[\dfrac{d}{d\tau} (\mathbf{F}_{(t)}(\mathbf{x}, \tau)) \right]_{\tau=t} \equiv \overline{[\mathbf{F}_{(t)}(\mathbf{x}, \tau)]}^{\,\cdot}_{\tau=t}$ we have

$$\left[\frac{d}{d\tau} (f^i_{(t)j}) \right]_{\tau=t} = \left[\dot{\xi}^i_{\ l} \frac{\partial \xi^l}{\partial x^j} \right]_{\tau=t} = \dot{x}^i_{,j} = v^i_{,j} \tag{5.43}$$

where v^i is the ith component of velocity of that material point which at time t is at co-ordinate $\mathbf{x}$. Symbolically one can write

$$\overline{[\mathbf{F}_{(t)}(\mathbf{x}, \tau)]}^{\,\cdot}_{\tau=t} \equiv \dot{\mathbf{F}}_{(t)}(t) = \nabla \dot{\mathbf{x}} = \nabla \mathbf{v} \tag{5.44}†$$

Thus $\dot{\mathbf{F}}_{(t)}(t)$ (the dependence on $\mathbf{x}$ being understood) is exactly what we call in elementary treatments the *velocity gradient*. We denote it here by $\mathbf{L}(t) \equiv \dot{\mathbf{F}}_{(t)}(t)$. This idea can be generalized to give the acceleration gradient

$$\mathbf{L}_2(t) = \nabla \ddot{\mathbf{x}} \tag{5.45}$$

or in general

$$\mathbf{L}_n(t) = \overset{(n)}{\mathbf{F}}_{(t)}(t) = \nabla \overset{(n)}{\mathbf{x}} \tag{5.46}$$

where

$$\mathbf{L}_0 = \mathbf{I}$$
$$\mathbf{L}_1 \equiv \mathbf{L}$$

---

† $\nabla \mathbf{v}$ is an equivalent expression for the tensor $v^i_{,j}\, \mathbf{e}_i \otimes \mathbf{e}^j$. Note that, according to our convention, a matrix representation of $\nabla \mathbf{v}$ with respect to rectangular Cartesian coordinates will have the form

$$\begin{bmatrix} \dfrac{\partial v^1}{\partial x^1} & \dfrac{\partial v^1}{\partial x^2} & \dfrac{\partial v^1}{\partial x^3} \\[2mm] \dfrac{\partial v^2}{\partial x^1} & \dfrac{\partial v^2}{\partial x^2} & \dfrac{\partial v^2}{\partial x^3} \\[2mm] \dfrac{\partial v^3}{\partial x^1} & \dfrac{\partial v^3}{\partial x^2} & \dfrac{\partial v^3}{\partial x^3} \end{bmatrix}$$

which is in contrast to the ordering implied by the dyadic notation $\nabla \mathbf{v}$. This convention, though widely used, is not universal.

5.9. RATE OF STRETCHING AND RATE OF SPIN

We now relate the quantities introduced in the last sections to well-known measures of rate of strain commonly used in fluid mechanics. We begin by applying the polar decomposition theorem (5.19) to the relative deformation gradient:

$$\mathbf{F}_{(t)}(\tau) = \mathbf{R}_{(t)}(\tau)\,\mathbf{U}_{(t)}(\tau) \tag{5.47}$$

At $\tau = t$ we know $\mathbf{F}_{(t)}(t) = \mathbf{I}$. From (5.47) it is then readily seen that

$$\mathbf{U}_{(t)}(t) = \mathbf{R}_{(t)}(t) = \mathbf{I} \tag{5.48}$$

Taking the derivative with respect to τ of (5.47) and evaluating at $\tau = t$, we find

$$\dot{\mathbf{F}}_{(t)}(t) = \mathbf{L}(t) = \dot{\mathbf{R}}_{(t)}(t) + \dot{\mathbf{U}}_{(t)}(t) \tag{5.49}$$

and note that the velocity gradient $\mathbf{L}$ can be composed from the sum of a *rate of spin* and a *rate of stretching* tensor. Furthermore, since $\mathbf{U}_{(t)}$ is symmetric, we know that $\dot{\mathbf{U}}_{(t)}(t)$ must also be symmetric. Also, since

$$\mathbf{R}_{(t)}(t)[\mathbf{R}_{(t)}(t)]^T = \mathbf{I} \tag{5.50}$$

we find that

$$\dot{\mathbf{R}}_{(t)}(t) + \overline{[\mathbf{R}_{(t)}(t)]^T} = \mathbf{0} \tag{5.51}$$

Equation (5.51) is the definition of an *antisymmetric* tensor.

Because of the importance of the rate of spin and rate of stretching tensors we give them the special symbols $\mathbf{W}$ and $\mathbf{D}$, respectively. Furthermore, from the symmetry and antisymmetry properties cited above one can write from (5.49):

$$\mathbf{W} = \dot{\mathbf{R}}_{(t)}(t) = \frac{1}{2}\,(\mathbf{L} - \mathbf{L}^T) \tag{5.52}$$

$$\mathbf{D} = \dot{\mathbf{U}}_{(t)}(t) = \frac{1}{2}\,(\mathbf{L} + \mathbf{L}^T) \tag{5.53}$$

5.10. PHYSICAL INTERPRETATION

A physical meaning of $\mathbf{D}$ and $\mathbf{W}$ is obtained by taking the material time derivative of (5.32):

$$\overline{d\xi} = \dot{\mathbf{F}}_{(t)}(\tau)d\mathbf{x} \tag{5.54}$$

Substitution of (5.49), (5.52), and (5.53) yields for $\tau = t$

$$\overline{d\xi} = (\mathbf{W} + \mathbf{D})d\mathbf{x} = \overline{d\mathbf{x}} \tag{5.55}$$

where $\overline{d\dot{\mathbf{x}}} \equiv (d\dot{\boldsymbol{\xi}})_{\tau=t}$. Since $\mathbf{D}$ is real and symmetric we know from Chapter 3 that it possesses orthogonal principal directions with real principal values, which we call d_i. Supposing for the moment that $\mathbf{W} = \mathbf{0}$ we can write, from (5.55),

$$\overline{d\dot{\mathbf{x}}_i} = d_i(d\mathbf{x}_i) \quad \text{(no sum)} \tag{5.56}$$

where $d\mathbf{x}_i$ is an element of length along the ith principal direction. Thus d_i represents the *rate* at which this element is being extended in the ith direction. Hence the designation *stretching* or *rate of stretching*. The term *rate of deformation* is also applied to $\mathbf{D}$.

If unit vectors $\mathbf{l}$ and $\mathbf{m}$ are taken along two orthogonal directions, Truesdell and Toupin [9] have shown that the angle between $\mathbf{l}$ and $\mathbf{m}$, when $\mathbf{l}$ is a fixed vector, changes at a rate

$$\dot{\phi}(m,l) = -\dot{x}_{i,j}\, l^i m^j \tag{5.57}$$

Thus the rotation rate of the ith vector with respect to a fixed unit vector in the jth direction is $-\dot{x}_{j,i}$, while the rotation rate of the jth unit vector with respect to a fixed unit vector in the ith direction is $-\dot{x}_{i,j}$. Then one-half of the sum of rotation rate of the ith and jth unit vectors in the same direction is

$$w_{ij} = \frac{1}{2}(\dot{x}_{i,j} - \dot{x}_{j,i}) \tag{5.58}$$

and we have a physical interpretation of $\mathbf{W}$, which is called the *spin*, or *rate of rotation*, or *vorticity* tensor.

From the notion of $\mathbf{D}$ as a rate of stretching, one can see that

$$\frac{1}{V}\frac{dV}{dt} = tr\,\mathbf{D} \tag{5.59}$$

where V is a volume. This result also follows directly from (4.13), since

$$\frac{\dot{\mathcal{J}}}{\mathcal{J}} = \frac{1}{V}\frac{dV}{dt}$$

and

$$tr\,\mathbf{D} = \nabla \cdot \dot{\mathbf{x}}$$

5.11. HIGHER TIME DERIVATIVES

The process of taking material time derivatives can be continued indefinitely. We set

$$\mathbf{D}_1 \equiv \mathbf{D}$$
$$\mathbf{D}_2 \equiv \dot{\mathbf{D}} = \ddot{\mathbf{U}}_{(t)}(t), \quad \text{etc.}$$
$$\mathbf{D}_n = \overset{(n)}{\mathbf{U}}_{(t)}(t) \tag{5.60}$$
$$\mathbf{W}_n = \overset{(n)}{\mathbf{R}}_{(t)}(t) \tag{5.61}$$

A special form of time derivative is the nth *Rivlin–Ericksen tensor* [10] defined by

$$\mathbf{A}_n(t) = \overset{(n)}{\mathbf{C}}_{(t)}(t) \tag{5.62}$$

The first Rivlin–Ericksen tensor is simply related to $\mathbf{D}$. Taking the material time derivative of the relative form of (5.25)

$$\overset{\bullet}{\mathbf{C}}_{(t)}(\tau) = [\mathbf{F}_{(t)}(\tau)]^T \overset{\bullet}{\mathbf{F}}_{(t)}(\tau) + \overline{[\mathbf{F}_{(t)}(\tau)]^T} \mathbf{F}_{(t)}(\tau)$$

and evaluating $\overset{\bullet}{\mathbf{C}}_{(t)}(\tau)$ at $\tau = t$, we obtain

$$\mathbf{A}_1 = \overset{\bullet}{\mathbf{C}}_{(t)}(t) = \mathbf{L} + \mathbf{L}^T = 2\mathbf{D} \tag{5.63}$$

Repeating the process, it can be shown that

$$\mathbf{A}_2 = \overset{\bullet\bullet}{\mathbf{C}}_{(t)}(t) = 2\mathbf{L}^T\mathbf{L} + \mathbf{L}_2 + \mathbf{L}_2{}^T \tag{5.64}$$

General expressions for $\mathbf{A}_n$ and $\mathbf{D}_n$ are

$$\mathbf{A}_n = \mathbf{L}_n + \mathbf{L}_n{}^T + \sum_{i=1}^{n-1} \binom{n}{i} \mathbf{L}_i{}^T \mathbf{L}_{(n-i)} \tag{5.65}$$

$$\mathbf{D}_n = \frac{1}{2}[\mathbf{A}_n - \sum_{i=1}^{n-1} \binom{n}{i} \mathbf{D}_i \mathbf{D}_{n-i}] \tag{5.66}$$

where $\binom{n}{i} = \dfrac{n!}{(n-i)!\,i!}$

REFERENCES

1. Truesdell, C. and Toupin, R. The classical field theories, in *Handbuch der Physik*, Band III/1, p. 270 (S. Flügge, ed.) (Berlin: Springer-Verlag, (1960).
2. Coleman, B. D., Markovitz, H. and Noll, W. *Viscometric Flows of Non-Newtonian Fluids*, p. 104 (New York: Springer-Verlag New York Inc., 1966).
3. Leigh, D. C. *Nonlinear Continuum Mechanics*, pp. 55, 103. (New York: McGraw-Hill, 1968).
4. Ericksen, J. L., Tensor fields, in *Handbuch der Physik*, Band III/1, p. 805, (S. Flügge, ed.) (Berlin: Springer-Verlag, 1960).
5. *Ibid.*, p. 841.
6. Hoffman, K. and Kunze, R. *Linear Algebra*, p. 279 (Englewood Cliffs, New Jersey: Prentice-Hall, 1961).
7. Sokolnikoff, I. S. *Tensor Analysis*, 2nd edn., p. 126 (New York: Wiley, 1964).
8. Truesdell, C. and Toupin, R. *loc. cit.*, p. 337.
9. *Ibid.*, p. 353.
10. *Ibid.*, p. 383.

PROBLEMS

5.1. Use (3.67) to prove that

$$(\mathbf{AB})^T = \mathbf{B}^T\mathbf{A}^T \tag{5.24a}$$

5.2. Prove that $\mathbf{U}_{(t)}(t) = \mathbf{R}_{(t)}(t) = \mathbf{I}$. Recall that $\mathbf{F}_{(t)}(t) = \mathbf{R}_{(t)}(t)\mathbf{U}_{(t)}(t) = \mathbf{I}$.

5.3. Show that the second Rivlin–Ericksen tensor $\mathbf{A}_2$ can be written

$$\mathbf{A}_2 = 2\mathbf{L}^T\mathbf{L} + \mathbf{L}_2 + \mathbf{L}_2{}^T \tag{5.64}$$

5.4. A plastic bar 10 cm long with a square cross-section of 4 cm² is deformed homogeneously so that the 10 cm side is extended to 12 cm and the cross-section remains square and is $3\frac{1}{3}$ cm². Define convenient coordinates and compute components of $\mathbf{F}_{(t)}(\tau)$, $\mathbf{C}_{(t)}(\tau)$, and $\mathbf{U}_{(t)}(\tau)$. Let τ refer to the time corresponding to the initial state and t the time for the present (deformed) state.

5.5. The bar described in Problem 4 is deformed from its initial state as shown below. The cross-section is not changed. Compute $\mathbf{F}_{(t)}(\tau)$, $\mathbf{C}_{(t)}(\tau)$, and $\mathbf{U}_{(t)}(\tau)$ for this deformation.

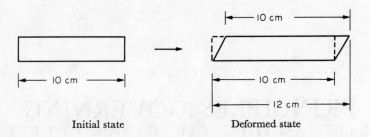

Initial state   Deformed state

CHAPTER 6

PRINCIPLES GOVERNING FORMULATION OF CONSTITUTIVE EQUATIONS

6.1. INTRODUCTION

Many topics covered in previous chapters, though essential for an exposition of non-Newtonian fluid mechanics, are also of primary importance to classical solid or fluid mechanics. In this brief chapter we discuss some basic principles which govern formulation of constitutive equations, thereby opening the way for treatment of fluids with response more complicated than that of the classical Newtonian fluid.

By the term *constitutive equation* we refer to an equation relating the stress on some material to the motion of the material. In so doing, all nonmechanical effects on the strets are being ignored. To isolate the stress in this way from thermodynamic, chemicas, and other experiences is obviously a restriction which may not always be justified. Part of the reason for this simplification rests on the fact that proper procedures for inclusion of these additional effects are still being sought. However, lest the reader infer that neglect of nonmechanical effects always introduces gross errors, it should be added that experimental evidence exists which shows nonmechanical effects to be negligible under many conditions.

In the sections below, the major requirements which we expect constitutive equations to satisfy are outlined. The exposition of these requirements follows closely the work of Noll, Truesdell, and Toupin [1–4].

6.2. COORDINATE INVARIANCE

This requirement merely states explicitly what has already been assumed in earlier chapters for all physical laws: their validity does not depend upon expression in a unique coordinate system. When this principle is applied to constitutive equations, the dynamic nature of the physical law requires that the principle of coordinate invariance apply to transformations between inertial coordinate systems (i.e., the coordinate systems are not undergoing relative acceleration) at any instant of time. The requirement insures that one will not obtain a new "law" every time a different coordinate system is used. The principle also establishes the validity of using coordinate-free notation to describe constitutive equations.

6.3. DETERMINISM OF THE STRESS

In essence one here proclaims that, though fluids may exhibit memory, they cannot possess predictive powers.[†] Consequently, the most general constitutive equation relates the stress in a material at, say, the present time, to the present and *previous* experience of the material. We say that the stress is determined by the *history* of the material.

6.4. PRINCIPLE OF MATERIAL OBJECTIVITY

This principle is the cornerstone of Noll's development of a general approach to formulation of constitutive equations. It is an expression of the belief that response of a material to a given experience or history of motion is independent of any motion of the person observing the response and the history. The principle itself has a long and somewhat confusing history [3]. It appears to have been enunciated near the turn of the century by Zaremba. However, it was first generally made known to rheologists through the important paper of Oldroyd [5] in 1950 and was expressed in a general form by Noll [1] in 1958.

In order to discuss the principle of material objectivity we must first define what is to be meant by a change of reference frame. By this phrase we mean a time dependent but spatially homogeneous transformation of physical space and time, i.e., a transformation of all ordered pairs $(\mathbf{x}, \tau)$ into corresponding pairs $(\mathbf{x}', \tau')$ according to the transformation rule

$$\begin{aligned} \mathbf{x}' &= \mathbf{c}(\tau) + \mathbf{Q}(\tau)\mathbf{x} \\ \tau' &= \tau - a \end{aligned} \tag{6.1}$$

where $\mathbf{c}(\tau)$ is an arbitrary vector-valued function of time, $\mathbf{Q}(\tau)$ is an arbitrary time dependent orthogonal transformation, and a is an arbitrary constant.

Entities which are invariant under the change of frame (6.1) are said to be *frame-indifferent* or *objective*. By invariant we mean that a scalar is unchanged, that a vector preserves its same physical meaning, i.e., if

$$\mathbf{v} = \mathbf{y} - \mathbf{x}$$

then

$$\mathbf{v}' = \mathbf{y}' - \mathbf{x}'$$

and that a (second-order) tensor is equivalent to a linear transformation which, when operating on an objective vector, yields an objective vector. These requirements lead one to conclude that under the change of frame (6.1), objective scalars, vectors, and tensors transform, respectively, according to

$$\begin{aligned} b' &= b \\ \mathbf{v}' &= \mathbf{Q}(\tau)\mathbf{v} \\ \mathbf{A}' &= \mathbf{Q}(\tau)\mathbf{A}\mathbf{Q}(\tau)^T \end{aligned} \tag{6.2}$$

Not all entities possessing physical significance transform objectively. For example, one can define a deformation gradient $\mathbf{F}'$ by

[†] A graduate student has observed: "If you are about to insert a plate into a free jet of fluid and you see the fluid flinch in anticipation, that's a violation of the principle of determinism of stress."

$$dx' = \mathbf{F}'d\mathbf{X}$$

and relate it to $\mathbf{F}$ through (6.1). The result is readily seen to be

$$\mathbf{F}' = \mathbf{Q}(\tau)\mathbf{F} \tag{6.3}$$

Note, however, that the *relative* deformation gradient transforms according to

$$\mathbf{F}'_{(t)}(\tau) = \mathbf{Q}(\tau)\mathbf{F}_{(t)}(\tau)\mathbf{Q}(t)^T \tag{6.4}$$

so that $\mathbf{F}$ and $\mathbf{F}_{(t)}$ transform differently under change of frame, and neither one transforms objectively.

The principle of material objectivity is an assertion that constitutive equations are to be frame-indifferent or, equivalently, must transform objectively under the change of frame (6.1). Physically, this principle codifies the belief that a constitutive equation expresses *material* behavior, and that this behavior must be indifferent to the motion of an observer. As noted by Truesdell and Noll [3], one can also state what is essentially the same result by requiring that the response of a material be indifferent to rigid body translations or rotations of the material.

Because of the indifference of constitutive equations we can regard any two motions

$$\mathbf{x} = \mathcal{X}_t(X) \quad \text{and} \quad \mathbf{x}' = \mathcal{X}'_{t'}(X)$$

where $\mathbf{x}$ and $\mathbf{x}'$ and t and t' are related through (6.1), to be equivalent motions in the sense that we must have

$$\mathbf{S}'(X, t') = \mathbf{Q}(t)\mathbf{S}(X, t)\mathbf{Q}^T(t) \tag{6.5}$$

6.5. APPLICATION

Following Noll we postulate the most general constitutive equation for a body to have the form

$$\mathbf{S}(X, t) = \mathcal{F}_{t,X}(\mathcal{X}) \tag{6.6}$$

where $\mathcal{F}_{t,X}(\mathcal{X})$ is a *functional*[†] of the motion of the body evaluated over all times up to and including the present time t. The statement (6.6) is more general than appears to be desirable or necessary. Consequently, we restrict (6.6) by postulating that the stress at material point X is only influenced by the motion of an arbitrarily small neighborhood around X (sometimes stated as the principle of *local action*). Then (6.6) becomes

$$\mathbf{S}(X, t) = \overset{\infty}{\underset{s=0}{\mathcal{F}}}(\mathcal{X}_{t-s}(X)) \tag{6.7}$$

We interpret the notation to mean that the motion $\mathcal{X}_{t-s}$ which affects $\mathbf{S}$ is in an arbitrarily small region around X. Note also that a time variable $0 \leqslant s \leqslant \infty$ has been introduced. Equation (6.7) is a symbolic statement that the stress at particle X and time t is determined by the motion, over all past times up to and including the present, of an arbitrarily small region around X. Furthermore, only those functionals $\mathcal{F}$ are admissible for which

$$\mathbf{Q}(t)\mathbf{S}(X, t)\mathbf{Q}^T(t) = \overset{\infty}{\underset{s=0}{\mathcal{F}}}(\mathcal{X}'_{t-s}(X)) \tag{6.8}$$

† See the appendix to this chapter.

when χ' and χ are equivalent processes. We shall see in the next chapter how this abstract result can be employed to make important inferences about the form of rather general constitutive relations.

6.6. OTHER PRINCIPLES

There are a number of other principles governing constitutive equations which could be included in this chapter [4], along with specific properties which one would intuitively expect constitutive functionals to possess. An example of the latter is the property of *fading memory*. By postulating that recent history of the motion should have a more dominant effect upon $\mathscr{F}$ than distant history one can show, for example, how the constitutive equation of a Newtonian fluid fits into the framework of general constitutive theory [6].

APPENDIX

The term *functional* employed in Section 6.4 may be new to the reader. For that reason we shall briefly discuss the concept. A thorough description is available elsewhere [7]. For the purposes employed in this book we merely need to consider a functional as an extremely useful notation for denoting dependence, in an unspecified way and over an interval of time, on a *function* in much the same way that a function indicates dependence, in an unspecified way, on a *variable*. Thus by writing $\mathscr{F}_{t,X}(\chi)$ in (6.6) we mean that the stress at time t and material point X is determined, in some unspecified way, by the history of the motion χ of the whole body. This notion is made more explicit in (6.7) where, for simplicity of notation, we understand the argument $\chi_{t-s}(X)$ to mean the motion in an arbitrarily small neighborhood around X and time $t-s$. By using functional notation we do not have to commit ourselves to any specific function such as one or more integrals over past time of one or more measures of the deformation. For a concise treatment of the relation between constitutive functionals and integral relations over past time the reader is referred to Lodge [8].

Virtues of functional notation include simplicity and generality. However, because of its generality a constitutive equation such as (6.7) is of little use in predicting material behavior until the nature of the functional is made more specific. We shall consider this problem at length in succeeding chapters.

REFERENCES

1. Noll, W. *Arch. Rational Mech. Anal.* **2**, 197 (1958).
2. Truesdell, C. *The Elements of Continuum Mechanics* (New York: Springer-Verlag New York, 1966).
3. Truesdell, C. and Noll W. The non-linear field theories of mechanics, in *Handbuch der Physik*, Band III/3, pp. 44–47 (S. Flügge, ed.), (Berlin: Springer-Verlag, 1965).
4. Truesdell, C. and Toupin, R. The classical field theories, in *Handbuch der Physik*, Band III/1, pp. 700–704 (S. Flügge, ed.) (Berlin: Springer-Verlag, 1960).
5. Oldroyd, J. G. *Proc. Roy. Soc.* (*London*) A, **200**, 523 (1950).
6. Coleman, B. D. and Noll, W. *Arch. Rational Mech. Anal.* **6**, 355 (1960); **9**, 273 (1962).
7. Volterra, V. *Theory of Functionals* (New York: Dover Publications, 1959).
8. Lodge, A. S. *Elastic Liquids*, pp. 174 *et seq.* (New York: Academic Press, 1964).

PROBLEMS

6.1. (a) Prove that the relation between second-order tensors which transform objectively is

$$\mathbf{A}' = \mathbf{Q}(\tau)\mathbf{A}\mathbf{Q}(\tau)^T \tag{6.2c}$$

(b) Prove (6.4).

6.2. Suppose one postulates a material such that the stress at a material point is a linear function of the velocity gradient at that material point. Thus

$$\mathbf{S}(X, t) = K\mathbf{L}(X, t)$$

where $\mathbf{L} = \nabla\mathbf{v}$. Show that the principle of material objectivity can only be obeyed if

$$\mathbf{S} = K\mathbf{D}$$

where $\mathbf{D} = \dfrac{1}{2}(\mathbf{L} + \mathbf{L}^T)$.

Hint: Write the equation as $\mathbf{S} = K(\mathbf{D} + \mathbf{W})$, where $\mathbf{W} = 1/2\,(\mathbf{L} - \mathbf{L}^T)$, and show that this must be reduced to $\mathbf{S} = K\mathbf{D}$ because, for two equivalent motions, $\mathbf{W}' = \mathbf{Q}\mathbf{W}\mathbf{Q}^T + \dot{\mathbf{Q}}\mathbf{Q}^T$.

CHAPTER 7

SIMPLE FLUIDS
AND SIMPLE FLOWS

7.1. INTRODUCTION

In this chapter we find ourselves on the horns of a dilemma that will recur throughout the rest of the book. We are faced with what might be called the cure of rheology: the search for some middle ground between generality and computational utility. At the one extreme is a functional statement of material response, such as (6.6), which includes all materials following the principle of determinism of stress. However, the statement is so general that it is of no value in providing predictions of how a fluid will respond to a given set of imposed conditions. At the other extreme is a highly specific constitutive equation for which the exact nature of the functional is specified. In the latter case, however, the probability that a real material will obey the stated constitutive equation is very small. Thus one makes a compromise, postulating a constitutive equation which should be appropriate to a large class of materials, but which is also specific enough to permit solution for interesting classes of flow problems.

In 1950 Oldroyd [1] showed how one could formulate properly invariant constitutive equations by working in a coordinate system embedded in the material. He then provided rules for transformation from the embedded coordinate system to one fixed in space, and illustrated the general technique by application to a few specific examples. Green, Rivlin, and Spencer [2–4] have shown how one can account for memory effects in materials by writing a constitutive functional as a series of integrals over past time. They showed how their results could be sufficiently truncated to permit application to certain flow problems. A third approach, employing similar physical ideas but a different mathematical formalism, is due to Noll [5]. In the following sections we shall see, using for the most part the formalism of Noll, how one can obtain a fruitful compromise between the generality and utility mentioned above.

7.2. THE SIMPLE FLUID

We begin by postulating a form of response functional which is more specific than that of (6.7) but is still general enough to encompass much of our experience with motion of materials.

Noll has defined a *simple material* as one for which the stress at a material point is determined by the history of the deformation gradient, evaluated at the material point in question, over all past times up to and including the present. Thus a simple material is defined by

$$\mathbf{S}(t) = \underset{s=0}{\overset{\infty}{\mathscr{F}}} \ (\mathbf{F}(t-s)) \tag{7.1}$$

where s is a variable over time. Note that, though not explicitly indicated, the form of the response functional $\mathscr{F}$ will in general depend upon the reference configuration $\mathbf{X}(X)$ used in the definition of $\mathbf{F}$. Also, specific notation showing the dependence of $\mathbf{S}$ upon X has been omitted.

Truesdell and Noll [6, 7] have shown that it is useful to write (7.1) in terms of the *relative* deformation gradient $\mathbf{F}_{(t)}(\tau)$. In the notation of Section 5.6, we may write in component form

$$\begin{aligned} d\xi^i(\tau) &= f^i{}_\alpha(\tau)dX^\alpha \\ d\xi^i(\tau) &= f_{(t)}{}^i{}_j(\tau)d\xi^j(t) = f_{(t)}{}^i{}_j(\tau)dx^j \end{aligned} \tag{7.2}$$

Recall that in this notation x^i is the coordinate ξ^i of material point X at reference time t. By the chain rule,

$$\frac{\partial \xi^i(\tau)}{\partial X^\alpha} = \frac{\partial \xi^i(\tau)}{\partial x^j} \frac{\partial x^j}{\partial X^\alpha} \tag{7.3}$$

or

$$\mathbf{F}(\tau) = \mathbf{F}_{(t)}(\tau)\mathbf{F}(t), \tag{7.4}$$

and we can write the defining equation (7.1) of a simple material

$$\mathbf{S}(t) = \underset{s=0}{\overset{\infty}{\mathscr{F}}} \ (\mathbf{F}_{(t)}(s)\mathbf{F}(t)) \tag{7.5}$$

Note that we shall often find it convenient to use the notation $\mathbf{F}_{(t)}(\tau) \equiv \mathbf{F}_{(t)}(t-s) \equiv \mathbf{F}_{(t)}(s)$, where $0 \leqslant s \leqslant \infty$. From the principle of material objectivity,

$$\mathbf{Q}(t)\mathbf{S}(t)\mathbf{Q}^T(t) = \underset{s=0}{\overset{\infty}{\mathscr{F}}}(\mathbf{Q}(t-s)\mathbf{F}_{(t)}(s)\mathbf{F}(t)) \tag{7.6}$$

for arbitrary orthogonal $\mathbf{Q}(t-s)$. In particular if we choose $\mathbf{Q}(t-s) = \mathbf{R}^T{}_{(t)}(s)$, where $\mathbf{F}_{(t)}(s) = \mathbf{R}_{(t)}(s)\mathbf{U}_{(t)}(s)$, (7.6) becomes, recalling (5.48),

$$\mathbf{S}(t) = \underset{s=0}{\overset{\infty}{\mathscr{F}}} \ (\mathbf{U}_{(t)}(s)\mathbf{R}(t)\mathbf{U}(t)) \tag{7.7}$$

The next step is to distinguish a *fluid* as a subclass of a *material*. There is no universally accepted definition of a fluid. For present purposes we follow Truesdell and Noll [6], who were guided by the idea that "a fluid should not alter its material response after an arbitrary deformation that leaves the density unchanged". A consequence of this property of a fluid is that the dependence of response functional $\mathscr{F}$ on reference configuration, to which we referred following (7.1), must disappear for all reference configurations which are related by a deformation at constant density. These include reference configurations related by a density-preserving rotation or reflection, i.e., orthogonal transformations. But invariance to reference configurations which are related by rotation or reflection is the defining characteristic of an isotropic material. Thus we conclude that a simple fluid is also an isotropic fluid. Hence in place of (7.7) one may write, recalling that $\mathbf{C}_{(t)}(s) = (\mathbf{U}_{(t)}(s))^2$,

$$\mathbf{S}(t) = \underset{s=0}{\overset{\infty}{\mathscr{G}}} (\mathbf{C}_{(t)}(s), \mathbf{U}(t)) \tag{7.8}$$

Furthermore, the definition of a fluid implies that though the response of a fluid is sensitive to the *history* of the deformation, the present state of deformation is important only to the extent that it affects the density. Thus in place of $\mathbf{U}(t)$ in (7.8) we may write the density $\rho(t)$. It is also convenient to isolate a scalar-valued function of the density $-p(\rho)$ as an isotropic portion of the response functional.

$$\mathbf{S}(t) = -p(\rho)\mathbf{I} + \underset{s=0}{\overset{\infty}{\mathscr{H}}} (\mathbf{C}_{(t)}(s), \rho) \tag{7.9}$$

We have already stated our intent to be concerned only with incompressible fluids. Consequently, it is necessary to alter somewhat the arguments given above. Rather than asserting that the influence of $\mathbf{U}(t)$ appears only in its effect on the density, we postulate, again in the light of experience with isochoric (i.e., volume preserving) motions, that $\mathbf{S}(t)$ is determined by the history of $\mathbf{C}_{(t)}(s)$ only to within an arbitrary isotropic stress. Then (7.9) becomes

$$\mathbf{S}(t) = -p\mathbf{I} + \underset{s=0}{\overset{\infty}{\mathscr{H}}} (\mathbf{C}_{(t)}(s)) \tag{7.10}$$

where p is now an arbitrary scalar-valued functional of $\mathbf{C}_{(t)}(s)$. The indeterminacy of p may be removed by defining an "extra stress"

$$\mathbf{T}(t) = \mathbf{S}(t) + p(t)\mathbf{I} = \underset{s=0}{\overset{\infty}{\mathscr{H}}} (\mathbf{C}_{(t)}(s)) \tag{7.11}$$

and requiring

$$p = -\frac{1}{3} \operatorname{tr} \mathbf{S} \tag{7.12}$$

whereupon

$$\operatorname{tr} \mathbf{T} = 0 \tag{7.13}$$

We could have proceeded immediately to (7.10) by using that equation as the defining relation for a simple incompressible fluid. Indeed, in one of their papers on the subject Coleman and Noll [8] have chosen just such a course. However, the more circuitous route chosen here has the advantage of relating the constitutive equation (7.10) to the deformation gradient, which admits a more immediate physical interpretation than $\mathbf{C}_t(s)$. Also, the distinction between a simple material, a simple fluid, and an incompressible simple fluid is made apparent.

7.3. THE REST HISTORY

A very special flow history is that for which the fluid has always been at rest. Then

$$\mathbf{F}_{(t)}(s) = \mathbf{C}_{(t)}(s) = \mathbf{I} \tag{7.14}$$

Let us apply the principle of material objectivity to the functional $\mathcal{H}$. Taking $\mathbf{Q}(\tau)$ in (6.1) to be a constant tensor, and recalling (6.4), one readily finds

$$\mathbf{Q} \underset{s=0}{\overset{\infty}{\mathcal{H}}} (\mathbf{I}) \mathbf{Q}^T = \underset{s=0}{\overset{\infty}{\mathcal{H}}} (\mathbf{I}) \tag{7.15}$$

But $\underset{s=0}{\overset{\infty}{\mathcal{H}}} (\mathbf{I})$ is just a constant symmetric tensor, say $\mathbf{A}$, so that

$$\mathbf{Q A Q}^T = \mathbf{A} \tag{7.16}$$

for any constant $\mathbf{Q}$. We note from (3.90) that $\mathbf{A}$ is an isotropic tensor as a consequence of (7.16). Now from the discussion of Section 3.17 we can see that if (7.16) holds, any direction will be a principal direction for $\mathbf{A}$. Hence we conclude that

$$\mathbf{A} = k\mathbf{I}$$

But from (7.13), tr $\mathbf{A}=0$, which requires $k=0$. Thus for a fluid at rest for all past time,

$$\mathbf{S} = -p\mathbf{I} \tag{7.17}$$

which is a justification for associating p with a scalar hydrostatic pressure. One can readily show [9, page 20] that (7.17) also obtains when the motion is a rigid rotation of the fluid.

Though it is customary to call p an arbitrary "hydrostatic" pressure, the term can be misleading. The value of p will of course, in general, change as one alters a fluid from a rest state to a state of motion. We shall in fact observe later that the arbitrary nature of p makes it appropriate to study *differences* between normal stress components rather than individual components themselves.

7.4. SOME STEADY LAMINAR SHEAR FLOWS

At this point the dilemma cited in Section 7.1 is apparent. We have made a hypothesis postulating a certain type of fluid behavior, and from this have determined some conditions which must be obeyed by the constitutive equation. However, the theory is of little use until one combines with the constitutive principles some information about the type of flow to which the constitutive principles will be applied. A class of flows possessing the happy combination of mathematical simplicity and practical significance is the class of *viscometric flows*, which we shall define in Section 7.6. Perhaps an appreciation of the significance of viscometric flows is best obtained by first discussing a common subclass of them [8, 10].

Consider those steady flows for which an orthogonal coordinate system can be found such that the contravariant components of the velocity $\mathbf{v}$ are of the form

$$v^1 = v(x^2)$$
$$v^2 = 0 \qquad (7.18)$$
$$v^3 = 0$$

Furthermore, the orthogonal coordinate system ($g_{ij}=0$ for $i \neq j$) must have a basis such that the covariant components of the metric tensor do not depend upon the coordinate in the flow direction, i.e.,

$$g_{ii} = [h_i(x^2, x^3)]^2 \quad \text{(no sum)} \qquad (7.19)$$

Thus the matrix formed from covariant components of the metric tensor is

$$[g_{ij}] = \begin{bmatrix} (h_1)^2 & 0 & 0 \\ 0 & (h_2)^2 & 0 \\ 0 & 0 & (h_3)^2 \end{bmatrix} \qquad (7.20)$$

We shall see that the usual types of viscometer flows are special cases of the class of steady laminar shear flows given by (7.18) and (7.19).

Our next task is to link the specification of the flow field to the specification of a constitutive equation such as (7.1) or, since we wish to deal with incompressible simple fluids, to equation (7.11). Hence we need to compute $\mathbf{C}_{(t)}$. It is a simple matter to show that the conditions (7.18) and (7.19) lead to the following covariant components for $\mathbf{C}_{(t)}$:

$$[c_{(t)ij}] = \begin{bmatrix} (h_1)^2 & -s(h_1)^2 v'(x^2) & 0 \\ -s(h_1)^2 v'(x^2) & (h_2)^2 + (h_1 s v'(x^2))^2 & 0 \\ 0 & 0 & (h_3)^2 \end{bmatrix} \qquad (7.21)$$

where $v'(x^2) \equiv \dfrac{dv}{dx^2}$ and we have used

$$c_{(t)ij} = f_{(t)}{}^l{}_i f_{(t)}{}^k{}_j \gamma_{lk} \qquad (5.36)$$

In the present system the coordinate metric is not time dependent; hence, $\gamma_{kl}=g_{kl}$.

Now define a quantity κ, the *shear rate*, by

$$\kappa = \frac{h_1}{h_2} v'(x^2) \qquad (7.22)$$

Then, using (3.57) we find the physical components corresponding to (7.21)

$$[c_{(t)}(ij)] = \begin{bmatrix} 1 & -\kappa s & 0 \\ -\kappa s & 1 + \kappa^2 s^2 & 0 \\ 0 & 0 & 1 \end{bmatrix} \qquad (7.23)$$

Let us recall the Rivlin–Ericksen kinematic tensors defined in Section 5.11. For the flows under consideration we find from (5.62) that

$$[a_1(ij)] = \begin{bmatrix} 0 & \kappa & 0 \\ \kappa & 0 & 0 \\ 0 & 0 & 0 \end{bmatrix} \qquad (7.24)$$

$$[a_2(ij)] = \begin{bmatrix} 0 & 0 & 0 \\ 0 & 2\kappa^2 & 0 \\ 0 & 0 & 0 \end{bmatrix} \tag{7.25}$$

so that the right relative Cauchy–Green tensor can be written

$$\mathbf{C}_{(t)}(s) = \mathbf{I} - s\,\mathbf{A}_1 + s^2\frac{1}{2}\,\mathbf{A}_2 \tag{7.26}$$

We note in passing that

$$\mathbf{A}_n = \mathbf{0} \quad (n > 2) \tag{7.27}$$

Equation (7.26) is significant because it provides an explicit form for $\mathbf{C}_{(t)}$ which is the same for *any* flow of the form (7.18), (7.19). Thus the functional $\mathscr{H}$ of (7.11) over all past time is reduced to a *function*

$$\mathbf{T} = \mathbf{h}(\mathbf{A}_1, \mathbf{A}_2) \tag{7.28}$$

or, in view of (7.24) and (7.25),

$$\mathbf{T} = \mathbf{f}(\kappa) \tag{7.29}$$

We can be even more specific about the form of (7.29) by considering symmetry properties of these laminar shear flows (see, for example, [11]), or by formally invoking the principle of material objectivity. Applying (6.8) to (7.28) and recalling (5.35) and (6.4) we have for the change of frame (6.1) with $\tau' = \tau$ and constant orthogonal tensor $\mathbf{Q}$,

$$\mathbf{Q}\,\mathbf{T}\,\mathbf{Q}^T = \mathbf{h}(\mathbf{Q}\,\mathbf{A}_1\,\mathbf{Q}^T, \mathbf{Q}\,\mathbf{A}_2\,\mathbf{Q}^T) \tag{7.30}†$$

In particular, if $\mathbf{Q}$ is taken to represent a reflection of the 3-axis

$$[q(ij)] = \begin{bmatrix} 1 & 0 & 0 \\ 0 & 1 & 0 \\ 0 & 0 & -1 \end{bmatrix}$$

one finds that

$$\mathbf{A}_1 = \mathbf{Q}\,\mathbf{A}_1\,\mathbf{Q}^T$$
$$\mathbf{A}_2 = \mathbf{Q}\,\mathbf{A}_2\,\mathbf{Q}^T$$

so that

$$\mathbf{Q}\,\mathbf{T}\,\mathbf{Q}^T = \mathbf{T} \tag{7.31}$$

The reader can quickly verify that this implies

$$[t(ij)] = \begin{bmatrix} t(11) & t(12) & 0 \\ t(21) & t(22) & 0 \\ 0 & 0 & t(33) \end{bmatrix} \tag{7.32}$$

---

† This result is also a consequence of the definition of a simple *fluid* given in Section 7.2. We noted there that a fluid will not alter its response after a deformation that leaves the density unchanged. Hence the form of the constitutive equation is unaffected by rotation or reflection with $\mathbf{Q}$, and if we have, say, $\mathbf{T} = \mathbf{h}(\mathbf{B}_1 \ldots \mathbf{B}_M;$ $\mathbf{u}_1 \ldots \mathbf{u}_N)$, then $\mathbf{QTQ}^T = \mathbf{h}(\mathbf{QB}_1\mathbf{Q}^T \ldots \mathbf{QB}_M\mathbf{Q}^T; \mathbf{Qu}_1 \ldots \mathbf{Qu}_N)$. This is the defining property of an *isotropic tensor function* $\mathbf{h}$ of tensor and vector arguments $\mathbf{B}_i$ and $\mathbf{u}_i$ [6, pp. 22 *et seq.*, p. 79).

Recalling that tr $\mathbf{T}=0$ and, from Section 4.7, that $t(12)=t(21)$, we see that $\mathbf{T}$ is completely determined by only three material functions of κ alone, which we choose to define by

$$
\begin{aligned}
t(12) &= t(21) = \tau(\kappa) \\
t(11) - t(22) &= \mathcal{N}_1(\kappa) \\
t(22) - t(33) &= \mathcal{N}_2(\kappa)
\end{aligned}
\tag{7.33}
$$

By choosing $\mathbf{Q}$ to represent a reflection of the 1-axis, the reader is urged to show that $\tau(\kappa)$ is an odd function of κ, while $\mathcal{N}_1(\kappa)$ and $\mathcal{N}_2(\kappa)$ are even functions.

At this point a word of warning is in order. Students of rheology are required to decode an amazing variety of conventions for expressions of stress and of stress differences. It is hoped that the hybrid scheme adopted here will reduce rather than multiply the confusion. By $\mathcal{N}_1(\kappa)$ we refer to the "primary" or "first" normal stress difference as it is usually called by experimentalists. This is a natural designation since $\mathcal{N}_1(\kappa)$ generally exceeds in magnitude the secondary or second normal stress difference $\mathcal{N}_2(\kappa)$. Also, experimentalists usually choose the 1-direction to be the direction of flow. These choices are in contrast to the convention employed by Coleman, Noll, and Truesdell, in whose notation the 1- and 2-directions are interchanged from those used here. Also, those writers describe normal stresses by $\sigma_1(\kappa)$ and $\sigma_2(\kappa)$, where $\sigma_1(\kappa) \equiv \mathcal{N}_2(\kappa)$ and $\sigma_2(\kappa) \equiv \mathcal{N}_1(\kappa) + \mathcal{N}_2(\kappa)$. Another source of confusion is the lack of a uniform sign convention for stress. Although most writers prefer to consider tensile stresses to be positive, in the work published by Bird and his students [12] stress is associated with momentum transport. Consequently, stresses are positive when compressive. A "conversion table" is supplied in Table 7.1.

TABLE 7.1. Various conventions for stress components $(\dot{\gamma}_{12} \equiv \kappa)$

This book	Bird and Marsh [12]	Coleman *et al.* [9]
$t(12) = \tau(\kappa)$	$-\tau_{12} = \eta(\dot{\gamma}_{12})\dot{\gamma}_{12}$	$T^{<12>} = \tau(\kappa)$
$t(11) - t(22) = \mathcal{N}_1(\kappa)$	$-(\tau_{11} - \tau_{22}) = \theta(\dot{\gamma}_{12})(\dot{\gamma}_{12})^2$	$T^{<22>} - T^{<11>} = \sigma_2(\kappa) - \sigma_1(\kappa)$
$t(22) - t(33) = \mathcal{N}_2(\kappa)$	$-(\tau_{22} - \tau_{33}) = \beta(\dot{\gamma}_{12})(\dot{\gamma}_{12})^2$	$T^{<11>} - T^{<33>} = \sigma_1(\kappa)$

Returning to equations (7.33), we see that they reveal exactly what is needed for a complete characterization of the laminar shear flows defined by (7.18) and (7.19). The task of the experimentalist is now clear: measure the functions τ, $\mathcal{N}_1$, and $\mathcal{N}_2$ in any *one* flow of the class (7.18), (7.19), and the behavior of a simple fluid in the *whole* class is completely determined.

7.5. A FEW EXAMPLES†

The equations developed in the previous section are sufficiently complete so that application to a specific flow is merely a matter of substitution of appropriate geometric factors and evaluation of equations. Nevertheless, we shall show the utility of (7.18) and

---

† These examples are also treated in several of the references cited earlier, as well as in the text by Fredrickson [13].

(7.19) by considering a few familiar examples. The results obtained will provide a basis for the discussion of viscometry in Chapter 8.

(i) PLANE COUETTE FLOW

We consider flow of a simple fluid between two horizontal parallel planes spaced a distance h apart. The flow is described with respect to rectangular Cartesian coordinates (x, y, z) as shown in Fig. 7.1. We assume that the flow field is given by

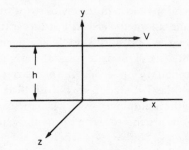

Fig. 7.1. Plane Couette flow.

$$\mathbf{v} = (v(y), 0, 0) \tag{7.34}$$

and the body force by the acceleration due to gravity

$$\mathbf{g} = (0, -g, 0) \tag{7.35}$$

where the 1-, 2-, and 3-directions are associated with x, y, and z of Fig. 7.1.

Cauchy's law of motion (4.26) becomes, in component form,

$$\frac{dt_{xy}}{dy} - \frac{\partial p}{\partial x} = 0$$

$$\frac{dt_{yy}}{dy} - \frac{\partial p}{\partial y} - \rho g = 0 \tag{7.36}$$

$$\frac{\partial p}{\partial z} = 0$$

where we have used the fact that the velocity, and hence components of the stress, can only change in the y-direction. Furthermore, we see from (7.36a) that $\partial p/\partial x$ cannot be a function of x. Thus, at most,

$$p = f(y)x + b(y)$$

But in order that $\partial p/\partial y$ in (7.36b) not be a function of x, $f(y) = c$, where c is a constant. Now suppose that we maintain a constant pressure p along the boundaries $y = 0, h$. Then $c = 0$ and the pressure is simply $p = b(y)$. Equations (7.36a) and (7.36b) are readily integrated to give

$$\begin{aligned} t_{xy} &= \tau(\kappa) = \text{const} \\ t_{yy} &= b(y) + \rho g\, y + \text{const} \end{aligned} \tag{7.37}$$

But the fact that the three material functions are solely dependent upon κ implies from (7.37a) that $\kappa = dv/dy = \text{const}$ and that therefore

$$b(y) = -\rho g y + \text{const} \tag{7.38}$$

Then if the lower plate is held fixed and the upper one is moved at velocity V, we have the linear velocity profile familiar from Newtonian fluid mechanics

$$v = \frac{y}{h} V \tag{7.39}$$

where we have assumed, as we shall throughout, that the fluid does not slip past solid boundaries (the no-slip condition). This example shows how, when a fluid is subjected to an extremely uncomplicated flow, one can be rather specific about the flow description even though the constitutive behavior of the fluid is limited only by the general formulation of (7.11). The result tells us that for any simple fluid (in the sense of Noll's definition) the familiar "linear velocity profile" of plane Couette flow is preserved.

(ii) LAMINAR FLOW THROUGH A TUBE

Consider fully developed laminar flow through a stationary tube with circular cross-section of radius R. The centerline of the tube is inclined at an arbitrary angle to the horizontal, and the only body force acting on the fluid is that of gravity $\mathbf{g}$. Let us consider the flow between two arbitrary reference planes P and Q, as shown in Fig. 7.2.

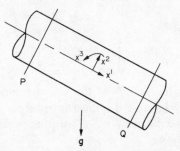

Fig. 7.2. Laminar flow through a tube.

The velocity field is assumed to have the form

$$\mathbf{v} = (v(x^2), 0, 0) \tag{7.40}$$

with respect to the coordinates shown. The flow is taken to be in the direction of increasing x^1. It is natural to associate the flow with a cylindrical coordinate system

$$x^1 = z; \quad x^2 = r; \quad x^3 = \phi$$

so that (7.19) is determined by

$$h_1 = h_2 = 1; \quad h_3 = r \tag{7.41}$$

and

$$\kappa = v'(r)$$

We express the gravitational force in terms of a scalar potential ψ by

$$\mathbf{g} = -\nabla\psi$$

and define

$$\Phi = p + \rho\psi \tag{7.42}$$

Then the components of the equation of motion, written in terms of physical components of the stress tensor, are

$$\frac{1}{r}\frac{d}{dr}(rt(rr)) - \frac{1}{r}t(\phi\phi) - \frac{\partial\Phi}{\partial r} = 0$$

$$\frac{1}{r}\frac{d}{dr}(rt(rz)) - \frac{\partial\Phi}{\partial z} = 0 \tag{7.43}†$$

$$\frac{\partial\Phi}{\partial\phi} = 0$$

since we know from (7.40) and our postulate of a simple fluid that the stress can only be a function of r. Then it is easy to show from (7.43) that the most general form for Φ is

$$\Phi = f(r) - az \tag{7.44}$$

where a is a constant representing the effective pressure gradient, corrected for the effect of gravity, which drives the fluid down the tube. To see this, consider the net driving force F in the direction of flow between planes P and Q. Recalling that our convention is to assign positive numbers to stresses which are tensile, one readily obtains for the appropriate force balance

$$F = \int_{Q} s(zz)dA - \int_{P} s(zz)dA - \int_{V_{P-Q}} \frac{\partial\psi}{\partial z}\rho\,dz\,dA \tag{7.45}$$

where the notation indicates integration of the first two integrals over the cross-sections at planes Q and P, respectively, and the third integration is over the volume of fluid enclosed between the bounding planes. We can integrate the last term over z while holding r and ϕ constant, the result being

$$-\rho\int\frac{\partial\psi}{\partial z}dz\,dA = -\rho\left[\int_{A_Q}\psi(r,\phi,z_Q)dA - \int_{A_P}\psi(r,\phi,z_P)dA\right]$$

so that (7.45) can be written

$$F = \int_{A_Q}(s(zz) - \rho\psi)dA - \int_{A_P}(s(zz) - \rho\psi)dA \tag{7.46}$$

But since $s(zz) = t(zz) - p$, $\rho\psi = \Phi - p$, and $\Phi = f(r) - az$, the integrands in (7.46) reduce to

$$t(zz) - f(r) + az$$

---

† See appendix to this chapter.

Now, $t(zz) - f(r)$ depends solely on r and its contribution will therefore be cancelled by the integration over A_Q and A_P. Hence we finally obtain the simple but significant result

$$a = \frac{F}{A(z_Q - z_P)} \tag{7.47}$$

We see that a has exactly the same physical significance that it holds for a Newtonian fluid; namely, the driving force per unit cross-section per unit length. In the absence of gravity a is of course merely the magnitude of the pressure gradient.

Let us now return to the equations of motion (7.43). We can readily integrate (7.43b) and obtain

$$t(rz) = -\frac{ar}{2} \tag{7.48}$$

where we have required the stress to be bounded at the centerline. This is another result familiar from elementary mechanics of Newtonian fluids.

Since we have postulated

$$t(rz) = \tau(\kappa) \tag{7.49}$$

we can make the physically reasonable assumption that there is a unique inverse to (7.49), i.e., that

$$\kappa = v'(r) = -\tau^{-1}\left(\frac{ar}{2}\right) \tag{7.50}$$

(where we have applied the earlier conclusion that $\tau(\kappa)$ is an odd function). Thus once the form of $\tau(\kappa)$ is known the velocity profile in the tube is, in principle, determined.

We note in passing that the development followed above can readily be modified to treat flow between parallel planes.

(iii) CYLINDRICAL COUETTE FLOW

We briefly consider the flow between vertical coaxial rotating cylinders, in the idealization of which we assume (7.18) to apply in the cylindrical coordinate system

$$x^1 = \phi; \quad x^2 = r; \quad x^3 = z \tag{7.51}$$

so that $h_1 = r$, $h_2 = h_3 = 1$, and

$$\kappa = r\,v'(r) \tag{7.52}$$

The equation of motion, expressed in terms of physical components of the stress, is

$$\frac{dt(rr)}{dr} + \frac{1}{r}(t(rr) - t(\phi\phi)) - \frac{\partial \Phi}{\partial r} = -\rho r(v)^2$$

$$\frac{dt(r\phi)}{dr} + \frac{2t(r\phi)}{r} - \frac{1}{r}\frac{\partial \Phi}{\partial \phi} = 0 \tag{7.53}$$

$$\frac{\partial \Phi}{\partial z} = 0$$

Following the reasoning employed in the previous examples, one notes that the stress can only be dependent upon r, and that Φ is of the form $\Phi = f(r) + c\phi$, where c is a constant. However, since we expect the pressure to be a single-valued function over the flow space we conclude that $c = 0$. Then, from (7.53b),

$$\frac{d \ln(t(r\phi))}{d \ln r} = -2$$

or

$$t(r\phi) = \tau(\kappa) = \frac{T}{2\pi r^2} \tag{7.54}$$

where T is a constant of integration. Assuming a unique inverse to exist for the function $\tau(\kappa)$, we write

$$\kappa = rv'(r) = r\frac{d\omega}{dr} = \tau^{-1}\left[\frac{T}{2\pi r^2}\right] \tag{7.55}$$

ω being the angular velocity of fluid rotation. Again we see that once the material function $\tau(\kappa)$ is known, the velocity in the flow field is determined in terms of boundary conditions which fix T.

These examples by no means exhaust the supply of laminar shear flows nor have we obtained all possible useful information from each example. We shall return to these flows in the next chapter. Right now, having seen something of the generality which is possible when one effects a proper combination of flow and fluid, we see how the definition of a laminar shear flow can be extended.

7.6. VISCOMETRIC FLOWS

Following Coleman [14] we define as *viscometric* any flow for which, at all times and at every material point X, the right relative Cauchy–Green tensor $\mathbf{C}_{(t)}(s)$ can be expressed in the form (7.26), and the components of $\mathbf{A}_1$ and $\mathbf{A}_2$ can be expressed in the form (7.24) and (7.25) relative to an orthogonal basis $\mathbf{e}_i$. The basis may itself also depend upon X and t.

Clearly, the laminar shear flows discussed in the previous section are viscometric flows. Note also that the arguments used to proceed from (7.26) to (7.33) are unaffected by this broadened interpretation of $\mathbf{C}_t(s)$. Thus we can conclude that the extra stress is completely determined for a viscometric flow by the three material functions of (7.33), where κ may be a function of position and time.

One additional important class of flows which is admitted by virtue of our broadened classification is that which Coleman and others have called "curvilineal flows" [9]. These are defined by the criterion that contravariant components of the velocity field have the form

$$\mathbf{v} = (v(x^2), 0, \omega(x^2)) \tag{7.56}^\dagger$$

when the components are expressed with respect to a set of orthogonal base vectors $\mathbf{a}_i$, the magnitudes of which

---

† Truesdell and Noll [6] have included the possibility of time dependence in their definition of curvilineal flows.

$$\sqrt{\mathbf{a}_i \cdot \mathbf{a}_i} = \sqrt{g_{ii}} \quad \text{(no sum)} \tag{7.57}$$

do not change as one follows the motion of any particular material point. We immediately see that all of the laminar shear flows discussed in the previous section are curvilineal flows since $v^2 = v^3 = 0$ and $g_{ii} = g_{ii}(x^2, x^3)$ (no sum).

It is straightforward to show that the flow given by (7.56) and (7.57), though distinct from laminar shear flow ((7.18), (7.19)), also satisfies (7.24), (7.25), and (7.26). Hence it is a viscometric flow. The reader can readily verify that physical components of $\mathbf{C}_{(t)}(s)$ are

$$[c_{(t)}(ij)(s)] = \begin{bmatrix} 1 & -sl & 0 \\ -sl & 1+s^2(l^2+m^2) & -sm \\ 0 & -sm & 1 \end{bmatrix} \tag{7.58}$$

where

$$l = \frac{h_1}{h_2} v'; \quad m = \frac{h_3}{h_2} \omega'$$

the prime denoting differentiation with respect to x^2. Equation (7.58) follows from the definition of $\mathbf{C}_{(t)}(s)$ and computation of covariant components by integration of components of

$$\frac{d\boldsymbol{\xi}(\tau)}{d\tau} = \mathbf{v}(\mathbf{x}) \tag{7.59}$$

with the condition $\boldsymbol{\xi} = \mathbf{x}(t)$ at $\tau \equiv t - s = t$, and application of (5.36) and (3.57). In (5.36) we can again equate $\gamma_{ij}(\xi^k)$ to $g_{ij}(x^k)$ because of the invariant magnitude of the base vectors $\mathbf{a}_i$ along the pathline of any particular material point. Furthermore, from the definitions of Section 5.11 it follows that components of $\mathbf{A}_1$ and $\mathbf{A}_2$ are given by

$$[a_1(ij)] = \begin{bmatrix} 0 & l & 0 \\ l & 0 & m \\ 0 & m & 0 \end{bmatrix}$$

$$[a_2(ij)] = 2(l^2+m^2)\begin{bmatrix} 0 & 0 & 0 \\ 0 & 1 & 0 \\ 0 & 0 & 0 \end{bmatrix} \tag{7.60}$$

so that a curvilineal flow satisfies (7.26). However, the physical components of $\mathbf{A}_1$ with respect to a basis in the directions $\mathbf{a}_i$ do not have the form of (7.24). The reader can readily verify that components of $\mathbf{A}_1$ and $\mathbf{A}_2$ are transformed into the form (7.24) and (7.25), respectively, by effecting a change of basis from $\mathbf{a}_i$ to the orthogonal system $\mathbf{e}_i$, defined by

$$\mathbf{e}_1 = \frac{l}{\kappa}\bar{\mathbf{a}}_1 + \frac{m}{\kappa}\bar{\mathbf{a}}_3$$

$$\mathbf{e}_2 = \bar{\mathbf{a}}_2 \tag{7.61}$$

$$\mathbf{e}_3 = -\frac{m}{\kappa}\bar{\mathbf{a}}_1 + \frac{l}{\kappa}\bar{\mathbf{a}}_3$$

where $\kappa = (l^2 + m^2)^{1/2}$ and the $\bar{\mathbf{a}}_i$ are unit vectors $\mathbf{a}_i / \sqrt{\mathbf{a}_i \cdot \mathbf{a}_i}$ (no sum). Thus we have proved that a curvilineal flow belongs to the class of viscometric flows.

The transformation (7.61) is an interesting one. The direction $\mathbf{a}_2$ must be retained to ensure that the transformation does not affect the form of $\mathbf{A}_2$, which already satisfies the requirement (7.25). The local velocity vector $\mathbf{v}$ is in the plane containing $\mathbf{e}_1$ and $\mathbf{e}_3$, and the orientation of these bases with respect to $\mathbf{v}$ is determined by the requirement that $d_{32} = 0$ with respect to the basis $\mathbf{e}_i$. That is to say, the component of $\mathbf{v}$ in the direction $\mathbf{e}_3$ can have no spatial gradient. Note that, in general, $\mathbf{e}_1$ is *not* parallel to $\mathbf{v}$.

In summary we see that all of the flows earlier called "laminar shear flows" are a subclass of curvilineal flows, which in turn form a subclass of viscometric flows. When we speak of a viscometric flow we usually mean one of the idealized laminar shear flows associated with common laboratory viscometers. However, the technical term "viscometric flow" is considerably broader in scope than the name may imply.

Our chief reason for describing curvilineal flows is to provide a point of departure for a special case; viz., *generalized steady helical flow* [14]. This is a curvilineal flow in which covariant components of the metric tensor take the more restricted form

$$g_{ii} = \mathbf{a}_i \cdot \mathbf{a}_i = [h_i(x^2)]^2 \quad \text{(no sum)} \tag{7.62}$$

the $\mathbf{a}_i$ being base vectors in an orthogonal coordinate system. In this coordinate system contravariant components of the velocity are again given by (7.56), so it is clear that generalized steady helical flows constitute a subclass of curvilineal flows.

As a final example we wish in particular to consider the steady helical flow of a simple fluid in the annular space between two coaxial circular cylinders. Contravariant components of the velocity are given by (7.56) with respect to the cylindrical coordinate system, and components of the metric tensor correspond to (7.62).

$$\begin{aligned} x^1 &= z; \, x^2 = r; \, x^3 = \phi \\ h_1 &= h_2 = 1; \, h_3 = r \end{aligned} \tag{7.63}$$

We envision such a flow to be generated by imposition of an axial driving force, characterized as before by a, and by steady rotation of one or both of the infinitely long circular cylinders bounding the flow. Thus the pathline of any material point is a helix. Components of the equation of motion (see appendix) must now include terms from both (7.43) and (7.53).

$$\frac{1}{r} \frac{d}{dr} (rt(rr)) - \frac{1}{r} t(\phi\phi) - \frac{\partial \Phi}{\partial r} = -\rho r \omega^2$$

$$\frac{1}{r} \frac{d}{dr} (rt(rz)) - \frac{\partial \Phi}{\partial z} = 0 \tag{7.64}$$

$$\frac{1}{r} \frac{d}{dr} [r^2 t(r\phi)] - \frac{\partial \Phi}{\partial \phi} = 0$$

The same reasoning which was previously applied to tube flow and cylindrical Couette flow leads to the conclusion that Φ can at most be of the form

$$\Phi = f(r) + c_1 z + c_2 \phi \tag{7.65}$$

where the c_i are constants. Then integration of (7.64b) and (7.64c) results in

$$t(rz) = \frac{c_1 r}{2} + \frac{1}{r} c_3$$

$$(7.66)$$

$$t(r\phi) = \frac{1}{2} c_2 + \frac{1}{r^2} c_4$$

where the constants are of course determined by application of suitable boundary conditions.

Our next task is to relate the stress components given above to the material functions for a simple fluid $\tau(\kappa)$, $\mathcal{N}_1(\kappa)$, and $\mathcal{N}_2(\kappa)$. This is readily done by noting that components of **T** have the "viscometric form" (7.32) and (7.33) when expressed with respect to the basis $\mathbf{e}_i$, which in turn is related to the unit vectors $\bar{\mathbf{a}}_i$ of a cylindrical cooordinate system by (7.61). In the present example

$$l = v'; \quad m = r\omega'; \quad \kappa = \sqrt{(v')^2 + (r\omega')^2} \qquad (7.67)$$

Recalling from Chapter 3 that a tensor is invariant to any change in coordinates, we write

$$\bar{\mathbf{a}}_i t(ij) \bar{\mathbf{a}}_j = \mathbf{e}_i \tilde{t}(ij) \mathbf{e}_j \qquad (7.68)$$

where the notation $\tilde{t}(ij)$ is used momentarily to indicate components as shown in (7.32) and (7.33). Using (7.61) it is now a simple matter to solve (7.68) for $t(21) = t(12) \equiv t(rz)$ and $t(32) = t(23) \equiv t(r\phi)$. We obtain

$$t(rz) = \frac{v'}{\kappa} \tau(\kappa); \quad t(r\phi) = \frac{r\omega'}{\kappa} \tau(\kappa) \qquad (7.69)$$

Note especially how the degenerate cases of axial-annular flow (flow in an annulus with fixed boundaries), tube flow, or cylindrical Couette flow result when one sets $\omega = 0$ in the first two cases or $v = 0$ in the third.

From combination of (7.69) with (7.66) there results a pair of differential equations for the velocity components

$$v'(r) = \left(\frac{c_1 r}{2} + \frac{c_3}{r} \right) \frac{\kappa}{\tau(\kappa)}$$

$$(7.70)$$

$$\omega'(r) = \left(\frac{c_2}{2r} + \frac{c_4}{r^3} \right) \frac{\kappa}{\tau(\kappa)}$$

where, from (7.67), one may write

$$\tau(\kappa) = \left[\left(\frac{c_1 r}{2} + \frac{c_3}{r} \right)^2 + \left(\frac{c_2}{2} + \frac{c_4}{r^2} \right)^2 \right]^{1/2} = f(r) \qquad (7.71)$$

$$\kappa = \tau^{-1}(f(r)) \qquad (7.72)$$

Thus in principle we can solve for the velocity profile from (7.70) once the boundary conditions, and hence the values of c_i, are known. As with all of the flows discussed earlier, we see that the velocity field is completely determined, for a given set of boundary conditions, by the single material function $\tau(\kappa)$.

Let us now consider the relation between differences of normal stress components in the cylindrical coordinate system and the normal stress material functions $\mathcal{N}_1(\kappa)$ and $\mathcal{N}_2(\kappa)$ of (7.33). The connection is again readily established through (7.68). The results can be cast in many different forms, one being

$$t(zz) - t(rr) = (v'/\kappa)^2(\mathcal{N}_1+\mathcal{N}_2) - \mathcal{N}_2$$

$$t(rr) - t(\phi\phi) = \mathcal{N}_2 - (r\omega'/\kappa)^2(\mathcal{N}_1+\mathcal{N}_2)$$

$$(7.73)$$

where we have used (7.61) along with $(v'/\kappa)^2 + (r\omega'/\kappa)^2 = 1$.

Because generalized helical flow includes many special cases which are of interest to viscometry, we shall have occasion in the next chapter to look further into the implications of the results which have just been developed.

7.7. OTHER APPROACHES TO SIMPLE FLOWS

Our development of viscometric flows of simple fluids has, to some extent, paralleled the evolution of the subject in the original literature. For a more direct treatment, in which explicit use of the right relative Cauchy–Green strain tensor is avoided, the reader is referred to the book of Coleman *et al.* [9]. Alternatively, one can formulate the kinematics of viscometric flows with heavy reliance on geometrical arguments, an example being the work of Yin and Pipkin [15].

We have restricted our examples to time-independent viscometric flows. However, we have noted that a general treatment of the subject includes the possibility of time as an independent variable [6, pp 427 *et seq.*]. Some applications to time-dependent problems have appeared [16, 17].

Another approach to the subject can be made by consideration of the class of flows with constant stretch history, a class from which viscometric flows form a special case [18; 6, pp. 427 *et seq.*]. We shall return to flows with constant stretch history later since some of them, though not viscometric in the technical sense, have utility as a means of rheological characterization.

APPENDIX:
Components of the Equation of Motion

We list below the continuity equation for flows with constant density

$$\operatorname{div} \mathbf{v} = 0$$

and components of the equation of motion

$$\rho\ddot{\mathbf{x}} = \rho g - \operatorname{grad} p + \operatorname{div} \mathbf{T} \tag{4.26}$$

for several common coordinate systems. Symbols in parentheses denote the particular physical component of a vector or tensor.

(I) RECTANGULAR CARTESIAN COORDINATES

$$\frac{\partial v(x)}{\partial x} + \frac{\partial v(y)}{\partial y} + \frac{\partial v(z)}{\partial z} = 0 \tag{7A1}$$

$$\rho \left[\frac{\partial v(x)}{\partial t} + v(x)\frac{\partial v(x)}{\partial x} + v(y)\frac{\partial v(x)}{\partial y} + v(z)\frac{\partial v(x)}{\partial z} \right]$$

$$= \rho g(x) - \frac{\partial p}{\partial x} + \frac{\partial t(xx)}{\partial x} + \frac{\partial t(yx)}{\partial y} + \frac{\partial t(zx)}{\partial z} \tag{7A2}$$

$$\rho \left[\frac{\partial v(y)}{\partial t} + v(x)\frac{\partial v(y)}{\partial x} + v(y)\frac{\partial v(y)}{\partial y} + v(z)\frac{\partial v(y)}{\partial z} \right]$$

$$= \rho g(y) - \frac{\partial p}{\partial y} + \frac{\partial t(xy)}{\partial x} + \frac{\partial t(yy)}{\partial y} + \frac{\partial t(zy)}{\partial z} \tag{7A3}$$

$$\rho \left[\frac{\partial v(z)}{\partial t} + v(x)\frac{\partial v(z)}{\partial x} + v(y)\frac{\partial v(z)}{\partial y} + v(z)\frac{\partial v(z)}{\partial z} \right]$$

$$= \rho g(z) - \frac{\partial p}{\partial z} + \frac{\partial t(xz)}{\partial x} + \frac{\partial t(yz)}{\partial y} + \frac{\partial t(zz)}{\partial z} \tag{7A4}$$

(II) CYLINDRICAL COORDINATES

$$\frac{1}{r}\frac{\partial}{\partial r}(rv(r)) + \frac{1}{r}\frac{\partial v(\phi)}{\partial \phi} + \frac{\partial v(z)}{\partial z} = 0 \tag{7A5}$$

$$\rho \left[\frac{\partial v(r)}{\partial t} + v(r)\frac{\partial v(r)}{\partial r} + \frac{v(\phi)}{r}\frac{\partial v(r)}{\partial \phi} - \frac{(v(\phi))^2}{r} + v(z)\frac{\partial v(r)}{\partial z} \right]$$

$$= \rho g(r) - \frac{\partial p}{\partial r} + \frac{1}{r}\frac{\partial}{\partial r}(rt(rr)) + \frac{1}{r}\frac{\partial t(r\phi)}{\partial \phi} - \frac{t(\phi\phi)}{r} + \frac{\partial t(rz)}{\partial z} \tag{7A6}$$

$$\rho \left[\frac{\partial v(\phi)}{\partial t} + v(r)\frac{\partial v(\phi)}{\partial r} + \frac{v(\phi)}{r}\frac{\partial v(\phi)}{\partial \phi} + \frac{v(r)v(\phi)}{r} + v(z)\frac{\partial v(\phi)}{\partial z} \right]$$

$$= \rho g(\phi) - \frac{1}{r}\frac{\partial p}{\partial \phi} + \frac{1}{r^2}\frac{\partial}{\partial r}(r^2 t(r\phi)) + \frac{1}{r}\frac{\partial t(\phi\phi)}{\partial \phi} + \frac{\partial t(\phi z)}{\partial z} \tag{7A7}$$

$$\rho \left[\frac{\partial v(z)}{\partial t} + v(r)\frac{\partial v(z)}{\partial r} + \frac{v(\phi)}{r}\frac{\partial v(z)}{\partial \phi} + v(z)\frac{\partial v(z)}{\partial z} \right]$$

$$= \rho g(z) - \frac{\partial p}{\partial z} + \frac{1}{r}\frac{\partial}{\partial r}(rt(rz)) + \frac{1}{r}\frac{\partial t(\phi z)}{\partial \phi} + \frac{\partial t(zz)}{\partial z} \tag{7A8}$$

(III) SPHERICAL COORDINATES

$$\frac{1}{r^2}\frac{\partial}{\partial r}(r^2 v(r)) + \frac{1}{r\sin\theta}\frac{\partial}{\partial \theta}(v(\theta)\sin\theta) + \frac{1}{r\sin\theta}\frac{\partial v(\phi)}{\partial \phi} = 0 \tag{7A9}$$

$$\rho\left\{\frac{\partial v(r)}{\partial t} + v(r)\frac{\partial v(r)}{\partial r} + \frac{v(\theta)}{r}\frac{\partial v(r)}{\partial \theta} + \frac{v(\phi)}{r\sin\theta}\frac{\partial v(r)}{\partial \phi} - \frac{1}{r}[(v(\theta))^2 + (v(\phi))^2]\right\}$$

$$= \rho g(r) - \frac{\partial p}{\partial r} + \frac{1}{r^2}\frac{\partial}{\partial r}(r^2 t(rr)) + \frac{1}{r\sin\theta}\frac{\partial}{\partial \theta}(t(r\theta)\sin\theta)$$

$$+ \frac{1}{r\sin\theta}\frac{\partial t(r\phi)}{\partial \phi} - \frac{1}{r}(t(\theta\theta) + t(\phi\phi)) \tag{7A10}$$

$$\rho\left[\frac{\partial v(\theta)}{\partial t} + v(r)\frac{\partial v(\theta)}{\partial r} + \frac{v(\theta)}{r}\frac{\partial v(\theta)}{\partial \theta} + \frac{v(\phi)}{r\sin\theta}\frac{\partial v(\theta)}{\partial \phi} + \frac{v(r)v(\theta)}{r} - \frac{(v(\phi))^2\cot\theta}{r}\right]$$

$$= \rho g(\theta) - \frac{1}{r}\frac{\partial p}{\partial \theta} + \frac{1}{r^2}\frac{\partial}{\partial r}(r^2 t(r\theta)) + \frac{1}{r\sin\theta}\frac{\partial}{\partial \theta}(t(\theta\theta)\sin\theta)$$

$$+ \frac{1}{r\sin\theta}\frac{\partial t(\theta\phi)}{\partial \phi} + \frac{t(r\theta)}{r} - \frac{\cot\theta}{r}t(\phi\phi) \tag{7A11}$$

$$\rho\left[\frac{\partial v(\phi)}{\partial t} + v(r)\frac{\partial v(\phi)}{\partial r} + \frac{v(\theta)}{r}\frac{\partial v(\phi)}{\partial \theta} + \frac{v(\phi)}{r\sin\theta}\frac{\partial v(\phi)}{\partial \phi} + \frac{v(\phi)v(r)}{r} + \frac{v(\theta)v(\phi)}{r}\cot\theta\right]$$

$$= \rho g(\phi) - \frac{1}{r\sin\theta}\frac{\partial p}{\partial \phi} + \frac{1}{r^2}\frac{\partial}{\partial r}(r^2 t(r\phi)) + \frac{1}{r}\frac{\partial t(\theta\phi)}{\partial \theta}$$

$$+ \frac{1}{r\sin\theta}\frac{\partial t(\phi\phi)}{\partial \phi} + \frac{t(r\phi)}{r} + \frac{2\cot\theta}{r}t(\theta\phi) \tag{7A12}$$

REFERENCES

1. Oldroyd, J. G., *Proc. Roy. Soc. (London)* A, **200**, 523 (1950).
2. Green, A. E., and Rivlin, R. S. *Arch. Rational Mech. Anal.* **1**, 1 (1957/58).
3. Green, A. E., Rivlin, R. S. and Spencer, A. J. M. *Arch. Rational Mech. Anal.* **3**, 82 (1959).
4. Green, A. E., and Rivlin, R. S., *Arch. Rational Mech. Anal.* **4**, 387 (1959/60).
5. Noll, W., *Arch. Rational Mech. Anal.* **2**, 197 (1958).
6. Truesdell, C., and Noll, W., The non-linear field theories of mechanics, in *Handbuch der Physik*, Band III/3 (S. Flügge, ed.), (Berlin: Springer-Verlag, 1965).
7. Truesdell, C., *The Elements of Continuum Mechanics* (New York: Springer-Verlag New York, 1966).
8. Coleman, B. D., and Noll, W., *Annals of the NY Acad. of Sci.* **89**, 672 (1961).
9. Coleman, B. D., Markovitz, H., and Noll, W., *Viscometric Flows of Non-Newtonian Fluids*, Chapter 3 (New York: Springer-Verlag New York, 1966).

10. Coleman, B. D., and Noll, W., *Arch. Rational Mech. Anal.* **3**, 289 (1959).
11. Astarita, G., *Ind. Eng. Chem. Fund.* **6**, 257 (1967).
12. Bird, R. B., and Marsh, B. D., *Trans. Soc. Rheol.* **12**, 479 (1968).
13. Fredrickson, A. G., *Principles and Applications of Rheology*, Chapter 7 (Englewood Cliffs, New Jersey: Prentice-Hall, 1964).
14. Coleman, B. D., *Arch. Rational Mech. Anal.* **9**, 273 (1962).
15. Yin, W.-L., and Pipkin, A. C., *Arch. Rational Mech. Anal.* **37**, 111 (1970).
16. Johns, L. E., Jr., *AIChE Jl.* **14**, 275 (1968).
17. Wankat, P. C., *AIChE Jl.* **15**, 150 (1969).
18. Huilgol, R. R., On the construction of motions with constant stretch history: Parts I and II, Mathematics Research Center, The University of Wisconsin, Reports 954 (1968) and 975 (1969); also *Quart. Appl. Math.* **29**, 1 (1971).

PROBLEMS

7.1. In section 7.2 a rationalization was put forward for replacing $\mathbf{U}(t)$ as an argument in the functional of (7.8) by $\rho(t)$ in the functional of (7.9). Use (4.11) to justify this substitution.

7.2. Use the principle of material objectivity to show that, for the rest history,

$$\mathbf{Q} \underset{s=0}{\overset{\infty}{\mathscr{H}}} (\mathbf{I}) \mathbf{Q}^T = \underset{s=0}{\overset{\infty}{\mathscr{H}}} (\mathbf{I}) \tag{7.15}$$

7.3. Derive (7.24) and (7.25).

7.4. Verify (7.32).

7.5. A Newtonian fluid (defined by $\tau(\kappa) = \mu\kappa$) is in laminar flow in the annular space between two long coaxial cylinders sketched below. Fluid viscosity $\mu = 4$ poise, and the fluid is being driven both axially and tangentially. Axial motion is due to a pressure gradient $dp/dz = -820$ dyne/cm³. Gravitational effects are not important. $R_i = 3.8$ cm and $R_o = 5.1$ cm. The inner cylinder is stationary, but the outer cylinder is rotated at an angular speed $\Omega = 100$ rpm.

a. Plot the angle, relative to the horizontal, of the velocity vector as a function of $\delta = (r - R_i)/(R_o - R_i)$.

b. Plot the angle, relative to the horizontal, of $\mathbf{e}_3$ as a function of δ. Note that $\mathbf{e}_3$ is normal to the plane containing $\mathbf{e}_1$ and $\mathbf{e}_2$, and that in the basis $\{\mathbf{e}_i\}$ $\partial v(3)/\partial x(2) = 0$ at a given value of δ.

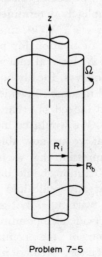

Problem 7-5

CHAPTER 8

MEASUREMENT OF THE VISCOMETRIC FUNCTIONS

8.1. INTRODUCTION

In the last two chapters we have described some of the cornerstones of rheology, the study of deformation and flow of matter. It has been shown how one achieves progress in the description of flows of non-Newtonian fluids by performing a sort of balancing act between constitutive equation and flow field. A case in point is the concept of a simple fluid which, of itself, is of little utility to one interested in solving flow problems. However, the definition of a simple fluid when joined with the specification of a viscometric flow offers great rewards. We saw in the preceding chapter how the extra stress field and hence also the velocity field are completely determined for any given viscometric flow as soon as the three material functions $\tau(\kappa)$, $\mathcal{N}_1(\kappa)$, and $\mathcal{N}_2(\kappa)$ of (7.33) are known for a given fluid. This result is a strong motivation for performing experiments which will permit one to determine τ, $\mathcal{N}_1$, and $\mathcal{N}_2$ for real materials, assuming of course that these behave as simple fluids in viscometric flows. Ideally one would wish for a catalogue of the viscometric functions to be tabulated in some convenient reference for all non-Newtonian materials of interest. Then we could indeed say that prediction of any viscometric flow of a simple fluid is within our grasp. We would be in a position to compute viscometric flow behavior to the same extent that we are now able for Newtonian fluids with known viscosity. Lest the reader gain an impression that we are at or are approaching the rheologist's Utopia, several statements of caution should be made.

First, we have no assurance that any real fluid *is* a "simple" fluid in the sense of Noll's definition. However, that warning is not as serious as it may sound. There is a collection of experimental evidence pointing to the fact that once $\tau(\kappa)$ has been determined in any one of the common viscometric flows, that information is often sufficient for satisfactory prediction of the velocity field in other common viscometric flows [1, 2]. The second warning is more serious: though we appear to be able to measure $\tau(\kappa)$ with relative ease and with high accuracy, the normal stress functions are *much* more elusive. We shall see that there has been conflicting evidence regarding the magnitude and even the sign of $\mathcal{N}_2(\kappa)$ for several polymer solutions. Third, it is not at all clear how the viscometric functions may be used, even if they are known, when one wishes to make predictions about the behavior of a simple fluid in a general *nonviscometric* flow.

In this chapter we shall develop the theory that underlies experimental measurement of the viscometric functions in a number of common viscometric devices. It is important to remember that *all* viscometric flows are idealizations of flow in some real configuration. Thus in a tube viscometer one must recognize the possible effect of the entrance region on the flow, and in flow between coaxial cylinders it is all too easily forgotten that the flow is bounded, generally by a stationary or rotating member at the bottom surface and by air at the top. These factors, as well as others which are more subtle, make meaningful measurement of the viscometric functions a difficult task. This is not the proper place to discuss the fine points of rheological experimentation. However, the reader is asked to remember that an understanding of the equations which are developed in this text constitutes only the beginning of an expertise in the subject.

The number of devices employed to make rheological measurements is limited only by experimenters' ingenuity and courage, both of which have been considerable. Though no systematic compilation exists for measurement of normal stresses,[†] a good discussion of devices for measurement of $\tau(\kappa)$ has been provided by Van Wazer *et al.* [3]. Our own treatment will be limited to a few of those experiments employed to measure τ, $\mathcal{N}_1$, and $\mathcal{N}_2$, and we shall restrict ourselves to flow systems which at least approximate viscometric flows. An exception is the discussion of measurements made with jet devices (Section 8.7).

The chief objective is to develop equations from which the viscometric functions may be obtained by making suitable measurements. A natural byproduct is specification of the velocity field when the assumption of a viscometric flow is valid. In Section 8.8 we briefly consider some of the more pervasive experimental problems which attend certain of the measurements.

8.2. PLANE COUETTE FLOW

There have been few attempts, with either Newtonian or non-Newtonian fluids, to generate flow between parallel planes in relative motion. However, plane Couette flow is such an important and uncomplicated limiting case that we include it in passing. Referring back to Section 7.5 and Fig. 7.1 we suppose again that the lower plane is at rest and the upper plane is moving at steady velocity V. The velocity field in the space $0 \leqslant y \leqslant h$ is immediately obtained from (7.39). The function $\tau(\kappa)$ follows directly from the measurement of the drag force exerted on either the upper or lower plane. By a simple force balance we see that if the drag force exerted in the positive x-direction, by the fluid *on* the upper plate is $-F$ over a plate surface of area A, then from (7.37a)

$$t_{xy}(h) = t_{xy}(y) = F/A = \tau(V/h) \tag{8.1}$$

since $dv(y)/dy = \kappa = V/h$ from (7.39).[‡] Plane Couette flow is an especially simple example since (7.39) provides the velocity field once V and h are specified, irrespective of $\tau(\kappa)$.

[†]*Note added in proof.* A recent book on viscometry includes descriptions of normal stress measurement: Walters, K. *Rheometry* (London: Chapman and Hall, 1975).

[‡]Recall that our sign convention for components of the stress tensor follows (4.23) and Fig. 4.4, with the unit normal **n** pointing *out* of the fluid. It is useful in this context to note that if one considers the "positive" y-face of an elemental cube of fluid (outer normal **n** pointing in the positive y-direction) then a stress component, e.g. t_{xy}, is positive when it exerts a force on the fluid in the positive x-direction. When considering the "negative" y-face of the elemental cube, t_{xy} at that face is positive if it exerts a force on the fluid in the negative x-direction. This result is in accord with the convention that "positive" stresses are tensile.

We remark briefly about the normal stress functions. Since κ is a constant $\mathcal{N}_1$ and $\mathcal{N}_2$ are everywhere constant. Furthermore, though we should expect to be able to measure t_{yy} by sensing the stress normal to the plate surface, we would actually be measuring the total stress s_{yy}, which is of little significance by itself since it contains an arbitrary function p. Thus if we collected $s_{yy}(\kappa)$ over a range of κ, this of itself would be of no use in evaluating $\mathcal{N}_1(\kappa)$ or $\mathcal{N}_2(\kappa)$. It is now clear why we desire to measure normal stress *differences*. In the present flow there appear to be no unambiguous means for measuring $s_{xx}(\kappa)$ or $s_{zz}(\kappa)$. Hence the results cannot be reduced to the meaningful quantities $\mathcal{N}_1(\kappa)$ and $\mathcal{N}_2(\kappa)$.

8.3. LAMINAR FLOW THROUGH TUBES

One of the most popular of simple viscometric devices is a cylindrical tube often called, with little justification, a "capillary" viscometer. The fluid to be tested is made to flow through a cylindrical tube at a sufficiently low value of Reynolds number so that the flow is laminar. Volumetric flow of the test fluid is measured over a known period of time and a measure of the "pressure drop" is obtained by a manometer or other suitable device which senses the difference in radial thrust against the tube wall at two axial positions a known distance apart. In some cases the "pressure drop" is inferred from the pressure difference between a reservoir at one end of the tube and the ambient pressure at the tube exit, a known axial distance from the reservoir. In this case one of course must deal with the complication of departure from viscometric flow near the tube entrance and exit. The interested reader is urged to consult Van Wazer *et al.* [3] or Fredrickson [4, ch. 9] for further details.

For an analysis of tube flow we may again draw upon the results of Section 7.5, keeping in mind that the data collected by the experimenter permit direct calculation of volumetric flow rate Q and driving force a for flow of a fluid with density ρ through a tube of radius R. If the radial thrust measurements P_P and P_Q are made at axial positions z_P and z_Q, respectively (Fig. 7.2), we have

$$P_P - P_Q = (s(rr))_{R,z_Q} - (s(rr))_{R,z_P} = p_{R,z_P} - p_{R,z_Q} \tag{8.2}$$

since the extra stress $(t(rr))_R$ is, by (7.29), a function only of κ and hence does not vary with axial position. From (7.42) and (7.44) we see that

$$a = \frac{\Phi_{R,z_P} - \Phi_{R,z_Q}}{z_Q - z_P} = \frac{(P_P - P_Q) + \rho(\psi_P - \psi_Q)}{z_Q - z_P} \tag{8.3}$$

where $(\psi_P - \psi_Q)$ is readily calculated from a knowledge of the difference in elevation of the two points at which radial thrust is measured. Since a is known, the distribution of shear stress t_{rz} is found from (7.48).

In the event that $\tau(\kappa)$ is known, we can compute the velocity field by integration of (7.50). Simple change of variable and application of the no-slip boundary condition give us the quadrature

$$v(r) = \int_r^R \tau^{-1}(a\xi/2)\,d\xi \tag{8.4}$$

However, in viscometry our problem is to determine the function τ, if possible without any *a priori* specification of its form, from data which relate Q to a. This may be done by application of one of any number of equivalent forms which is usually dubbed the "Rabinowitsch equation" [5].[†] We have

$$Q = \int_0^R 2\pi r v(r)\,dr \tag{8.5}$$

which is conveniently integrated by parts to give

$$Q = -\int_0^R \pi r^2 \frac{dv(r)}{dr}\,dr = \frac{8\pi}{a^3}\int_0^{(aR)/2} \xi^2 \tau^{-1}(\xi)\,d\xi \tag{8.6}$$

Now we rearrange (8.6) and apply Leibniz' rule for differentiation under the integral [7, p. 348], with the result

$$\tau\left\{\frac{1}{\pi a^2 R^3}\frac{d}{da}\left[a^3 Q(a)\right]\right\} = \frac{aR}{2} \tag{8.7}$$

where the curly brackets indicate functional dependence for τ. Thus an experimental determination of $\tau(\kappa)$ is possible by applying (8.7) to a set of data relating Q and a. One can obtain a graphical or numerical estimate of the required derivative by recording the dependence of $a^3 Q(a)$ on a.

Whenever possible, (8.7) should be applied to data from more than one tube diameter. Absence of any dependence on R beyond that indicated in (8.7) is a good test of the assumptions leading to the equation. Suspensions, for example, may undergo particle segregation in tube flow. This heterogeneity can cause a diameter effect not accounted for in (8.7).

Equation (8.7) is *a posteriori* a straightforward result. Nevertheless, it is one of the more clever techniques developed for treatment of rheological data. Its careful study is bound to be rewarding.

One would of course next wish to prescribe means for measurement of the normal stress functions from tube flow data. However, this is not readily done, and the reasons are identical to those set forth in the preceding section. We have no way to isolate the extra stress $(t(rr))_R$ from measurements of $(s(rr))_{R,z}$, and there is no unequivocal way to measure $s(zz)$ or $s(\phi\phi)$.

8.4. THE PARALLEL-PLATE VISCOMETER

We now discuss the first of several devices from which normal as well as shear stress measurements can be made. Typical applications are available in the reports of Greensmith and Rivlin [8] and Kotaka *et al.* [9]. This flow, often referred to as torsional flow [10, p. 79], is shown in Fig. 8.1. We envision a fluid contained between two horizontal circular disks of radius R and separated by a distance h. The upper disk is stationary and the lower disk rotates with angular speed Ω. We assume a laminar shearing motion of the form (7.18) with contravariant components

[†]Apparently the technique was first employed by Herzog and Weissenberg [6].

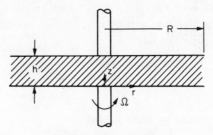

Fig. 8.1. Torsional flow.

$$\mathbf{v} = [\omega(z), 0, 0] \qquad (8.8)$$

where

$$x^1 = \phi; \quad x^2 = z; \quad x^3 = r$$
$$h_1 = r; \quad h_2 = h_3 = 1$$

so that, from (7.22),

$$\kappa = r\omega'(z)$$

For a simple fluid the extra stress can be dependent only upon κ so that physical components of the equation of motion take the form

$$\left.\begin{aligned}
\frac{1}{r}\frac{\partial}{\partial r}(rt(rr)) - \frac{1}{r}t(\phi\phi) - \frac{\partial \Phi}{\partial r} &= -\rho r\omega^2 \\[2mm]
\frac{\partial}{\partial z}(t(zz)) - \frac{\partial \Phi}{\partial z} &= 0 \\[2mm]
r\frac{\partial}{\partial z}t(z\phi) - \frac{\partial \Phi}{\partial \phi} &= 0
\end{aligned}\right\} \qquad (8.9)$$

A necessary condition for validity of the assumption (8.8) is of course the requirement that the velocity components satisfy any stated boundary conditions and are compatible with each component of the equation of motion. In previous examples, the latter requirement was trivially fulfilled. However, in the present case we meet the first of several important examples where a difficulty in this regard is encountered. We find that (8.8) is *not* compatible with (8.9) if we wish to impose only a gravitational body force and if we assume a simple pressure distribution on the flow boundaries. The difficulty is introduced through the "inertial term" $\rho r\omega^2$ in (8.9a). Our theoretical problems are overcome if we neglect that term. It is clear, however, that experimental problems may be more difficult to eliminate. For the present we shall proceed on the assumption that the term $\rho r\omega^2$ is of negligible importance.

(i) SHEAR STRESS MEASUREMENTS

Following earlier arguments one concludes that $\partial\Phi/\partial\phi = 0$ in (8.9c) and, consequently,

$$t(z\phi) = f(r) = \tau(r\omega'(z)) \qquad (8.10)$$

from which it follows that

$$\omega' = c_1; \quad \omega = c_1 z + c_2$$

From boundary conditions we obtain the values of constants c_1 and c_2. Then

$$\omega = \Omega \left(1 - \frac{z}{h}\right) \tag{8.11}$$

$$v = r\omega \tag{8.12}$$

$$\omega' = -\Omega/h \tag{8.13}$$

$$\kappa = -\Omega r/h \tag{8.14}$$

where v here signifies the physical component of the velocity in the ϕ-direction. As in the case of plane Couette flow, the velocity field is fixed by the boundary conditions without regard to the form of $\tau(\kappa)$.

To determine $\tau(\kappa)$ by experiment one measures the torque $\mathscr{T}z$ exerted by the fluid on, say, the upper disk over a range of rotational speeds Ω. The torque is related to $\tau(\kappa)$ by

$$\mathscr{T} = -\int_0^R t(z\phi) 2\pi r^2 dr = 2\pi \left(\frac{h}{\Omega}\right)^3 \int_0^{(R\Omega/h)} \tau(\xi)\xi^2 d\xi \tag{8.15}$$

Note the similarity to the result of the previous section. The function that we wish to determine is the kernel of an integral equation. Once again we can apply Leibniz' rule and obtain

$$\tau\left(\frac{R\Omega}{h}\right) = \frac{1}{2\pi R^3 \Omega^2} \frac{d(\mathscr{T}\Omega^3)}{d\Omega} \tag{8.16}$$

so that once the geometry of the system is known one may proceed to evaluate the function $\tau(\kappa)$ from a series of measurements which provides $\mathscr{T} = \mathscr{T}(\Omega)$.

(ii) NORMAL STRESS MEASUREMENTS

In addition to the limitations imposed by neglect of inertial effects, measurement of normal stresses is hampered by two further complications. First, one is not able from a single set of measurements to obtain N_1 and N_2 separately. Second, in order to relate thrust measurements to N_1 and N_2 it is necessary to specify a boundary condition at the rim of the apparatus. Because of uncertainties in the flow pattern at the rim, the stated boundary condition is only approximately correct.

From the requirement that Φ be single-valued over the flow space, and noting from (8.14) that the extra stress is only a function of spatial variable r, one obtains from (8.9)

$$\frac{d}{dr} t(rr) + \frac{1}{r} [t(rr) - t(\phi\phi)] = \frac{d\Phi}{dr}$$

where the term $-\rho r\omega^2$ of (8.9) has been neglected. Now although $\Phi = \Phi(r)$, its component parts p and ψ certainly depend upon z as well. In fact, for the configuration under discussion, $\psi = \psi(z)$. Suppose we are only interested in the stress distribution at a given elevation, say $z = h$. Then $d\Phi(r)/dr = dp(r)/dr$, and

$$\frac{d}{dr}[s(zz) - N_2] - \frac{1}{r}(N_1 + N_2) = 0 \tag{8.17}$$

since $N_1 = t(\phi\phi) - t(zz)$; $N_2 = t(zz) - t(rr)$

Integrating (8.17) we obtain

$$(s(zz))_r - (s(zz))_0 = N_2(\kappa) + \int_0^\kappa \frac{1}{\xi}[N_1(\xi) + N_2(\xi)]d\xi \tag{8.18}$$

where we have used $N_2(0) = 0$. We wish to employ (8.18) to derive an expression for the thrust F which the fluid exerts in an upward direction on the top disk. To do this we need to say something about the state of stress at the rim boundary. For simplicity we neglect any effects of interfacial tension or of motion of the fluid in the region outside of the rim. It is simply asserted that $(s(rr))_R = -p_s$, where p_s is a static pressure, e.g., atmospheric pressure plus any hydrostatic head which may exist if the apparatus is immersed into a reservoir to some distance below the liquid surface. The force F is computed relative to p_s. Then

$$F = -\int_0^R [s(zz) + p_s]2\pi r dr = -\left|\begin{array}{c} R \\ 0 \end{array}\right. \pi r^2[s(zz) + p_s] + \int_0^R \pi r^2 \frac{d}{dr}[s(zz) + p_s]dr \tag{8.19}$$

Before substituting (8.18) into (8.19) we eliminate $(s(zz))_0$ by applying the rim boundary condition

$$(s(zz))_R = (s(rr))_R + N_2(\kappa_R) = -p_s + N_2(\kappa_R)$$

where $\kappa_R = -\Omega R/h$. Then (8.18) becomes

$$(s(zz))_r = -p_s + N_2(\kappa) + \int_{\kappa_R}^\kappa \frac{1}{\xi}[N_1(\xi) + N_2(\xi)]d\xi \tag{8.20}$$

Substitution of (8.20) into (8.19) with some subsequent rearrangement yields

$$F = \frac{\pi R^2}{\kappa_R^2}\int_0^{\kappa_R}(N_1 - N_2)\kappa d\kappa \tag{8.21}$$

Applying Leibniz' rule, we obtain

$$\frac{1}{\pi R^2 \kappa_R}\frac{d}{d\kappa_R}(F\kappa_R^2) = N_1(\kappa_R) - N_2(\kappa_R) \tag{8.22}$$

Thus we see how the combination $(N_1 - N_2)$ may be obtained from torsional flow measurements in which the normal thrust is measured as a function of rotational speed. In order to isolate the effects of N_1 and N_2 it is clear that more data on this or some other apparatus are required. One possibility is measurement of the radial *distribution* of axial thrust. Equation (8.17) can be written

$$\frac{ds(zz)}{d\ln r} = \frac{d\mathcal{N}_2(\kappa)}{d\ln \kappa} + \mathcal{N}_1(\kappa) + \mathcal{N}_2(\kappa) \qquad (8.23)$$

and, of course, we have $\kappa = -\Omega r/h$. From combination of (8.22) and (8.23) and use of the boundary condition $\mathcal{N}_2(0) = 0$, we have, in principle, a means for experimental measurement of $\mathcal{N}_1$ and $\mathcal{N}_2$.

This device has been used by a number of experimenters. If consistent results are obtained from devices with different combinations of h and R, one can infer that the assumptions made in the theoretical development are valid. Greensmith and Rivlin have discussed means for making corrections for the effect of the centrifugal term which was dropped from (8.9a) [8].

8.5. CONE-AND-PLATE VISCOMETER

Though the configuration of the cone-and-plate viscometer, as well as the parallel-plate device, does not fulfill the requirements for a viscometric flow without introduction of some simplifying assumptions, the cone-and-plate apparatus has enjoyed wide popularity as a viscometer for non-Newtonian fluids. This is due partially to the fact that one can obtain information about both normal and shear stress behavior of test fluids, and partially to the ready availability of versatile commercial instruments manufactured specifically for use with rheologically complex fluids.

A schematic sketch of the cone-and-plate device is shown in Fig. 8.2a, which we describe

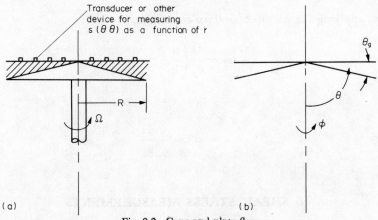

Fig. 8.2. Cone-and-plate flow.

in terms of a spherical coordinate system illustrated in Fig. 8.2b. Contravariant components of the velocity are assumed to be of the form

$$\mathbf{v} = (\omega(\theta), 0, 0) \qquad (8.24)$$

where

$$x^1 = \phi; \quad x^2 = \theta; \quad x^3 = r$$
$$h_1 = r\sin\theta; \quad h_2 = r; \quad h_3 = 1$$

so that

$$\kappa = \omega'(\theta) \sin \theta$$

If the flow postulated in (8.24) is to be compatible with physically reasonable boundary conditions for a simple fluid, it can be shown that the effect of inertia must be negligible, and that the gap angle θ_g must be sufficiently small so that terms containing $\cot \theta$ can be neglected. Then the equation of motion in spherical coordinates (see appendix to Chapter 7) reduces to the following physical components:

$$\left.\begin{array}{c} 2t(rr) - t(\theta\theta) - t(\phi\phi) - r\dfrac{\partial \Phi}{\partial r} = 0 \\[3mm] \dfrac{dt(\theta\theta)}{d\theta} - \dfrac{\partial \Phi}{\partial \theta} = 0 \\[3mm] \dfrac{dt(\theta\phi)}{d\theta} - \dfrac{\partial \Phi}{\partial \phi} = 0 \end{array}\right\} \qquad (8.25)$$

and

$$\kappa = \omega'(\theta) \qquad (8.26)$$

Assuming that boundary conditions require $\partial\Phi/\partial\phi = 0$, one immediately concludes that

$$t(\theta\phi) = \text{const} = \tau(\kappa) = \tau(\omega') \qquad (8.27)$$

Assuming also, as we have before, that τ has a unique inverse,

$$\omega' = \text{const}$$

whereupon, applying the no-slip boundary conditions,

$$\omega = \frac{\Omega}{\theta_g}\left(\frac{\pi}{2} - \theta\right) \qquad (8.28)$$

$$v_\phi = r\omega \qquad (8.29)$$

$$\kappa = \omega' = -\frac{\Omega}{\theta_g} \qquad (8.30)$$

(i) SHEAR STRESS MEASUREMENTS

The torque $\mathscr{T}$ exerted on the upper plate of the apparatus (or on the lower cone, to the order of approximation being used here) is easily found.

$$\mathscr{T} = -\int_0^R t(\theta\phi)2\pi r^2 dr = \frac{2}{3}\pi R^3\tau(\Omega/\theta_g) \qquad (8.31)$$

Thus the material function $\tau(\kappa)$ is readily evaluated from measurements of torque as a function of rotational speed. From extensive measurements one can conclude that the "small gap" approximation imposed in the development of (8.31) is quite reasonable for cone angles less than about 4°. Several commercial instruments employ $\theta_g = 2°$. A convenience of the cone-and-plate geometry is that the shear rate is essentially constant over the flow field. However, it shares with the parallel-plate device the limitation of values of κ below which centrifugal terms are dominant, and this is indeed a serious limitation.

(ii) NORMAL STRESS MEASUREMENTS

As with parallel-plate flow we have a means for separating $\mathcal{N}_1$ and $\mathcal{N}_2$ by measuring both total thrust and the variation of $s(\theta\theta)$ with r along a platen surface. To derive the relevant equations we apply (8.25a) along the plate at $\theta = \pi/2$. Then, since this plate is horizontal,

$$\frac{\partial \Phi}{\partial r} = \frac{\partial}{\partial r}(p + \rho\psi) = \frac{\partial}{\partial r}p(r, \theta) = -\frac{\partial}{\partial r}s(\theta\theta)$$

since $t(\theta\theta)$ is not a function of r. Recalling the definitions of $\mathcal{N}_1$ and $\mathcal{N}_2$ we find

$$\frac{\partial}{\partial \ln r}s(\theta\theta) = \mathcal{N}_1 + 2\mathcal{N}_2 \tag{8.32}$$

which we integrate along $\theta = \pi/2$ to obtain

$$(s(\theta\theta))_r = (s(\theta\theta))_R + (\mathcal{N}_1 + 2\mathcal{N}_2)\ln\left(\frac{r}{R}\right) \tag{8.33}$$

We eliminate $(s(\theta\theta))_R$ by assuming that along the rim $(r = R, \theta = \pi/2)$ there is a uniform known static pressure (such as atmospheric pressure) p_s. Then we write, for $\theta = \pi/2$,

$$(s(rr))_R = -p_s = (t(rr))_R - p(R, \pi/2) = (s(\theta\theta))_R - \mathcal{N}_2$$

Combining with (8.33) we have

$$(s(\theta\theta))_r = -p_s + \mathcal{N}_2 + (\mathcal{N}_1 + 2\mathcal{N}_2)\ln(r/R) \tag{8.34}$$

Let us use F to indicate the upward thrust, over and above that due to p_s, which is exerted by the fluid on the upper plate. Then

$$F = -\int_0^R [(s(\theta\theta))_r + p_s]2\pi r\,dr = -\int_0^R \left[\mathcal{N}_2 + (\mathcal{N}_1 + 2\mathcal{N}_2)\ln\left(\frac{r}{R}\right)\right]2\pi r\,dr$$

Since $\kappa = $ const throughout the flow field, we may integrate this equation directly and obtain

$$F = \frac{1}{2}\pi R^2 \mathcal{N}_1 \tag{8.35}$$

So from thrust measurements the first normal stress material function is readily obtained. However, note that from (8.32) we have means both to check consistency of the assumptions employed, and also to evaluate $\mathcal{N}_1 + 2\mathcal{N}_2$. We see from (8.32) that if our assumptions are valid, one expects the slope of a plot of $s(\theta\theta)$ vs. $\ln r$ to be a constant at any given value of κ. Also, we see that from the combination of total thrust and thrust-gradient measurements implied by (8.32) and (8.35), we should be able to obtain separate functional relationships for $\mathcal{N}_1(\kappa)$ and $\mathcal{N}_2(\kappa)$. The literature contains numerous reports of cone-and-plate experiments in which (8.32) has been used to estimate $(\mathcal{N}_1 + 2\mathcal{N}_2)$, and the separate functions $\mathcal{N}_1$ and $\mathcal{N}_2$ have been obtained by combination of thrust-gradient data with total thrust measurements or with measurements in other geometries [10, p. 75]. Validity of most of these results, however, has been questioned by Ginn and Metzner [11].

8.6. FLOW IN ANNULAR SPACES

In this section we consider two special cases of the form of helical flow which was discussed in Section 7.5. We could proceed to carry forward the discussion of helical flow from Section 7.5, and indeed specific helical flows have been studied experimentally [1, 2]. However, we shall consider separately the degenerate cases of cylindrical Couette flow and axial-annular flow, which are readily treated within the framework of laminar shear flows of Section 7.4.

(i) CYLINDRICAL COUETTE FLOW

This is a viscometric flow which has been used since the earliest days of rheology [3, ch. 2] to characterize the shear behavior of non-Newtonian fluids. More recently the configuration has also been utilized to characterize normal stress differences [10, p. 65; 12; 13].

(a) *Shear Stress Measurements.* Cylindrical Couette flow is characterized by the special form of (7.18), where we set $v(x^2) \equiv \omega(x^2)$, so that

$$\mathbf{v} = (\omega(x^2), 0, 0) \tag{8.36}$$

in the cylindrical coordinate system (7.51). The flow is approximated experimentally by inserting a test fluid in the annular space between two vertical coaxial cylinders, one of which is rotating with respect to the other. For the sake of definiteness we consider the flow generated by the apparatus pictured in Fig. 8.3 where, as is customary in Couette

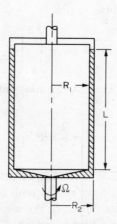

Fig. 8.3. Cylindrical Couette device.

viscometers, the inner cylinder is held stationary and the outer one is rotated at angular speed Ω. The contribution from end effects must of course be considered in any real system [3, ch. 2]. End effects can be minimized by ensuring that the instrument dimensions are such as to generate much more drag on the vertical cylinder walls than on the other surfaces.

Over any annular surface at radius r in the flow space we readily compute the torque $\mathcal{T}$ exerted on the fluid interior to r.

$$\mathcal{T} = t(r\phi)2\pi r^2 L$$

so that we can ascribe definite physical significance to the integration constant T of (7.54). It is the torque per unit height exerted on any annular surface, this torque being independent of r. If the material function $\tau(\kappa)$ is known, we may obtain the velocity field from integration of (7.55). One readily finds

$$\omega(r) = \int_{\frac{\mathcal{T}}{2\pi r^2 L}}^{\frac{\mathcal{T}}{2R_1^2 L}} \frac{\tau^{-1}(\xi)}{2\xi} \, d\xi \tag{8.37}$$

However, in order to evaluate $\tau(\kappa)$ from data which relate $\mathcal{T}$ to Ω we, as in previous cases, wish to isolate the kernel of the integral equation (8.37). If, as is frequently true, $(R_2 - R_1) \ll R_1$, the isolation and inversion are easily accomplished. The mean value theorem can be invoked [14], and we write for (8.37) between R_1 and R_2,

$$2\Omega \cong \tau^{-1}(\bar{\xi})\bar{\xi}^{-1}\Delta\xi$$

where $\bar{\xi}$ is the arithmetic mean value between limits of integration. Evaluating to $O(\Delta R/R)$ one obtains

$$\tau\left(\frac{R\Omega}{\Delta R}\right) = \frac{\mathcal{T}}{2\pi R^2 L} \tag{8.38}$$

where $\Delta R = R_2 - R_1$. Thus $\tau(\kappa)$ is readily found from measurements of torque and angular velocity Ω. When the assumption $\Delta R/R \ll 1$ is not valid, one must use an approximate procedure, apparently first proposed by Krieger and Elrod [15], to invert (8.37). Although the accuracy of (8.38) is dependent upon a small gap between cylinders, the Krieger and Elrod method converges most readily when the gap is appreciable, i.e., when $\lambda = (R_1/R_2)^2 \ll 1$.

Suppose that we attempt to proceed by the method used in earlier sections and differentiate (8.37). Taking limits of $\mathcal{T}/(2\pi R_2^2 L)$ and $\mathcal{T}/(2\pi R_1^2 L)$ in (8.37) and differentiating with respect to $\mathcal{T}$ for given apparatus constants, one obtains

$$2\frac{d\Omega}{d\ln\mathcal{T}} = \tau^{-1}\left(\frac{\mathcal{T}}{2\pi R_1^2 L}\right) - \tau^{-1}\left(\frac{\mathcal{T}}{2\pi R_2^2 L}\right) \tag{8.39}$$

In contrast to earlier cases, it is not apparent that $\tau(\kappa)$ can be isolated. However, following Krieger and Elrod, and others [10, p. 41; 16; 17], note that (8.39) can be written

$$2 \frac{d\Omega}{d \ln \mathscr{T}} \equiv F(\mathscr{T}) = \tau^{-1}\left(\frac{\mathscr{T}}{2\pi R_1^2 L}\right) - \tau^{-1}\left(\frac{\mathscr{T}\lambda}{2\pi R_1^2 L}\right) \tag{8.40}$$

from which it follows that

$$\sum_{n=0}^{N} F(\lambda^n \mathscr{T}) = \tau^{-1}\left(\frac{\mathscr{T}}{2\pi R_1^2 L}\right) - \tau^{-1}\left(\frac{\mathscr{T}\lambda^{N+1}}{2\pi R_1^2 L}\right)$$

But since $\lambda < 1$ and $\tau(0) = 0$, we can write

$$\tau\left\{\sum_{n=0}^{\infty} F(\lambda^n \mathscr{T})\right\} = \frac{\mathscr{T}}{2\pi R_1^2 L} \tag{8.41}$$

where the curly brackets represent functional dependence. From data relating $\mathscr{T}$ to Ω we readily relate $F(\Upsilon)$ to Υ from its definition in (8.40). Then the summation of (8.41) is calculable to the desired degree of accuracy, with the result that the function $\tau(\kappa)$ can be found. Clearly, this method is most easily applied for relatively wide gaps. Then the summation of (8.41) will converge in just a few terms.

Krieger [18] has shown how evaluation of $\tau(\kappa)$ from data in a coaxial cylinder apparatus can be greatly facilitated. He has rearranged the series solution (8.41) so that the dominant term corresponds to the result obtained for a power-law fluid, a particular model of rheological behavior which we consider in Chapter 10. The interested reader is referred to Krieger's paper for details.

(b) *Normal Stress Measurements.* Suppose that we now modify the apparatus of Fig. 8.3 in such a way that the difference in radial thrust $(s(rr))_{R_2} - (s(rr))_{R_1}$ is measured, in addition to previous measurements. A schematic drawing of such an apparatus is depicted in Fig. 8.4 (see, e.g., [12]). For a fluid with density ρ we may write

$$\rho g \Delta h = (s(rr))_{R_2} - (s(rr))_{R_1} \tag{8.42}$$

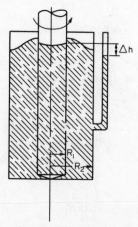

Fig. 8.4. Normal stress measurements in cylindrical Couette flow. The schematic diagram indicates measurement of $(s(rr))_{R_2} - (s((rr))_{R_1}$ by means of conventional pressure holes. This technique, which has been used in most instances, can lead to serious errors. A discussion of pressure-hole error is included in Section 8.8.

The stress difference may be used in conjunction with knowledge of $\tau(\kappa)$ to determine the primary normal stress function $\mathcal{N}_1(\kappa)$. The basic equation which we wish to integrate between R_1 and R_2 is (7.53a). Recalling that $\Phi = p + \rho\psi$ and that ψ is a function only of z, we may express the integration of (7.53a) over r at a fixed value of z as

$$\rho g \Delta h = \int_{R_1}^{R_2} \left[\frac{\mathcal{N}_1(\kappa)}{r} - \rho r (\omega(r))^2 \right] dr \qquad (8.43)$$

It is worth noting that with the Couette apparatus we were not required to drop the inertial contribution to the equation of motion. Thus, knowing $\tau(\kappa)$ from (8.41) we can include the term $\rho r(\omega(r))^2$ in (8.43) and evaluate its contribution to Δh. At sufficiently low values of Ω, the inertial contribution is negligible. In the event that its presence constitutes an important correction we acknowledge its presence by defining

$$\rho g (\Delta h)_c = \rho g \Delta h + \int_{R_1}^{R_2} \rho r (\omega(r))^2 dr \qquad (8.44)$$

Isolation of $\mathcal{N}_1(\kappa)$ proceeds essentially as with the development leading to (8.41). Define

$$G\left(\frac{\mathcal{T}}{2\pi r^2 L}\right) = \mathcal{N}_1\left\{\tau^{-1}\left(\frac{\mathcal{T}}{2\pi r^2 L}\right)\right\} = \mathcal{N}_1(\kappa)$$

Differentiation of (8.43) yields

$$2\rho g \frac{d(\Delta h)_c}{d \ln \mathcal{T}} \equiv H(\mathcal{T}) = G\left(\frac{\mathcal{T}}{2\pi R_1^2 L}\right) - G\left(\frac{\mathcal{T}\lambda}{2\pi R_1^2 L}\right)$$

from which it follows that

$$\sum_{n=0}^{\infty} H(\lambda^n \mathcal{T}) = \mathcal{N}_1\left\{\tau^{-1}\left(\frac{\mathcal{T}}{2\pi R_1^2 L}\right)\right\} \qquad (8.45)$$

and we see that the Couette apparatus can be used to determine $\mathcal{N}_1(\kappa)$. This is an extremely delicate experiment to perform, and the interested reader is urged to review with care the original experimental papers on the subject as well as the difficulties associated with pressure-hole error, a subject discussed in Section 8.8. We note that apparatuses designed for measurement of $\tau(\kappa)$ have usually employed very small gaps. However, in order to obtain substantial values of Δh, normal stresses are generally measured in a "wide-gap" instrument.

(ii) AXIAL-ANNULAR FLOW

Reliable measurement of the secondary normal stress difference $\mathcal{N}_2(\kappa)$ has proved to be an especially elusive goal. Apparently Rivlin [19] was the first to suggest that measurements of this function were, in principle, possible from data relating flow throughput to difference in radial thrust exerted on inner and outer cylinders in an axial-annular flow. The apparatus, shown schematically in Fig. 8.5, has, since the time of Rivlin's suggestion, been used by several investigators to measure $\mathcal{N}_2$ for various polymer solutions [20–22].

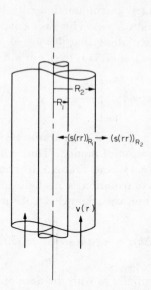

Fig. 8.5. Normal stress measurements in axial-annular flow.

Flow is taken to be in the upward vertical direction, but corrections can be made for any other orientation with respect to gravity. The flow is assumed to follow (7.40) in the cylindrical coordinate system (7.41), and components of the equation of motion are given by (7.43). From (7.43b) and (7.44) we obtain

$$t(rz) = \tau(\kappa) = -\frac{ar}{2} + \frac{c}{r}$$

where a is the driving force, defined as before by (7.47). It is convenient to relate the integration constant to the radius r_0 at which $t(rz) = 0$. Thus

$$t(rz) = \frac{a}{2}\left[\frac{r_0^2}{r} - r\right] = \tau(\kappa) \tag{8.46}$$

We assume that the function $\tau(\kappa)$ is known and we can now concentrate on evaluation of $\mathcal{N}_2(\kappa)$. Define

$$\Delta P = -\left[(s(rr))_{R_2} - (s(rr))_{R_1}\right] \tag{8.47}$$

Then, from (7.43a) we obtain

$$\Delta P = \int_{R_1}^{R_2} \mathcal{N}_2 \frac{dr}{r} \tag{8.48}$$

where the integration over r has been carried out at a fixed value of z. It is convenient to employ (8.46) to effect a change of variable in (8.48). Then

$$\Delta P = - \int_{\tau_1}^{\tau_2} \frac{\mathcal{N}_2}{[\tau^2 + (a\dot{r}_0)^2]^{1/2}} \, d\tau \tag{8.49}^\dagger$$

where we use subscripts 1 and 2 on τ to indicate values of $\tau(\kappa)$ at R_1 and R_2, respectively.

At this point a new difficulty arises. In contrast to procedures used in earlier sections, we can no longer solve the integral equation by merely differentiating (8.49) with respect to one of the wall shear stresses. In general both a and r_0 will depend upon wall shear stress. Because of this, early analyses of axial-annular flow were treated by assuming a functional form for $\mathcal{N}_2$ [20, 21]. However, a more satisfactory approximate procedure can be utilized [22, 23]. In this section we follow the method of Hoffman, which depends upon the requirement that $(\tau/ar_0)^2 < 1$ over the flow space. It is clear from (8.46) that $|\tau|$ will achieve its maximum values on each side of r_0 at the two walls, and that

$$4 \left(\frac{\tau_1}{ar_0} \right)^2 = \left[\frac{\alpha}{\beta} - \frac{\beta}{\alpha} \right]^2 ; \quad 4 \left(\frac{\tau_2}{ar_0} \right)^2 = \left[\frac{1}{\beta} - \beta \right]^2$$

where $\alpha = R_1/R_2$ and $\beta = r_0/R_2$. From these results it is readily shown that for any value of β one is assured that $(\tau/(ar_0))^2 < 1$ if

$$0.4142 < \beta < 1.0 \tag{8.50}$$

a requirement readily achieved in practice. Assuming then that $(\tau/(ar_0))^2 < 1$, we expand the integrand of (8.49) to give

$$\Delta P = \int_{\tau_1}^{\tau_2} \frac{\mathcal{N}_2}{ar_0} \left[1 - \frac{1}{2} \left(\frac{\tau}{ar_0} \right)^2 + \frac{3}{8} \left(\frac{\tau}{ar_0} \right)^4 - \ldots \right] d\tau \tag{8.51}$$

Equation (8.51) can be rewritten as

$$\Delta P = \int_{\tau_1}^{\tau_2} \frac{\mathcal{N}_2}{ar_0} \, d\tau + \int_{\tau_1}^{\tau_2} \frac{\mathcal{N}_2}{ar_0} \varepsilon(\tau, \tau_2) d\tau \tag{8.52}$$

where

$$\varepsilon(\tau, \tau_2) = - \frac{1}{2} \left(\frac{\tau}{ar_0} \right)^2 + \frac{3}{8} \left(\frac{\tau}{ar_0} \right)^4 - \ldots \tag{8.53}$$

Rearranging (8.52) and differentiating with respect to τ_2, one ultimately obtains

$$- ar_0 \frac{d\Delta P}{d\tau_2} + \Delta P \left\{ \frac{2\beta}{1-\beta^2} - \frac{(\beta^2+1)}{(1-\beta^2)} aR_2 \frac{d\beta_2}{d\tau_2} \right\} \equiv G(\tau_2)$$

$$= - (\mathcal{N}_2)_{\tau_2} + (\mathcal{N}_2)_{\tau_1} \left\{ \frac{(\alpha - \beta^2/\alpha)}{(1-\beta^2)} - \frac{aR_2\beta(\alpha - 1/\alpha)}{(1-\beta^2)} \frac{d\beta}{d\tau_2} \right\} + \frac{d}{d\tau_2} \int_{\tau_1}^{\tau_2} \mathcal{N}_2 \varepsilon(\tau, \tau_2) d\tau \tag{8.54}$$

The differentiation is of course being carried out with the understanding that one is to

---

†In this section it is convenient to employ τ as a variable denoting the local value of stress $t_{rz}(r)$. This use of τ is not to be confused with the functional designation $\tau(\kappa)$. Note also in (8.49) and equations which follow from it that $\mathcal{N}_2$ is considered to be a function of the local shear stress τ.

apply the result to a given fluid in a given apparatus. We wish to obtain an estimate of the magnitude of the last term in (8.54). We do this by examining the nature of $\varepsilon(\tau, \tau_2)$, defined by (8.53). Note that since (8.53) is an alternating series we can estimate the error from the magnitude of the first neglected term [7, p. 20]. Thus

$$-\frac{1}{2}\left[\frac{\tau}{ar_0}\right]^2 \leqslant \varepsilon \leqslant 0$$

It can be shown [22] that for any fluid which displays a Newtonian regime as $\kappa \to 0$ (and this behavior is typical of non-Newtonian materials), $|\tau_1/\tau_2| > 1$. Then the largest magnitude of ε is given by evaluating $\frac{1}{2}(\tau/ar_0)^2$ at $\tau = \tau_1$; viz.,

$$\varepsilon(\tau_1, \tau_2) = -\frac{1}{8}\left[\frac{\alpha}{\beta} - \frac{\beta}{\alpha}\right]^2$$

Given α, there will be some value of τ_2 for which $|\varepsilon|$ will achieve its largest value. Call this $\varepsilon_{\max}$. Substituting $\varepsilon_{\max}$ for ε in (8.52) we obtain an estimate of $\mathcal{N}_2$, which we call $\hat{\mathcal{N}}_2$. Then

$$G(\tau_2) = (1 + \varepsilon_{\max})\left[-(\hat{\mathcal{N}}_2)_{\tau_2} + (\hat{\mathcal{N}}_2)_{\tau_1}\,\delta(\tau_2)\right] \tag{8.55}$$

where

$$\delta(\tau_2) = \frac{d\tau_1}{d\tau_2}$$

One can easily obtain a good estimate of the value of $\varepsilon_{\max}$ by assuming that $r_0 = (R_1 + R_2)/2$, an assumption that is expected to be reasonably accurate for well-behaved fluids and for values of $\alpha > 0.5$ [24]. One finds the values of $\varepsilon_{\max}$ listed below in Table 8.1 [22].

TABLE 8.1. Value of $\epsilon_{\max}$ when $r_0 = (R_1 + R_2)/2$

α	$-100\epsilon_{\max}$
0.5	8.68
0.6	4.25
0.7	1.91
0.75	1.20
0.8	0.697
0.9	0.146

Thus in the region $\alpha \simeq 0.7$ the error is well within the limits of accuracy of measurement and can be neglected.

Because of the factor $\delta(\tau_2)$ on the right side of (8.55) the convenient summation procedure employed in, for example, (8.41) is not directly applicable here. Instead we use (8.55) to obtain an approximate solution to (8.54) by neglecting ε and applying an iteration technique. From data we know the function $G(\tau_2)$ over a range of τ_2. As a zeroth approximation to $\mathcal{N}_2$ we write

$$G(\tau_2) = \mathcal{N}_2^0(\tau_1)\,(\delta(\tau_2) - 1)$$

Then

$$\mathcal{N}_2^0(\tau_1) = \mathcal{N}_2^0(\sigma\tau_2) = G(\tau_2)/(\delta(\tau_2) - 1)$$

where

$$\sigma = \tau_1/\tau_2 = (\alpha - \beta^2/\alpha)/(1-\beta^2)$$

Thus

$$\mathcal{N}_2^0(\tau_2) = \frac{G(\tau_2/\sigma)}{(\delta(\tau_2/\sigma)-1)} \tag{8.56}$$

so that from experimental data one can obtain $\mathcal{N}_2^0$ as a function of τ_2 or, alternatively, as a function of shear rate κ. As the next step one obtains a new estimate $\mathcal{N}_2^1(\tau_1)$ by inserting (8.56) into (8.55) (again neglecting the contribution from ε). Then

$$\mathcal{N}_2^1(\tau_1) = \mathcal{N}_2^1(\sigma\tau_2) = \frac{G(\tau_2)}{\delta(\tau_2)} + \frac{G(\tau_2/\sigma)}{\delta(\tau_2)[\delta(\tau_2/\sigma)-1]}$$

or

$$\mathcal{N}_2^1(\tau_2) = \frac{G(\tau_2/\sigma)}{\delta(\tau_2/\sigma)} + \frac{G(\tau_2/\sigma^2)}{\delta(\tau_2/\sigma)[\delta(\tau_2/\sigma^2)-1]} \tag{8.57}$$

It is clear that one can now proceed to higher iterations $\mathcal{N}_2^n(\tau_2)$. Hoffman has shown that for any fluid exhibiting a Newtonian regime as $\kappa \to 0$ and possessing reasonable flow characteristics, the iteration procedure yields, for $n = \infty$,

$$\mathcal{N}_2(\kappa_1) \cong \mathcal{N}_2^\infty(\tau_1) = \sum_{n=0}^{\infty} \frac{G(\tau_2/\sigma^n)}{\prod\limits_{m=0}^{n} \delta(\tau_2/\sigma^m)} \tag{8.58}$$

The only limitation to the accuracy of the approximation (8.58) is the error incurred by neglect of the last term of (8.52). The series (8.58) is convergent, but may converge slowly. Hoffman has also indicated how one can rapidly achieve convergence through use of appropriate weighting functions.

Another approximate method for solution of (8.49) has been proposed by Tanner [23]. His result is

$$(\mathcal{N}_2)_{\tau_2} \cong \frac{(R_1+R_2)}{2(R_1-R_2)}\left[a\frac{d\Delta P}{da} + \Delta P \right] \tag{8.59}$$

Equation (8.59) can be obtained from (8.53) if one makes the following approximations:

(i) $\varepsilon = 0$.
(ii) $d\beta/d\tau_2 = 0$.
(iii) $(\mathcal{N}_2)_{\tau_1} = (\mathcal{N}_2)_{\tau_2}$.
(iv) $r_0 \cong (R_1+R_2)/2$.

(v) $\left[\dfrac{2(R_2-R_1)}{R_2+R_1}\right]^2 \ll 1$.

Though these conditions may appear to be highly restrictive, Tanner has shown that they are probably valid approximations in many cases of practical interest.

8.7. JET DEVICES

It has already been mentioned that normal stress measurements are often made with commercial instruments, such as the Weissenberg rheogoniometer, and that a cone-and-plate configuration is frequently used. It was also noted that secondary flow effects, such as those due to inertia, limit meaningful operation of the instrument to shear rates below about 10^3 s^{-1}. That is most unfortunate, since for technological applications one often wishes to have normal stress data to 10^4 s^{-1} or higher. This desire has led to development, by Metzner (1961) and coworkers at the University of Delaware, of a series of novel techniques which to date have been applied to a number of polymer solutions. In contrast to all of the configurations studied earlier in this chapter, the flow which we now wish to discuss is *not* an approximation to a viscometric flow. The apparent disadvantage is overcome by applying Cauchy's law in an integral form.

Consider a free jet exiting from a circular tube at the plane z_1 as shown schematically in Fig. 8.6. Note that the jet has been shown to exhibit an *increase* in diameter as it leaves

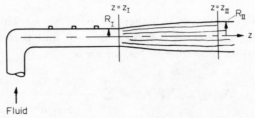

Fig. 8.6. Free jet.

the tube. This is in marked contrast to elementary treatments which consider only the adjustment due to change in axial momentum flux. The increase in diameter is a consequence of relaxation of the viscoelastic fluid as the confining boundary condition of the tube wall is removed. We shall see below how this relaxation can be connected to normal stress. In the analysis we assume that there is no interaction between the fluid jet and the surrounding atmosphere, and that at the plane z_{II} the jet has reached a "fully relaxed" state, i.e., that the jet velocity is uniform over the cross-section and the jet diameter is no longer changing with z. Interfacial forces are also assumed to be negligible. (For a more complete treatment of the jet momentum balance, the reader is referred to Slattery and Schowalter [25].) Additional assumptions include axisymmetric and steady flow. We also, as shown in Fig. 8.6, have neglected the effect of gravity. In any actual experiment [26] an initially horizontal jet will of course begin to droop, and one must hope that the conditions for z_{II} are reached before the effect of gravity becomes significant. (This difficulty is not readily obviated by using a vertical jet, since gravity may mask the effect of jet relaxation which one wishes to measure.)

(i) RELATION OF NORMAL STRESS TO JET EXPANSION

The restrictions cited above permit us to write a relatively simple form for the z-component of the equation of motion when it is integrated over the macroscopic volume bounded by vertical planes at z_I and z_II and the jet–air interface. One readily obtains from (4.21) (see, for example, Fredrickson [4, p. 214])

$$\int_0^{R_\mathrm{I}} 2\pi r \rho (v(r))_\mathrm{I}^2 dr - \int_0^{R_\mathrm{I}} 2\pi r (s(zz))_\mathrm{I} dr = \rho \pi R_\mathrm{II}^2 v_\mathrm{II}^2 \tag{8.60}$$

where $(v(r))_\mathrm{I}$ is the axial velocity at the tube exit and v_II is the (uniform) axial velocity at z_II. For convenience we have taken the ambient pressure as zero. This merely means that all stresses (or pressures) are measured with respect to atmospheric pressure. We shall also assume that, in addition to the condition $\partial/\partial z = 0$ at z_II, the flow at z_I is that of fully developed laminar tube flow. Then we can apply (7.43) at these two planes. Integrating (7.43a) at z_I from the centreline to any value r,

$$(s(rr))_r - (s(rr))_0 + \int_0^r \frac{1}{\xi} (t(rr) - t(\phi\phi)) d\xi = 0 \tag{8.61}$$

the subscript r referring of course to evaluation at radial position r. Subsequent analysis can be greatly simplified if at this point we invoke what has come to be known as the "Weissenberg hypothesis"; viz., that

$$\mathcal{N}_2(\kappa) = 0 \tag{8.62}$$

We shall have more to say about the approximation (8.62) in a later section. At this point it is merely noted that convincing evidence exists which indicates that $\mathcal{N}_2 \neq 0$. However, it is generally found that $|\mathcal{N}_2| < |\mathcal{N}_1|$, and that at low shear rates $\mathcal{N}_2 \cong 0$. Proceeding on the assumption that (8.62) is valid, we simplify (8.61) to

$$(s(rr))_r = (t(rr))_r - p_r = \mathrm{const} = 0 \tag{8.63}$$

This is true because we postulate that right at the tube exit $(z = z_\mathrm{I}, r = R_\mathrm{I})$ the radial thrust must balance atmospheric pressure, which we have taken to be zero. Note that this argument precludes any discontinuity in stress or in $\partial R/\partial z$ at the tube exit, even though we do admit a discontinuity in the wall boundary condition. Now from (8.63) and the definition of extra stress,

$$(s(zz))_r = (t(zz))_r - p_r = (t(zz))_r - (t(rr))_r = (\mathcal{N}_1)_r \tag{8.64}$$

so that the momentum balance (8.60) becomes

$$\int_0^{R_\mathrm{I}} 2r (\mathcal{N}_1)_r dr = \int_0^{R_\mathrm{I}} 2r \rho (v(r)_\mathrm{I})^2 dr - \rho R_\mathrm{II}^2 v_\mathrm{II}^2 \tag{8.65}$$

Standard techniques employed in previous sections lead to

$$\mathcal{N}_1 \left(\frac{aR_\mathrm{I}}{2} \right) = \frac{\rho}{R_\mathrm{I}^2 a} \frac{d}{da} \left\{ 4 \int_0^{(-aR_\mathrm{I})/2} v_\mathrm{I}^2 \tau d\tau - \frac{1}{2\pi^2} \left[\frac{aQ}{R_\mathrm{II}} \right]^2 \right\} \tag{8.66}$$

The driving force a is measured by pressure taps indicated in Fig. 8.6. Volumetric throughput Q is measured directly, and jet radius R_II is easily measured photographically. Knowing $\tau(\kappa)$ and assuming fully developed tube flow at the exit plane z_I, one can evaluate the remaining integral in (8.66) and thus obtain $\mathcal{N}_1(\kappa)$. The equation can be cast in various forms which are convenient for computation [26; 27, p. 42].

(ii) JET THRUST MEASUREMENTS

The method outlined above is desirable in at least two respects: the apparatus is relatively inexpensive to construct and easy to operate. Also one may, with dilute polymer solutions, obtain shear rates in the range of 10^5 s^{-1}, an achievement beyond the reach of cone-and-plate instruments. However, an obvious drawback is dependence upon the Weissenberg hypothesis. In subsequent work with jet devices the Delaware group has overcome this objection by modifying the jet experiment.

Consider a new jet configuration shown schematically in Fig. 8.7. Test fluid enters a

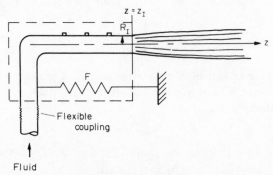

Fig. 8.7. Jet thrust device.

tube through a flexible coupling, is directed from vertical to horizontal flow, and exits into the atmosphere as a free jet. Among the measured quantities is the reaction force F which must be applied in the z-direction to hold the jet device stationary. Application of a momentum balance over the control volume (dashed lines in Fig. 8.7) yields the following z-component:

$$F = \int_0^{R_I} [\rho(v(r))^2 - (s(zz))]_{z_I} 2\pi r dr \tag{8.67}$$

We again take atmospheric pressure as the zero reference for stress. Differentiation of (8.67) leads to

$$(s(zz))_{R_I} = \frac{1}{aR_I^2} \frac{d}{da} \left\{ \int_0^{(-aR_I)/2} 4\rho(v(r))^2 \tau d\tau - \frac{Fa^2}{2\pi} \right\} \tag{8.68}$$

where τ is again used to denote $t_{rz}(r)$. We see that one is now in a position to relate $s(zz)$ to the wall shear stress and, knowing $\tau(\kappa)$, to κ. The stress $(s(rr))_{R_I}$ is obtained directly from extrapolation of radial thrust measurements, available from, for example, pressure transducers along the tube, to the plane $z = z_I$. Then we have, by definition,

$$(s(zz) - s(rr))_{R_I} = (\mathcal{N}_1)_{R_I}$$

from which we obtain $\mathcal{N}_1(\kappa)$, assuming fully developed flow up to the exit plane so that $\tau(\kappa)$ may be used to find $(\kappa)_{R_I}$. Note that the Weissenberg hypothesis was not required.

One may also employ a jet device to obtain $\mathcal{N}_2(\kappa)$. Differentiation of (8.61) leads to the result

$$(\mathcal{N}_2)_{R_{\mathrm{I}}} = [t(rr) - t(\phi\phi)]_{R_{\mathrm{I}}} = \frac{d}{d \ln a} [(s(rr))_{R_{\mathrm{I}}} + p_0] \qquad (8.69)$$

where p_0 is the value of p at z_1 and $r = 0$. *If* one can assume that the contribution of p_0 to the right-hand side of (8.69) is negligible, then measurement of $\mathcal{N}_2(\kappa)$ is straightforward. However, there is no *a priori* justification for making such an assumption. One can attempt to account for p_0 by an iterative procedure as follows: from (8.61) we can write

$$(s(rr))_{R_{\mathrm{I}}} = - p_0 + \int_0^{(-aR_{\mathrm{I}})/2} (\mathcal{N}_2)_\tau \frac{d\tau}{\tau} \qquad (8.70)$$

As a first approximation we assume p_0 to be zero and find $(\mathcal{N}_2)_\tau$ from (8.69). Then the integral of (8.70) can be estimated for a range of $(-aR_{\mathrm{I}}/2)$ and an estimate of p_0 over a range of a obtained from (8.70). These results are then used to recompute $\mathcal{N}_2$ from (8.69). If desired, the iteration may then be repeated.

The use of jet devices described above is by no means the only way in which they may be employed, nor are the results obtained by different investigators entirely compatible [28]. It is probable that measurement of radial thrust by extrapolation of pressure measurements along the tube axis is subject to errors not originally recognized by those doing the experiments (see Section 8.8). Experiments with polymer melts as opposed to polymer solutions have yielded anomalous results when die swell measurements were analyzed according to the momentum balance principles used above [29]. Thus though the jet device appears to be useful under certain conditions, we still do not understand the theory behind the experiment in sufficient detail to obtain unequivocal measurements of normal stress differences. Even Newtonian jet behavior is not completely understood [30].

8.8. EXPERIMENTAL DIFFICULTIES

It is not the purpose of this section to provide a critical review of the data which have been collected and reported for the viscometric functions of polymer melts and solutions. However, to offer some balance to the largely theoretical orientation of the chapter we include a description of some of those problems which are bound to face any experimenter. These problems become, in general, increasingly severe as one proceeds from measurement of $\tau(\kappa)$ to $\mathcal{N}_1(\kappa)$ to $\mathcal{N}_2(\kappa)$. A summary of experimental methods and results is contained in the book by Coleman *et al.* [10]. In spite of its brevity, the coverage is rather thorough for the period up to 1965.

The discussion here is confined to three particular sources of difficulty: end effects, viscous heating, and pressure-hole errors. The first two are probably the most common obstacles faced in rheometry, while the last has, since the time of publication of the book by Coleman *et al.*, become recognized as an important source of error in certain normal stress measurements.

(i) END EFFECTS

Discussion of magnitude and importance of end effects is of course dependent upon the geometry in which viscometric measurements are being made. Consider, for example, measurement of $\tau(\kappa)$ in the coaxial cylinder device discussed in Section 8.6. The torque measured at a given rotational speed contains a contribution from that portion of the flow which is influenced by solid surfaces at the bottom of the cylinders and by the air/liquid interface at the top. The obvious way to account for such end effects is to employ an apparatus of sufficient length L (Fig. 8.3) so that the contribution of the ends to the measured torque $\mathscr{T}$ is negligible. If that is not practical, one might account for end effects by comparing the torques in two viscometers which differ only in the length L of the test section. The difference between the two torque measurements, each taken for the same rotational speed, is then a measure of the torque over the distance $(L_2 - L_1)$, exclusive of end effects. In practice, however, it is not always practical to effect such comparisons. To avoid a large error due to a small difference $(\mathscr{T}_2 - \mathscr{T}_1)$, $(L_2 - L_1)$ must be appreciable. This implies considerable flexibility in drive apparatus and associated equipment. If the material is highly viscous and perhaps is also being measured at an elevated temperature, as would be the case with polymer melts, appropriate heating jackets may not be available for the two lengths even if sufficient driving power can be provided. Thus one must often resort to less direct, more empirical means. These are well documented in the specialized literature on the subject [3].

To give an example of the kinds of empiricism which have been applied with some success we turn to the capillary apparatus of Section 8.3. It is not always practical to nullify the contributions from ends by using an extremely long tube or by comparing results in tubes of different lengths [4, ch. 9]. In fact, because of temperature and pressure requirements, viscometric characterization of polymer melts is often accomplished by forcing the melt through a capillary die with a length/diameter ratio in the neighborhood of unity. Because of the extremely high viscosity of the melt (often O (10^6 poise)), pressures in the neighborhood of 1000 psi may be required to achieve appreciable shear rates.

The most widely used end correction for tube flow is that due to Bagley [31]. Imagine a polymer being forced through a capillary tube in a device such as that shown in Fig. 8.8. From the force applied to the plunger one can estimate the reservoir pressure p_1. Neglecting for the moment the complications cited in the previous section with respect to pressure

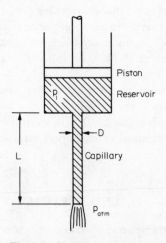

Fig. 8.8. Capillary viscometer.

at the tube exit, we write for the total pressure change which drives the fluid out of the reservoir and through the capillary tube

$$\Delta p = p_1 - p_{atm}$$

From Δp one wishes to obtain a value for $(dp/dz)_{fd}$, the pressure gradient in that portion of the tube over which the flow is fully developed. Then the value of a to be used in, for example, (8.7) can be ascertained. Bagley proposed an end-correction parameter n defined by

$$\left| \left(\frac{dp}{dz} \right)_{fd} \right| = \frac{\Delta p}{L + nD} \tag{8.71}$$

where n is found from a series of experiments with tubes of different L/D but with fluid flowing at the same value of wall shear rate κ_w in the fully developed portion of the capillary. This is readily accomplished by maintaining constant throughput in a series of capillaries with the same diameter but different lengths. Assuming that in each case L is sufficient to generate a region of fully developed flow, one expects curves such as those found by Bagley for polyethylene and shown in Fig. 8.9. Note that the values of wall

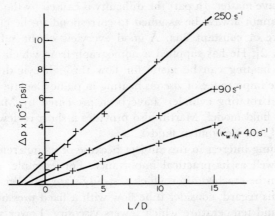

Fig. 8.9. Entry-length data for a polyethylene melt (from Bagley [31]).

shear rate are those which correspond to a *Newtonian* fluid flowing through the tube with volumetric flow rate Q. From the parabolic velocity profile one finds

$$(\kappa_w)_N = \frac{32Q}{\pi D^3} \tag{8.72}$$

Values of n are found from intercepts of the curves shown in Fig. 8.9 with the L/D axis. The correction factor n will be a function of the material being tested and of the wall shear rate. Typically n varies between 1 and 10 and exhibits a shear-rate dependence of the form shown in Fig. 8.10.

The analysis presented here is only one of a number of attempts to account for entrance effects. A more complete analysis which includes both entrance and exit behavior has been presented by LaNieve and Bogue [32], who sought a correlation between capillary pressure drop and normal stresses.

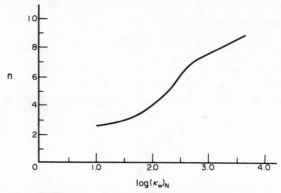

Fig. 8.10. Bagley end correction n for a polyethylene melt [31].

(ii) VISCOUS HEATING

Discussion of the role of the energy equation has been conspicuous by its absence (cf. Section 4.8). Quantitative assessment of the importance of viscous heating in viscometry is a difficult and elusive matter. In part the difficulty is caused by the apparatus boundary conditions, which cannot always be assumed to correspond to the classical conditions of constant temperature or constant flux. A good overview of the subject is provided by Middleman [27, ch. 2]. He has supplied a nomograph from which estimates of the importance of viscous heating can be made for flow through cylindrical tubes. Detailed solutions showing the importance of viscous heating in plane Couette flow, tube flow, and flow between coaxial rotating cylinders have been presented by Martin [33], who employed a power-law fluid model. Martin also provides a short review of prior work with both Newtonian and non-Newtonian fluids.

There is a continuing interest in this subject because of its inherent mathematical and physical appeal as well as its practical importance. For example, it is possible that a critical condition can be reached after which a steady-state solution to the flow problem is not possible. In this regard, consider tube flow with a fixed pressure gradient. Viscous heating raises the fluid temperature which lowers viscosity. Lower viscosity leads to an increased flow rate, which in turn increases viscous heating [34]. Turian [35] has computed critical conditions for the nonexistence of steady plane Couette flow of a specific (Ellis model) fluid. Even under conditions for which steady-state solutions are possible, physical interpretation of solutions to the governing differential equations is not obvious, The existence of multiple steady-state solutions for Newtonian fluids has been discussed. for example, by Joseph [36] and by Gavis and Laurence [37].

(iii) PRESSURE-HOLE ERRORS

Thus far, our discussion of errors has been in the context of measurement of $\tau(\kappa)$. However, we have noted that normal stresses are considerably more difficult to measure. (The reader can obtain a ready appreciation of these difficulties by consulting the paper of Adams and Lodge [38], in which measurements with a cone-and-plate apparatus are described in great detail.) There has been particular confusion over the meaning of reported values for $\mathcal{N}_2(\kappa)$ [38–40], a confusion which now appears to have been resolved.

Recall from earlier sections that deduction of $N_2(\kappa)$ has often involved measurements of local thrust on bounding surfaces. In practice these measurements are frequently made with standard manometric techniques whereby one observes the height of fluid in a manometer tube. The fluid in the tube is attached, by means of a pressure tap, to fluid adjacent to the bounding surface. It is well known that the presence of a hole in the bounding surface is a source of error, and the subject has been studied in detail for Newtonian fluids [41]. However, the source and magnitude of the error for elastic fluids are evidently entirely different, and the error has been sufficient to have caused investigators to report even the wrong sign for N_2. Instruments most vulnerable to a "pressure-hole error" are those in which pressures are compared between positions at which different values of shear stress obtain. A prime example is the measurement of $\Delta P = -[(s(rr))_{R_2} - (s(rr))_{R_1}]$ of (8.47) for axial-annular flow (Fig. 8.5).

That pressure taps may be a serious source of error was first documented with some certainty by Broadbent *et al.* [42] in 1968, though evidently Schultz-Grünow had mentioned the possibility somewhat earlier [43]. Subsequent work by Kaye *et al.* [44] reinforced the concept of an appreciable error due to the presence of pressure taps. These workers, and others [45], have reported that the magnitude of the error incurred with a manometer is, within broad limits, independent of the size of the hole but is a function of the value of the shear stress at the bounding surface in question. When this error was accounted for, values of N_2 were obtained which were different in sign from N_1 (N_1 is positive in our convention) but appreciably smaller in magnitude. Most previous results had indicated identical signs for N_1 and N_2. Results confirming the existence of pressure-hole error have been supplied by many sources. In fact, if one is willing to postulate certain specific types of constitutive behavior, pressure-hole error can be used to advantage to measure viscometric functions [46, 47]. For example, Kearsley [48] has measured the pressure exerted on slots aligned perpendicular and parallel to gravity flow down an inclined channel. By postulating that the fluid is appropriately modeled by a "second-order fluid" (see (10.16)), one can obtain both N_1 and N_2 from the deviation in recorded pressure from that expected for slow flow of a Newtonian fluid. At low shear rates, experimental results are consistent with predictions based on the second-order model.

8.9. NORMAL STRESSES BY THE METHOD OF JACKSON AND KAYE

In 1966 Jackson and Kaye reported an ingenious method for measuring both N_1 and N_2 from total thrust measurements on a cone-and-plate apparatus [49]. Since only one apparatus is necessary, and since *local* thrust measurements (for which one would normally employ manometric techniques) are not necessary, their work took on an added importance following the disclosure of Broadbent *et al.* [42]. That is not to say that the Jackson–Kaye method is the final solution to measurement of normal stress differences. Serious experimental difficulties remain, although at least some of these seem to be capable of resolution [50].

As a final example of measurement of N_1 and N_2 we describe the principles which lead to the technique of Jackson and Kaye. We follow primarily the treatment presented by Marsh and Pearson [51]. By choosing an appropriate coordinate system and applying a form of the lubrication approximation, these authors have formulated a generalized description of flows which includes, as limiting cases, the parallel plate flow of Section 8.4 and cone-and-plate flow of Section 8.5.

In essence the suggestion of Jackson and Kaye is the following: operate a cone-and-plate viscometer with a small but finite gap between the apex of the cone and the plate (Fig. 8.11a). Measure the thrust F on the plate (or cone) for different gaps c in order to find an approximation to dF/dc as $c \to 0$. From these data one can arrive at values for $\mathcal{N}_1$ and $\mathcal{N}_2$. In the approach of Marsh and Pearson, one avoids the necessity for going to the limit $c \to 0$.

In contrast to Section 8.5, we now choose to describe the flow of Fig. 8.11a in terms of

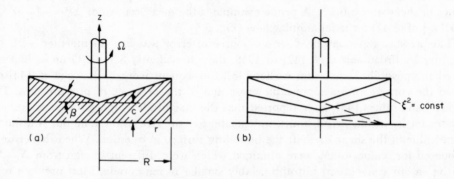

Fig. 8.11. Cone-and-plate viscometer with finite gap.

a *cylindrical* coordinate system with the origin for z taken at the intersection of the center-line with the plate. In this (r, ϕ, z) system, physical components of the velocity are postulated to have the form

$$\mathbf{v} = \left\{ 0, \frac{\Omega r z}{c + r \tan \beta}, 0 \right\} \tag{8.73}$$

From (8.73) one can readily find the nonzero components of $\mathbf{D}$ (Section 5.9), the rate of stretching or rate of deformation tensor

$$d(\phi z) = d(z\phi) = \frac{\Omega r}{2(c + r \tan \beta)}; \quad d(\phi r) = d(r\phi) = \frac{\Omega r z \tan \beta}{2(c + r \tan \beta)^2} \tag{8.74}$$

One must of course show that the velocity profile (8.73) is consistent with Cauchy's equation of motion and any stated assumptions and boundary conditions. Then, if one is willing to forego questions of uniqueness, the solution is the valid solution. In any event satisfaction of the motion equation at least demonstrates consistency.

We now show that the postulated flow (8.73) is viscometric. Note that, from (8.73), conical surfaces from the family

$$z = k(c + r \tan \beta) \quad (0 \leqslant k \leqslant 1)$$

are lines of constant angular velocity. Using these surfaces to define a new coordinate system (ξ^1, ξ^2, ξ^3) (Fig. 8.11b) we have

$$\xi^1 = \phi; \quad \xi^2 = \frac{z}{c + r \tan \beta} \tag{8.75}$$

$$\xi^3 = (c + r \tan \beta)^2 + (z \tan \beta)^2 - c^2$$

Considering the plane of the paper to be a representative surface of constant ξ^1, one notes that the solid lines shown in Fig. 8.11b correspond to intersections of planes $\xi^1 = \text{const}$ with the conical surfaces $\xi^2 = \text{const}$. It is readily shown that the extensions indicated by the dashed lines of Fig. 8.11b intersect at a common point. This point serves as an origin for circles which are lines of constant ξ^3 and which intersect the lines $\xi^1 = \text{const}$, $\xi^2 = \text{const}$ at right angles. Since the representation of Fig. 8.11b applies at any angle ϕ, we conclude that the (ξ^1, ξ^2, ξ^3) system is orthogonal. Furthermore, from the definition of the metric tensor one finds

$$\left. \begin{array}{l} h_1 = \dfrac{1}{\tan \beta} \left\{ \left[\dfrac{\xi^3 + c^2}{1 + (\xi^2)^2 \tan^2 \beta} \right]^{1/2} - c \right\} \\[3ex] h_2 = \dfrac{(\xi^3 + c^2)^{1/2}}{1 + (\xi^3)^2 \tan^2 \beta} \\[3ex] h_3 = \dfrac{1}{2 \tan \beta (\xi^3 + c^2)^{1/2}} \end{array} \right\} \tag{8.76}$$

Transformation of (8.73) to the (ξ^1, ξ^2, ξ^3) system yields

$$\mathbf{v} = \{v^1(\xi^2), 0, 0\} = \{\Omega \xi^2, 0, 0\}$$

$$\kappa = \frac{h_1}{h_2} \Omega \tag{8.77}$$

so that the flow is a laminar shearing flow and hence viscometric. Note that along the plate $(\xi^2 = z = 0)$, $\kappa = \Omega r / (c + r \tan \beta)$.

We now seek a solution to the equation of motion (4.26) subject to the following important restrictions:

(i) inertial and gravitational effects are negligible;
(ii) boundary conditions at $r = R$ are compatible with the assumed velocity field (8.73).

We continue to assume, as we have with the flows analyzed earlier, that viscous heating is negligible. Then Cauchy's equation becomes

$$\nabla \cdot \mathbf{S} = 0 \tag{8.78}$$

which is compatible with a simple fluid following the velocity field (8.73). Consider now the ξ^3-component of the equation over the surface $\xi^2 = 0$, i.e., along the surface of the plate. Then lines of constant ξ^3 correspond to lines of constant r, and one obtains from (8.78)

$$\frac{\partial s(33)}{\partial r} = \frac{t(11) - t(33)}{r} + \frac{\tan \beta (t(22) - t(33))}{c + r \tan \beta} \tag{8.79}$$

For convenience we measure stresses with respect to a zero reference pressure at $r = R$. Thus

$$s(33)(R) = 0$$

and (8.79) can be integrated to yield

$$(s(22)(r))_{z=0} = \int_R^r \left[\frac{\tan \beta}{c + \zeta \tan \beta} N_2 + \frac{N_1 + N_2}{\zeta} \right] d\zeta + (N_2)_r \qquad (8.80)$$

The total downward thrust against the plate is

$$F = - \int_0^R (s(22)(r))_{z=0} 2\pi r dr \qquad (8.81)$$

Substitution of (8.80) into (8.81) and subsequent reversal of order of integration yields

$$F = - 2\pi \int_0^R r N_2 dr + \pi \int_0^R r^2 \left[\frac{\tan \beta}{c + r \tan \beta} N_2 + \left\{ \frac{N_1 + N_2}{r} \right\} \right] dr \qquad (8.82)$$

The limiting cases of a parallel-plate (8.21) or a cone-and-plate instrument (8.35) can be recovered by setting, respectively, $\beta = 0$ or $c = 0$. The important new result is obtained by differentiating (8.82) with respect to gap distance c. It is useful first to change the variable of integration for (8.82) to shear rate. Then

$$\frac{F}{\pi c^2 \Omega} = \int_0^{\kappa_R} \frac{\kappa(N_1 - N_2) d\kappa}{(\Omega - \kappa \tan \beta)^3} + \int_0^{\kappa_R} \frac{\kappa^2 N_2 \tan \beta \, d\kappa}{\Omega(\Omega - \kappa \tan \beta)^3} \qquad (8.83)$$

where $\kappa_R = \Omega R / (c + R \tan \beta)$. Differentiation leads to

$$N_2(\kappa_R) = \left[\frac{-F}{\pi R^2} (2 + m) + N_1(\kappa_R) \right] \frac{\Omega}{\Omega - \kappa_R \tan \beta} \qquad (8.84)$$

where $m = - \partial \ln F / \partial \ln c$.

The experimental opportunity latent in (8.84) is evident. One computes $N_1(\kappa)$ from conventional operation with $c = 0$ and use of the limiting form (8.35). Then, from data obtained at various values of c, (8.84) can be used to find N_2. Experimental data which are available [44–51] tend to corroborate the erroneous nature of early measurements of N_2 because of errors introduced by pressure holes. It now appears that it is customary for N_1 and N_2 to have *opposite* signs and for N_2 to be $O(10^{-1})$ of N_1. However, one should emphasize that reliable measurements of N_2 are still scant, and it is premature to attempt any generalizations about the value of N_2/N_1 [50].

Given the current unsettled state of normal stress measurements, one readily appreciates the need for further development of convenient and inexpensive means for measurement of normal stresses. Progress is being made in this direction [52], including measurement of the effect of normal stresses on secondary flow [53].

REFERENCES

1. Dierckes, A. C., Jr. and Schowalter, W. R., *Ind. Eng. Chem. Fund.* **5**, 263 (1966).
2. Rea, D. R. and Schowalter, W. R., *Trans. Soc. Rheol.* **11**, 125 (1967).
3. Van Wazer, J. R., Lyons, J. W., Kim, K. Y. and Colwell, R. E., *Viscosity and Flow Measurement* (New York: Interscience 1963).
4. Fredrickson, A. G., *Principles and Applications of Rheology* (Englewood Cliffs, New Jersey: Prentice-Hall, 1964).
5. Rabinowitsch, B., *Z. physik. Chemie* A **145**, 1 (1929).

6. Herzog, R. O. and Weissenberg, K., *Kolloid Z.* **46**, 277 (1928); see also, *Physics Today* **21** (8), 13 (1968).
7. Sokolnikoff, I. S., and Redheffer, R. M., *Mathematics of Physics and Modern Engineering*, 2nd edn. (New York: McGraw-Hill, 1966).
8. Greensmith, H. W., and Rivlin, R. S., *Proc. Roy. Soc. (London)*, A **245**, 399 (1953).
9. Kotaka, T., Kurata, M., and Tamura, M., *J. Appl. Phys.* **30**, 1705 (1959).
10. Coleman, B. D., Markovitz, H., and Noll, W., *Viscometric Flows of Non-Newtonian Fluids* (New York: Springer-Verlag New York Inc. 1966).
11. Ginn, R. F., and Metzner, A. B., *Trans. Soc. Rheol.* **13**, 429 (1969).
12. Padden, F. J., and DeWitt, T. W., *J. Appl. Phys.* **25**, 1086 (1954).
13. Markovitz, H., *Proc. 4th Int. Cong. on Rheology*, Part 1, p. 189 (E. H. Lee and A. L. Copley, eds.), (New York: Interscience, 1965).
14. Fulks, W., *Advanced Calculus*, p. 94 (New York: J. Wiley, 1961).
15. Krieger, I. M., and Elrod, H., *J. Appl. Phys.* **24**, 134 (1953).
16. Pawlowski, J., *Koll. Z.* **130**, 129 (1953).
17. Coleman, B. D., and Noll, W., *Arch. Rational Mech. Anal.* **3**, 289 (1959).
18. Krieger, I. M., *Trans. Soc. Rheol.* **12**, 5 (1968).
19. Rivlin, R. S., *J. Rat. Mech. Anal.* **5**, 179 (1956).
20. Huppler, J. D., *Trans. Soc. Rheol.* **9**, 273 (1965).
21. Hayes, J. W., and Tanner, R. I., *Proc. 4th Int. Cong. on Rheology*, Part 3, p. 389 (E. H. Lee and A. L. Copley, eds.), (New York: Interscience, 1965).
22. Hoffman, R. L., PhD thesis, Princeton University, 1968.
23. Tanner, R. I., *Trans. Soc. Rheol.* **11**, 347 (1967).
24. Fredrickson, A. G., and Bird, R. B., *Ind. Eng. Chem.* **50**, 347 (1958).
25. Slattery, J. C., and Schowalter, W. R., *J. Appl. Pol. Sci.* **8**, 1941 (1964).
26. Metzner, A. B., Houghton, W. T., Sailor, R. A., and White, J. L., *Trans. Soc. Rheol.* **5**, 133 (1961).
27. Middleman, S., *The Flow of High Polymers* (New York: Interscience, 1968).
28. Powell, R. L., and Middleman, S., *Trans. Soc. Rheol.* **13**, 111 (1969).
29. Graessley, W. W., Glasscock, S. D., and Crawley, R. L., *Trans. Soc. Rheol.* **14**, 519 (1970).
30. Richardson, S., *Rheol. Acta* **9**, 193 (1970).
31. Bagley, E. B., *J. Appl. Phys.* **28**, 624 (1957).
32. LaNieve, H. L., and Bogue, D. C., *J. Appl. Pol. Sci.* **12**, 353 (1968).
33. Martin, B., *Int. J. Nonlinear Mech.* **2**, 285 (1965).
34. Gruntfest, I. J., *Trans. Soc. Rheol.* **7**, 195 (1963).
35. Turian, R. M., *Chem. Eng. Sci.* **24**, 1581 (1969).
36. Joseph, D. D., *Phys. Fluids* **8**, 2195 (1965).
37. Gavis, J., and Laurence, R. L., *Ind. Eng. Chem. Fund.* **7**, 232 (1968).
38. Adams, N., and Lodge, A. S., *Phil. Trans. Roy. Soc. (London)* A, **256**, 149 (1964).
39. Weissenberg, K., *Nature* **159**, 310 (1947).
40. Markovitz, H., and Brown, D. R., *Trans. Soc. Rheol.* **7**, 137 (1963).
41. Shaw, R., *J. Fluid Mech.* **7**, 550 (1960).
42. Broadbent, J. M., Kaye, A., Lodge, A. S. and Vale, D. G., *Nature* **217**, 55 (1968).
43. Schultz-Grünow, F., *Ver. Deut. Ing.-Z.* **97**, 409 (1955).
44. Kaye, A., Lodge, A. S., and Vale, D. G., *Rheol. Acta* **7**, 368 (1968).
45. Tanner, R. I., and Pipkin, A. C., *Trans. Soc. Rheol.* **13**, 471 (1969).
46. Pipkin, A. C., and Tanner, R. I., *Mechanics Today* **1**, 262 (1972).
47. Higashitani, K., and Pritchard, W. G., *Trans. Soc. Rheol.* **16**, 687 (1972).
48. Kearsley, E. A., *Trans. Soc. Rheol.* **17**, 617 (1973).
49. Jackson, R., and Kaye, A., *Br. J. Appl. Phys.* **17**, 1355 (1966).
50. Cowsley, C. W., Improvement to total thrust methods for the measurement of second normal stress differences, University of Cambridge, Department of Chemical Engineering, Polymer Processing Research Center, Report No. 6, 1970.
51. Marsh, B. D., and Pearson, J. R. A., *Rheol. Acta* **7**, 326 (1968).
52. Tanner, R. I., *Trans. Soc. Rheol.* **14**, 483 (1970).
53. Denn, M. M., and Roisman, J. J., *AIChE Jl.* **15**, 454 (1969).

PROBLEMS

8.1. The shear dependent viscosity of many fluids is adequately described by the empirical "power law" (see Section 10.3). In the case of laminar flow through a tube with circular cross-section, the power law is expressed by $\tau(\kappa) = K \left| \frac{\partial v_z}{\partial r} \right|^n$, where K and n are empirical constants. The quantity $K \left| \frac{\partial v_z}{\partial r} \right|^{n-1}$ is often called the "effective viscosity".

The data presented below were taken for a polyisobutylene solution flowing between two points in a capillary tube viscometer.† Tests were carried out with two different capillaries, tubes A and B. The flow was fully developed.

From the data:

(a)   estimate the power-law constants over the range $25 \text{ s}^{-1} < \kappa < 250 \text{ s}^{-1}$;
(b)   plot the effective viscosity as a function of shear rate.

NB: if graphical means are employed to evaluate the argument of τ in (8.7), it is convenient to plot Q vs. a and to use

$$\frac{d}{da}[a^3 Q(a)] = 3a^2 Q(a) + a^3 \frac{dQ(a)}{da}$$

I. Tube A		II. Tube B	
$-\Delta P$	Q	$-\Delta P$	Q
3,590	0.374	29,760	0.189
46,800	5.906	64,980	0.506
18,360	1.861	109,120	1.028
24,610	2.543	74,900	0.599
33,900	3.730	122,760	1.239
41,960	4.754	143,840	1.579
43,690	5.054	172,360	2.126
6,250	0.645	155,000	1.786
17,150	1.774	184,760	2.415
7,670	0.696	213,280	3.089
11,700	1.189	235,600	3.711
22,590	2.327	260,400	4.497
9,480	0.906	302,560	5.900
27,030	2.870	333,560	7.195
35,100	3.872	364,560	8.459
55,470	6.832		
4,840	0.442		
13,520	1.355		
4,640	0.425		

	Radius (cm)	Length over which ΔP was measured (cm)
Tube A	0.4953	49.530
Tube B	0.1930	19.300

Units are $Q=\text{cm}^3/\text{s}$; $\Delta P=\text{dynes/cm}^2$.

8.2.   Prove that the assumption (8.8) leads to (8.10) and (8.11). Next, show that the result is incompatible with (8.9a). This result implies that secondary flow, which is due to the effect of inertia, can induce a non-viscometric flow between parallel plates.

8.3.   Show how the equations of motion in the appendix of Chapter 7 reduce to (8.25).

8.4.   One wishes to estimate the first and second normal stress differences $N_1(\kappa)$ and $N_2(\kappa)$ for a polymeric material. To do this, total normal thrust data are obtained in parallel-plate and cone-and-plate rheometers.

(a)   Parallel plate instrument: gap=0.11 cm; plate radius=3 cm.

Rotational speed (rpm)	Force against plate (g)
0.8	13.8
1.9	65.5
2.5	117
4.6	321
7.3	515
10.0	843
14.0	975

† Dierckes, A. C., Jr., PhD thesis, Princeton University, 1965.

(b)   Cone-and-plate instrument. The cone-and-plate instrument has a 3 cm radius. The cone angle is 2°, and the following data were obtained:

Rotational speed	Force against plate
(rpm)	(g)
0.75	27.5
1.50	101
4.75	465
7.30	810
10.10	1150
13.10	1290

Comment on the value of this procedure as a means for measuring $N_1(\kappa)$ and $N_2(\kappa)$.

8.5. Shear stress measurements in a cone-and-plate viscometer are often made by supporting the non-rotating member (the plate, for example) on a torsion bar. The fluid stress transmitted to the plate causes a small rotation of the torsion bar which is sensed by the motion of the end of a rod suitably attached to the torsion bar. This arrangement is shown below.

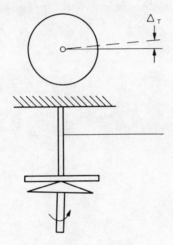

The manufacturer of a cone-and-plate viscometer gives the following equation for obtaining the shear stress from measured variables:

$$t(12) = \frac{3.82 \, k_t \Delta_t}{d^3}$$

where d is the platen diameter (cm), k_t is the torsion constant (dyne cm/0.001 in deflection), and Δ_t is the rod displacement (in $\times 10^3$).

(a)   Prove the correctness of the general form of this equation.

(b)   Plot the effective viscosity $\tau(\kappa)/\kappa$ for a polymer solution with which the data given below were obtained.

Instrument constants:

plate diameter, 7.5 cm
cone angle,   1.533°
torsion bar constant, $k_t = 3.41 \times 10^4$ dyne cm/0.001 in deflection

Data:

Ω (rev/sec)	Δ_t(in $\times 10^3$)
7.50	2.03
2.38	1.31
0.750	0.88
0.238	0.59
0.0750	0.40
0.0238	0.28
7.50×10^{-3}	0.172
2.38×10^{-3}	0.097
7.50×10^{-4}	0.049

8.6. The following data were obtained from a polymer solution in a coaxial cylinder viscometer:[†] instrument constants: inner radius, 2.253 cm; outer radius, 2.508 cm; length, 7.25 cm.

Rotational speed Ω of outer cylinder (rad/sec)	Torque on inner cylinder (dyne cm $\times 10^{-4}$)
0.0217	0.569
0.0969	1.008
0.391	1.293
1.384	2.54
3.26	3.59
8.48	5.61
15.16	7.58
22.0	9.40
28.1	10.90
35.2	12.61
41.7	14.11

Compare plots of the effective viscosity $\tau(\kappa)/\kappa$ computed from (8.38) and (8.41).

[†] Adapted from Krieger, I. M., and Maron, S. H., *J. Appl. Phys.* **25**, 72 (1954).

CHAPTER 9

NONVISCOMETRIC FLOWS OF SIMPLE FLUIDS

9.1. INTRODUCTION

Just as the word "non-Newtonian" fails to bound a subject until some agreement is reached over those particular non-Newtonian materials which are to be discussed, so "nonviscometric" poses similar difficulties because of the infinity of nonviscometric flows. The scope of this chapter is much more modest than might be inferred from the title.

In the last two chapters we dealt at some length with viscometric flows of simple fluids and the description of those flows in terms of the three viscometric functions. We noted the failure, in general, of the approach taken there to be helpful for flows failing to qualify for viscometric status. Much research in rheology is a groping for means to connect, in a productive way, flows and constitutive equations which together might yield some of the generality that evolved from the union of simple fluids and viscometric flows. The engineering significance of this search and the description of specific applications are deferred to a later chapter. Our intent here is to show how, first of all, one can productively classify flows of simple fluids in a manner different from that employed in Chapters 7 and 8. There will be emphasis on the description of a flow characterized by "stretching" rather than "shearing". Since in many applications, for example fiber spinning and film blowing of polymers, the stretching nature of the flow is predominant, there is strong practical motivation for study of these flows. Second, we shall describe an attempt to extend our knowledge of viscometric flows to apply to flows which are in some sense "slightly non-viscometric". Use of the viscometric functions as a base from which one makes small excursions into nonviscometric regimes can be very helpful in providing information necessary for both fundamental theory and, one might hope, for applications.

9.2. MOTIONS WITH CONSTANT STRETCH HISTORY

In this section the concept of a steady viscometric flow will be broadened to include a much larger class of possible motions. Recall that in Section 7.6 a viscometric flow was defined by the requirements that $\mathbf{C}_{(t)}(s)$ be expressible in the form (7.26) and that $\mathbf{A}_1$ and $\mathbf{A}_2$ have components given, respectively, by (7.24) and (7.25) with respect to an orthogonal basis $\mathbf{e}_i$. One of the features of a viscometric flow is that the strain history for a fluid element is an extremely simple one. This is made clear by the choice of $\mathbf{e}_i$. If the

117

flow is steady $(\kappa \neq f(t))$, then one sees from (7.24) to (7.26) that components of $\mathbf{C}_{(t)}(s)$ depend only on the time interval s from the present time t into the past. This feature of steady viscometric flows can be extended to provide a criterion by which we define flows with constant stretch history. These are all flows for which

$$\mathbf{C}_{(t)}(t-s) \equiv \mathbf{C}_{(t)}(s) = \mathbf{Q}(t)\mathbf{C}_{(0)}(0-s)\mathbf{Q}^T(t) \tag{9.1}†$$

where $\mathbf{Q}(0) = \mathbf{I}$.

The significance of (9.1) is clear if we consider an orthonormal basis $\{\mathbf{a}_i(t)\} = \{\mathbf{Q}(t)\mathbf{a}_i(0)\}$, $\mathbf{Q}(t)$ being the same as that used in (9.1). With respect to this basis components of $\mathbf{C}_{(t)}(t-s)$ are seen to depend upon the time interval s from the present into the past but not upon t. Recalling from (5.23) that $\mathbf{C}_{(t)} = \mathbf{U}_{(t)}^2$ and from (7.10) that a functional over past time of $\mathbf{C}_{(t)}(s)$ can be a useful measure of the history of the motion, one sees the justification for the term *constant stretch history* [1, p. 438].‡ A constant stretch history, in turn, implies a stated value for the constitutive functional (7.11) of a simple fluid. Hence the stress at a material point in a simple fluid is constant in a motion with constant stretch history.

Noll [1, p. 438; 4] has provided an interesting classification of these flows. To develop his result we use a notation involving tensor exponents. By $e^{\mathbf{A}}$ we mean

$$e^{\mathbf{A}} \equiv \mathbf{I} + \mathbf{A} + \frac{1}{2!}\mathbf{A}^2 + \frac{1}{3!}\mathbf{A}^3 + \cdots \tag{9.2}$$

We wish to demonstrate that an alternative form for a flow with constant stretch history, equation (9.1), is

$$\mathbf{C}_{(t)}(t-s) = e^{-s\kappa \mathbf{N}^T}e^{-s\kappa \mathbf{N}} \tag{9.3}$$

where κ is a scalar constant, and $\mathbf{N}$ is defined in (9.10) and the definitions following (9.5) and (9.9). To show the equivalence of (9.1) and (9.3) recall (5.25), (5.35), and (7.4). From them we immediately obtain

$$\mathbf{C}(\tau) = \mathbf{F}^T(t)\mathbf{C}_{(t)}(\tau)\mathbf{F}(t)$$

Clearly, we can specialize this equation to some specific reference configuration for $\mathbf{C}$ and $\mathbf{F}$, say the configuration at t'. Then

$$\mathbf{C}_{(t')}(\tau) = \mathbf{F}^T_{(t')}(t)\mathbf{C}_{(t)}(\tau)\mathbf{F}_{(t')}(t) \tag{9.4}$$

Interpreting τ as $t-s$ we can write, from (9.1), for motions with constant stretch history,

$$\mathbf{C}_{(0)}(t-s) = \mathbf{F}^T_{(0)}(t)\mathbf{Q}(t)\mathbf{C}_{(0)}(0-s)\mathbf{Q}^T(t)\mathbf{F}_{(0)}(t) \tag{9.5}$$

where we have set $t' = 0$. For convenience define $\mathbf{M}(t) = \mathbf{Q}^T(t)\mathbf{F}_{(0)}(t)$. Equation (9.5) can now be written

$$\mathbf{C}_{(0)}(t-s) = \mathbf{M}^T(t)\mathbf{C}_{(0)}(0-s)\mathbf{M}(t) \tag{9.6}$$

with $\mathbf{M}(0) = \mathbf{I}$ since $\mathbf{Q}(0) = \mathbf{F}_{(0)}(0) = \mathbf{I}$. Let us differentiate (9.6) with respect to t:

$$\frac{\partial}{\partial t}[\mathbf{C}_{(0)}(t-s)] = \frac{d\mathbf{C}_{(0)}(t-s)}{d(t-s)} \equiv \dot{\mathbf{C}}_{(0)}(t-s)$$

---

† Recall that it was agreed, following (7.5), to abbreviate $\mathbf{C}_{(t)}(t-s)$ by writing $\mathbf{C}_{(t)}(t-s) \equiv \mathbf{C}_{(t)}(s)$. In the present section, however, it is convenient to express the full argument of $\mathbf{C}_{(t)}$.

‡ The flows described by (9.1) were first delineated by Coleman [2, 3], who called them "substantially stagnant" motions.

$$= \dot{\mathbf{M}}^T(t)\mathbf{C}_{(0)}(0-s)\mathbf{M}(t) + \mathbf{M}^T(t)\mathbf{C}_{(0)}(0-s)\dot{\mathbf{M}}(t) \qquad (9.7)$$

where the dot refers to differentiation with respect to the respective arguments. Now suppose we agree to fix t at some set value, say $t=0$. Then $\dot{\mathbf{C}}_{(0)}(t-s)$ can be written as $-d[\mathbf{C}_{(0)}(0-s)]/ds$, and (9.7) becomes

$$- \frac{d}{ds}[\mathbf{C}_{(0)}(0-s)] = \kappa\mathbf{N}_0^T\mathbf{C}_{(0)}(0-s) + \mathbf{C}_{(0)}(0-s)\kappa\mathbf{N}_0 \qquad (9.8)$$

a differential equation with the solution

$$\mathbf{C}_{(0)}(0-s) = e^{-s\kappa\mathbf{N}_0^T} e^{-s\kappa\mathbf{N}_0} \qquad (9.9)$$

where $\kappa\mathbf{N}_0$ is the constant tensor $\left\{\dfrac{d\mathbf{M}(t)}{dt}\right\}_{t=0}$ and κ is defined by the requirement $|\mathbf{N}_0|=1$.[†]

From (9.1) and (9.9) we obtain

$$\mathbf{C}_{(t)}(t-s) = \mathbf{Q}(t)e^{-s\kappa\mathbf{N}_0^T}e^{-s\kappa\mathbf{N}_0}\mathbf{Q}^T(t)$$

But from the definition (9.2) of an exponential with tensor argument and from the fact that $(\mathbf{Q}\mathbf{A}\mathbf{Q}^T)^n = \mathbf{Q}\mathbf{A}^n\mathbf{Q}^T$ we have, finally, the demonstration of (9.3):

$$\mathbf{C}_{(t)}(t-s) = e^{-s\kappa\mathbf{N}^T}e^{-s\kappa\mathbf{N}} \qquad (9.3)$$

where

$$\mathbf{N} = \mathbf{Q}(t)\mathbf{N}_0\mathbf{Q}^T(t) \qquad (9.10)$$

The alternative definition (9.3) and (9.10) of constant stretch history flows allows an interesting subdivision used by Noll. We note three cases:

(i) $\mathbf{N} \neq \mathbf{0}$ but $\mathbf{N}^2 = \mathbf{0}$.

If $\mathbf{N}^2=\mathbf{0}$ we know that $\mathbf{N}^n=\mathbf{0}$ for $n\geqslant 3$, so expansion of (9.3) yields

$$\mathbf{C}_{(t)}(t-s) = \mathbf{I} - s\kappa(\mathbf{N}+\mathbf{N}^T) + (s\kappa)^2\mathbf{N}^T\mathbf{N} \qquad (9.11)$$

and since from (5.62) we know that $\mathbf{A}_n(t) = \overset{n}{\overline{[\mathbf{C}_{(t)}(t-s)]}}_{s=0}$ one finds

$$\left.\begin{aligned} \mathbf{A}_1 &= 2\mathbf{D} = \kappa(\mathbf{N}+\mathbf{N}^T) \\ \mathbf{A}_2 &= 2\kappa^2\mathbf{N}^T\mathbf{N} \\ \mathbf{A}_n &= 0 \qquad\qquad\qquad \text{for } n \geqslant 3 \end{aligned}\right\} \qquad (9.12)$$

so that (9.11) corresponds to a portion of the definition of a viscometric flow, (7.24)–(7.26). Also, since $|\mathbf{N}_0|=1$ it follows that $|\mathbf{N}|=1$. Now, if $|\mathbf{N}|=1$ and $\mathbf{N}^2=\mathbf{0}$, one can show (see appendix) that it is possible to put $\mathbf{N}$ into the form

$$[n(ij)] = \begin{bmatrix} 0 & 1 & 0 \\ 0 & 0 & 0 \\ 0 & 0 & 0 \end{bmatrix} \qquad (9.13)$$

and we see from (9.12) that $\mathbf{A}_1$ and $\mathbf{A}_2$ have the forms (7.24) and (7.25), respectively.

[†] The magnitude of a tensor $\mathbf{A}$ is defined by $|\mathbf{A}|=[\text{tr}(\mathbf{A}\mathbf{A}^T)]^{1/2}$.

Thus we conclude that all constant stretch history flows with $\mathbf{N}^2 = \mathbf{0}$ are *viscometric flows*. It is a simple matter to show that this is a necessary as well as a sufficient condition for steady viscometric flows.

(ii) $\mathbf{N}^3 = \mathbf{0}$ but $\mathbf{N}^2 \neq \mathbf{0}$.

$\mathbf{N}^3 = \mathbf{0}$ implies $\mathbf{N}^m = \mathbf{0}$, and conversely, for $m \geqslant 3$. Huilgol [5] has shown how a number of flows in this class can be generated by superposing viscometric flows.

(iii) $\mathbf{N}^m \neq \mathbf{0}$ for any value of m ($\mathbf{N}$ is not *nilpotent*).

An example of this classification is *steady extension*, a flow which is discussed in Section 9.4.

Because of the particularly simple history of these flows one might expect that some generalizations could be made about the rheological information required to characterize flows with constant stretch history. Though that is true to some extent, specific features of the constant stretch history flow under discussion are important. We shall see this more clearly in some of the examples given below.

9.3. EXTENSIONAL MOTIONS OF SIMPLE FLUIDS

In 1962 Coleman and Noll [6] analyzed a class of motions which they called *steady extensions*. That work was later clarified and extended by Coleman [7], and we provide below a description which is based upon portions of Coleman's paper. We restrict our discussion to *rectilinear* extensions, in which case the coordinates employed below are Cartesian and we do not need to distinguish between covariant and contravariant components.

Let us consider the history of a material point X and suppose that at least one ortho-normal basis $\{\mathbf{a}_i\}$ can be found such that components of the right relative stretch tensor $\mathbf{U}_{(t)}(s)$ (cf. Section 5.6) have the form

$$[(u_{(t)}(s))_{ij}] = \begin{bmatrix} \alpha_{(t)1}(s) & 0 & 0 \\ 0 & \alpha_{(t)2}(s) & 0 \\ 0 & 0 & \alpha_{(t)3}(s) \end{bmatrix} \tag{9.14}$$

where the basis $\{\mathbf{a}_i\}$ is time independent and (9.14) holds for all values of $s(s \geqslant 0)$. Such a motion constitutes an *extension*. We wish to consider some of the properties of extensions of (incompressible) simple fluids.

From (7.7) and the discussion leading to (7.10) we write for the constitutive equation of a simple fluid

$$\mathbf{S}(t) + p\mathbf{I} = \mathop{\mathscr{S}}_{s=0}^{\infty} (\mathbf{U}_{(t)}(s)) \tag{9.15}$$

p being an arbitrary isotropic pressure.

Our first task is to identify those components $s_{ij}(t)$ which, with respect to the basis $\{\mathbf{a}_i\}$, must be zero. This can be done in a manner analogous to the method employed in Section 7.4 to ascertain the character of the viscometric functions. From a restricted form of the principle of material objectivity it is clear that

$$\mathbf{Q}(\mathbf{S}(t)+p\mathbf{I})\mathbf{Q}^T = \mathop{\mathscr{S}}_{s=0}^{\infty} (\mathbf{Q}\mathbf{U}_{(t)}(s)\mathbf{Q}^T) \tag{9.16}$$

for any constant orthogonal $\mathbf{Q}$. Then if one applies various combinations of the rotations and reflections

$$[q_{ij}] = \begin{bmatrix} \pm1 & 0 & 0 \\ 0 & \pm1 & 0 \\ 0 & 0 & \pm1 \end{bmatrix}$$

to the constitutive equation (9.15) for the extension (9.14), it is readily seen that a contradiction is avoided only if, with respect to the basis $\{\mathbf{a}_i\}$, the stress has the form

$$[t(t)_{ij}] = [s(t)_{ij} + p\,\delta_{ij}] = \begin{bmatrix} t_1(t) & 0 & 0 \\ 0 & t_2(t) & 0 \\ 0 & 0 & t_3(t) \end{bmatrix} \tag{9.17}^{\dagger}$$

Then we can write (9.15) in the component form

$$\begin{aligned} t_1(t) &= \mathscr{S}_1(\alpha_{(t)1},\,\alpha_{(t)2},\,\alpha_{(t)3}) \\ t_2(t) &= \mathscr{S}_2(\alpha_{(t)1},\,\alpha_{(t)2},\,\alpha_{(t)3}) \\ t_3(t) &= \mathscr{S}_3(\alpha_{(t)1},\,\alpha_{(t)2},\,\alpha_{(t)3}) \end{aligned} \tag{9.18}$$

where the $\mathscr{S}_i$ are of course functionals over s, and the dependence of each $\alpha_{(t)i}$ on s is understood. The functionals of (9.18) cannot be independent, however, as one can readily deduce by again applying (9.16). Let us interchange the 1- and 2-principal axes of stretch by the reflection

$$[q_{ij}] = \begin{bmatrix} 0 & 1 & 0 \\ 1 & 0 & 0 \\ 0 & 0 & 1 \end{bmatrix}$$

From (9.16) and (9.18) we find

$$\begin{aligned} \mathscr{S}_2(\alpha_{(t)1},\,\alpha_{(t)2},\,\alpha_{(t)3}) &= \mathscr{S}_1(\alpha_{(t)2},\,\alpha_{(t)1},\,\alpha_{(t)3}) \\ \mathscr{S}_3(\alpha_{(t)1},\,\alpha_{(t)2},\,\alpha_{(t)3}) &= \mathscr{S}_3(\alpha_{(t)2},\,\alpha_{(t)1},\,\alpha_{(t)3}) \end{aligned} \tag{9.19}$$

A similar exchange between the 1- and 3-axes with

$$[q_{ij}] = \begin{bmatrix} 0 & 0 & 1 \\ 0 & 1 & 0 \\ 1 & 0 & 0 \end{bmatrix}$$

leads to

$$\begin{aligned} \mathscr{S}_3(\alpha_{(t)1},\,\alpha_{(t)2},\,\alpha_{(t)3}) &= \mathscr{S}_1(\alpha_{(t)3},\,\alpha_{(t)2},\,\alpha_{(t)1}) \\ \mathscr{S}_2(\alpha_{(t)1},\,\alpha_{(t)2},\,\alpha_{(t)3}) &= \mathscr{S}_2(\alpha_{(t)3},\,\alpha_{(t)2},\,\alpha_{(t)1}) \end{aligned} \tag{9.20}$$

Equations (9.19) and (9.20) indicate that

$$\begin{aligned} t_1(t) &= \mathscr{S}_1(\alpha_{(t)1},\,\alpha_{(t)2},\,\alpha_{(t)3}) \\ t_2(t) &= \mathscr{S}_1(\alpha_{(t)2},\,\alpha_{(t)1},\,\alpha_{(t)3}) \\ t_3(t) &= \mathscr{S}_1(\alpha_{(t)3},\,\alpha_{(t)1},\,\alpha_{(t)2}) \end{aligned} \tag{9.21}$$

and the functional $\mathscr{S}_1$ has the symmetry property

---

† Note that it is not essential for the tensor $\mathbf{T}$ which results from the components $(t(t))_{ij}$ to be the *extra* stress, such that tr $\mathbf{T}=0$.

$$\mathscr{S}_1(a, b, c) = \mathscr{S}_1(a, c, b) \tag{9.22}$$

where a, b, c refer to any allowable combination of functions $a_{(t)i}(s)$. Thus we conclude that extensions, as defined in this section, are characterized by *one* scalar-valued material functional of the histories of the three principal relative stretches. Furthermore, the material functional possesses the symmetry property (9.22).

9.4. STEADY EXTENSIONS

By far the greatest attention, both theoretical and experimental, has been directed toward situations where the stretch, ideally, is homogeneous and the stretch *rates* do not depend on time. How one approaches realization of this flow in the laboratory is far from obvious. We shall discuss extension experiments in some detail in Chapter 12.

Steady extension is defined, with respect to the orthogonal basis discussed in the previous section, by the kinematics

$$\frac{d\xi^i}{ds} = -l_i\xi^i \quad \text{(no sum on } i\text{)} \tag{9.23}$$

with the initial conditions $\xi^i(s=0) = x^i$. Solving, we obtain

$$\xi^i = x^i e^{-l_i s} \tag{9.24}$$

Then

$$\mathbf{\xi} = \mathbf{x} \cdot e^{-\mathbf{L}s}$$

describes the trajectory of a material point which at the present time t is at position $\mathbf{x}$. $\mathbf{\xi}$ is the position at time $\tau = t - s$. Equation (9.23) has been written for an orthonormal basis with respect to which $\mathbf{L}$ is diagonalized and has constant components l_i ($i = 1, 2, 3$). It is clear that

$$\mathbf{F}_{(t)}(\tau) = e^{-\mathbf{L}s}$$

and therefore

$$\mathbf{C}_{(t)}(\tau) = \mathbf{F}_{(t)}^T(\tau)\mathbf{F}_{(t)}(\tau) = e^{-\mathbf{L}^T s}e^{-\mathbf{L}s}$$

We find that steady extensions belong to the class of flows with constant stretch history. Here $\mathbf{Q}(t) = \mathbf{I}$ in (9.10), $\mathbf{N} = \mathbf{N}_0 = \mathbf{L}/\kappa$, and $\kappa = |\mathbf{L}|$. Note that the symbol $\mathbf{L}$ being used here is consistent with the definition given in Section 5.8.

The physical nature of the flow (9.23) can be inferred from

$$-\frac{d\xi^i}{ds}\bigg|_{s=0} = v^i = l_i x^i \quad \text{(no sum on } i\text{)}$$

Since $\mathbf{D} = (\dot{\mathbf{U}}_{(t)}(s))_{s=0} = \frac{1}{2}(\mathbf{L} + \mathbf{L}^T)$ (Section 5.9) we can write, in the basis of (9.23),

$$[l_{ij}] = [(\dot{u}(s))_{ij}]_{s=0} = [d_{ij}] = \begin{bmatrix} l_1 & 0 & 0 \\ 0 & l_2 & 0 \\ 0 & 0 & l_3 \end{bmatrix} \tag{9.25}$$

so that the steady extension (9.23) corresponds to a pure stretching motion with the constant *rates* of stretch given by (9.25). Since we are restricting ourselves to incompressible fluids, continuity requires

$$\sum_i l_i = 0 \tag{9.26}$$

which implies, from (9.24),

$$\xi^1 \xi^2 \xi^3 = \text{const} \tag{9.27}$$

Note that, except for the trivial case $l_i = 0$, $(\mathbf{L})^n \geqslant 0$. Thus steady extension is an example of the classification (iii) of flows with constant stretch history, since the tensor $\mathbf{N}$ of (9.3) is not nilpotent. Other examples are discussed by Huilgol [5, 8], and some of these will be considered later.

Since the dependence of $\mathbf{U}_{(t)}(s)$ on s is known, we can now interpret $\mathcal{S}_1$ in (9.21) as a function instead of a functional. In fact it is possible to write

$$t_i = f(l_i, l_j, l_k) \tag{9.28}$$

where i, j, k can be permuted over 1, 2, 3, and f is a single material function. We have the further constraint (9.26).

An alternative expression for the stresses is available. Recalling the tensor origin of (9.28) we can write, for the flow under discussion,

$$\mathbf{T} = \mathbf{F}(\mathbf{D})$$

Now let us apply our knowledge of the most general form of $\mathbf{F}(\mathbf{D})$, $\mathbf{F}$ being an isotropic tensor-valued function of $\mathbf{D}$.[†] It is noted in the appendix to this chapter that the Cayley–Hamilton theorem or, in the present case, the properties of an isotropic tensor function, may be employed to show that the most general form of $\mathbf{F}(\mathbf{D})$ is given by

$$\mathbf{T} = \lambda_0 \mathbf{I} + \lambda_1 \mathbf{D} + \lambda_2 \mathbf{D}^2 \tag{9.29}$$

where the λ_i are functions of the principal invariants of $\mathbf{D}$ (Section 3.16). Because of the familiar difficulty of the undetermined isotropic pressure, especially severe in the present instance where our concern is with normal stress components, we deal again with stress differences. Then, from (9.29), we may write

$$\left.\begin{aligned} s_1 - s_2 &= (l_1 - l_2)\lambda_1 + (l_1^2 - l_2^2)\lambda_2 \\ s_2 - s_3 &= (l_2 - l_3)\lambda_1 + (l_2^2 - l_3^2)\lambda_2 \end{aligned}\right\} \tag{9.30}$$

where $s_i = t_i - p$ and $\lambda_i = \lambda_i(I_2, I_3)$, since incompressibility ensures that the first invariant $I_1 = 0$. Using the notation of the preceding section $I_2 = -\frac{1}{2}\kappa^2$ and $I_3 = l_1 l_2 l_3$.

Let us consider three examples of steady extension:

(i) Extension of a circular cylinder.

Take the axis of the cylinder to coincide with the x^1 coordinate direction. The stretching is taken to be axisymmetric so that, in a rectangular Cartesian system of coordinates,

$$l_1 = l; \quad l_2 = l_3 = -\tfrac{1}{2}l \tag{9.31}$$

Converting to cylindrical coordinates

$$x^1 = z$$
$$[(x^2)^2 + (x^3)^2]^{1/2} = r$$

[†] An isotropic tensor function of a tensor is one for which $\mathbf{Q}\mathbf{F}(\mathbf{D})\mathbf{Q}^T = \mathbf{F}(\mathbf{Q}\mathbf{D}\mathbf{Q}^T)$. Objectivity requirements ensure this property. (See (7.30) and the accompanying footnote.)

$$\tan^{-1}\left[\frac{x^3}{x^2}\right] = \theta$$

and (9.23) becomes

$$\frac{d}{dt}\ln L = l; \quad -2\frac{d}{dt}\ln R = l \tag{9.32}$$

where $L=L(t)$ and $R=R(t)$ are the length and radius, respectively, of the cylinder at time t. Note that, according to (9.32), an experimental realization of this form of steady extension requires a stretching $L(t)=L(0)e^{lt}$.

It is instructive to combine the kinematics and constitutive behavior given above with the equation of motion (4.26). The form of the solution to (4.26) can be written down immediately from a theorem for homogeneous motions [1, p. 74], or can be inferred from the following components of (4.26):

$$\left.\begin{array}{l}(s_1-\rho\psi)_{,1} = \rho(l_1)^2x^1\\(s_2-\rho\psi)_{,2} = \rho(l_2)^2x^2\\(s_3-\rho\psi)_{,3} = \rho(l_3)^2x^3\end{array}\right\} \tag{9.33}$$

where $\mathbf{g}=-\nabla\psi$. Since each equation has the same form we can integrate the individual equations partially and combine the results to give

$$s_i = \tfrac{1}{2}\rho[(l_1x^1)^2+(l_2x^2)^2+(l_3x^3)^2]+\rho\psi+l_i\lambda_1+l_i^2\lambda_2+g(t) \tag{9.34}$$

$g(t)$ being an arbitrary function of time. Now apply (9.34) to the cylindrical geometry of the present example in order to compute those tensions which must be applied to each end of the cylinder (s_0, s_L at $z=0, L$) and at the walls (s_R at $r=R$) to achieve the motion (9.32):

$$\left.\begin{array}{l}s_0 = \tfrac{1}{8}\rho(lr)^2+\rho\psi+l(\lambda_1+l\lambda_2)+g(t)\\s_L = \tfrac{1}{2}\rho l^2(L^2+\tfrac{1}{4}r^2)+\rho\psi+l(\lambda_1+l\lambda_2)+g(t)\\s_R = \tfrac{1}{2}\rho l^2(z^2+\tfrac{1}{4}R^2)+\rho\psi-\tfrac{1}{2}l(\lambda_1-\tfrac{1}{2}l\lambda_2)+g(t)\end{array}\right\} \tag{9.35}$$

Before dropping several terms in (9.35) which one might hope are negligible in cases of interest, it is instructive to consider some of the consequences of the full equations. In particular the terms arising from the presence of a body force and from inertia indicate that the extension (9.32) cannot be maintained without a continuous adjustment to the (constant) ambient pressure at the fluid/surroundings interface. Hence the flow is not realizable if one merely stretches a fluid cylinder by moving the ends in accord with (9.32a). If, however, we can neglect effects of gravitation and inertia, spatial dependence of the stresses is eliminated. In fact, since $g(t)$ is arbitrary, one can measure thrusts with respect to the value of s_R at the cylinder wall. Then

$$\left.\begin{array}{l}s_R= 0\\s_0 = s_L = \tfrac{3}{2}l\lambda_1+\tfrac{3}{4}l^2\lambda_2\end{array}\right\} \tag{9.36}$$

If the fluid is Newtonian one immediately concludes from (9.29) that $\lambda_1=2\mu$, $\lambda_2=0$, μ being the fluid viscosity. Then (9.36) becomes

$$s_0 = s_L = 3\mu l \tag{9.37}$$

We shall return to this celebrated result later. The ratio of (s_L-s_R) to stretch rate in extension experiments is often referred to as the *extensional* or *Trouton viscosity* [9, p. 97]. Though it is clear that the Trouton viscosity of a Newtonian fluid is three times the ordinary viscosity when body and inertial forces can be ignored, the significance of this ratio in general is by no means assured. In particular note that we have made absolutely no connection between the viscometric functions for a simple fluid and the function relevant to steady extensions.

Naturally, one can consider extension of a cylinder in terms of the single scalar-valued function of (9.28). In fact, because of the connections between the l_i (9.31) we can write

$$s_L - s_R = \mu_T(l)l \qquad (9.38)$$

and concern ourselves with evaluation of the Trouton viscosity μ_T in lieu of explicit evaluations of the two scalar invariant functions λ_1 and λ_2. As long as we are concerned with cylindrical extension of the type described, use of (9.38) is perfectly suitable.

(ii) Extension of a sheet in one dimension and contraction in another (pure shearing).

In this case,

$$l_1 = l; \quad l_2 = -l; \quad l_3 = 0$$

Associating 1, 2, 3 with a rectangular Cartesian coordinate system we have for the two finite dimensions of the sheet which at $t=0$ are X_0 and Y_0,

$$x = X_0 e^{lt}; \quad y = Y_0 e^{-lt} \qquad (9.39)$$

One finds, after invoking the simplifications used in the preceding example,

$$s_x - s_y = 2l\lambda_1 \qquad (9.40)$$

The effect of λ_2 cancels in this case, a result that suggests a combination of configurations (i) and (ii) to ascertain the separate contributions of λ_1 and λ_2. Note also that in this case the extensional viscosity $(s_x - s_y)/l = 2\lambda_1$, which is 4μ for a Newtonian fluid.

(iii) Biaxial extension of a sheet.

Imagine a thin sheet which is extended at equal rates l in two orthogonal directions in the plane of the sheet. Then, in contrast to the case (ii) above, we have

$$l_1 = l; \quad l_2 = l; \quad l_3 = -2l \qquad (9.41)$$

The extensional viscosity in this case is usually defined as $(s_x - s_z)/l$. We find

$$s_x - s_z = 3l(\lambda_1 - l\lambda_2) \qquad (9.42)$$

and the extensional viscosity for a Newtonian fluid is

$$\frac{s_x - s_z}{l} = 6\mu \qquad (9.43)$$

Further examples of extensional motions are discussed by Coleman [7] and by Huilgol [8]. We note in passing that Coleman [7] has developed properties of a hybrid class of motions, not necessarily steady, which he calls *sheared extensions*. The interested reader is referred to his paper for details.

9.5. NEARLY VISCOMETRIC FLOWS

The reader has been reminded on several occasions that rather sweeping statements about broad classes of materials, such as simple fluids, were possible for viscometric flows and steady extensions because these flows provide each material point with extremely simple histories. It is natural to ask what the value of information obtained for a given fluid in one of these special flows might be when one wishes to describe a more complicated flow. We have already seen that, within the framework of the continuum approach used thus far, there is no apparent connection between the material functions relevant to viscometric and to steady extensional flow. However, we have also seen that although they are both flows with constant stretch histories their structure, in terms of the nature of $\mathbf{N}$ in (9.3), is very different.

There are numerous ways by which one may perturb a flow (or a fluid) from a condition (or a response) which is readily characterized to one which is in some sense "near" to the basic condition. Joseph and his coworkers have analyzed the behavior of free surfaces of non-Newtonian fluids by perturbation from a rest state [10]. In a review of various kinds of approximating procedures, Pipkin [11] has referred to small motions, short motions, and slow motions about a basic state.

In this section we develop a formalism appropriate to flows of simple fluids which are "nearly viscometric". The presentation follows closely an important paper on the subject by Pipkin and Owen [12]. Our purpose is to show how the results of Pipkin and Owen can be cause to the practitioner for hope as well as anxiety. Hope because one finds that indeed the viscometric functions are important for description of nonviscometric flows; anxiety, however, because several new material functionals occur which can, in principle, have an important effect upon the fluid response. Furthermore, unambiguous means for experimental measurement of these new functions are not at all clear.

Points for departure are the constitutive functional of a simple incompressible fluid (as usual we restrict ourselves to isochoric, i.e., constant density, motions)

$$\mathbf{T}(t) = \mathbf{S}(t) + p(t)\mathbf{I} = \underset{s=0}{\overset{\infty}{\mathscr{H}}} [\mathbf{C}_{(t)}(s)] \tag{7.11}$$

and the particularly simple form of the right relative Cauchy–Green strain tensor in a viscometric flow

$$\bar{\mathbf{C}}_{(t)}(s) = \mathbf{I} - s\mathbf{A}_1 + s^2 \tfrac{1}{2}\mathbf{A}_2 \tag{7.26}$$

where $\mathbf{A}_1$ and $\mathbf{A}_2$ can be put in the forms (7.24) and (7.25). It is convenient here to use the overbar to designate a viscometric history. Pipkin and Owen's basic assumption is that the constitutive functional $\mathscr{H}(\mathbf{C}_{(t)}(s))$ for a nearly viscometric flow can be expanded about the functional for a *related* viscometric history $\bar{\mathbf{C}}_{(t)}(s)$, so that

$$\mathbf{T}(t) = \underset{s=0}{\overset{\infty}{\mathscr{H}}} [\bar{\mathbf{C}}_{(t)}(s)] + \delta \underset{s=0}{\overset{\infty}{\mathscr{H}}} [\bar{\mathbf{C}}_{(t)}(s), \mathbf{E}_{(t)}(s)] + \mathbf{O}(\|\mathbf{E}_{(t)}\|^2) \tag{9.44}$$

where $\mathbf{E}_{(t)}(s) = \mathbf{C}_{(t)}(s) - \bar{\mathbf{C}}_{(t)}(s)$ is a small parameter which is the expansion variable and accounts for the effect of the *difference* history between the actual flow and the related viscometric flow.

Some explanation is in order regarding the last two terms on the right-hand side of

(9.44). It is clear that the degree of departure of the flow from viscometric is measured by $\mathbf{E}_{(t)}$. Over the whole history $s \geqslant 0$ we must measure the size of the departure by defining some appropriate scalar-valued *norm* of the difference history, for example

$$\left[\frac{1}{t_E} \int_{s=0}^{t_E} \mathrm{tr}(\mathbf{E}_{(t)} \mathbf{E}_{(t)}^T) \, ds \right]^{1/2}$$

where t_E is a time interval appropriate to the flow being considered. The exact form of the norm is not central to the development. The functional $\delta\mathscr{H}$ is taken to be *linear* in the difference history $\mathbf{E}_{(t)}(s)$. Thus in using (9.44) we are assuming that there is a region of interest over which the departure of $\mathscr{H}$ from its value for a viscometric flow can be expressed in terms of a linear functional in $\mathbf{E}_{(t)}(s)$, and that higher-order terms (terms $O(\|\mathbf{E}\|^2)$ and smaller) are sufficiently small to be neglected. Note that the basic theory says nothing about the proper definition of $\|\mathbf{E}\|$ and hence the exact extent to which (9.44) is applicable. That will depend upon the details of the constitutive equation for the fluid under study and the precise nature of the nonviscometric flow. In spite of this degree of ambiguity, which can only be resolved on an individual basis for fluids and flows, some important generalizations are possible.

From the theory of functionals there is good reason to expect that an expansion of the type (9.44) is legitimate (without answering the question of the degree to which it may be useful in treating nonviscometric flows). Inherent in (9.44) is the assumption that $\mathscr{H}$ is Fréchet-differentiable and that $\delta\mathscr{H}$ is therefore a Fréchet differential. Simply stated, we are assuming that $\mathscr{H}$ can be expanded in the functional analog to a Taylor-series expansion of functions, and that $\delta\mathscr{H}$ is the first term of this expansion. As with a Taylor-series, we are assuming in (9.44) that $\mathscr{H}$ possesses sufficient smoothness properties so that an expansion in Fréchet differentials is possible. For further detail the reader is referred to the brief discussion by Truesdell and Noll [1, p. 104] or to more specialized treatments [13, ch. 1].

Our task now is of course to see how the various constraints which we have imposed upon fluid and flow can be used to infer something about the nature of $\delta\mathscr{H}$. Since we know that $\overline{\mathbf{C}}_{(t)}(s)$ can be reduced to the form (7.26), it is clear that we can write $\delta\mathscr{H}$ in terms of different arguments

$$\delta\mathscr{H} = \underset{s=0}{\overset{\infty}{\delta\mathscr{H}}} [\{\mathbf{e}^i\}, \kappa, \mathbf{E}_{(t)}(s)] \tag{9.45}$$†

where $\{\mathbf{e}^i\}$ is the orthonormal basis with respect to which $\mathbf{A}_1$ and $\mathbf{A}_2$ have the form (7.24) and (7.25). We can go considerably beyond (9.45) by making the following observations:

(i) As a consequence of material objectivity $\delta\mathscr{H}$ is an isotropic tensor-valued function of the orthonormal vectors $\mathbf{e}^i$. From the footnote following (7.30) we see that the $\mathbf{e}^i$ must occur in $\delta\mathscr{H}$ so that

$$\delta\mathscr{H} = \mathbf{e}^i \otimes \mathbf{e}^j \underset{s=0}{\overset{\infty}{\delta\mathscr{H}}}{}_{ij}(\mathbf{e}^k \cdot \mathbf{e}^l, \kappa, \mathbf{E}_{(t)}(s))$$

where $\mathbf{e}^k \cdot \mathbf{e}^l = \delta^{kl}$. Otherwise, the isotropy condition will be violated.

---

† Recall that both $\{\mathbf{e}^i\}$ and κ can, in general, be functions of t (Section 7.6). For simplicity we omit showing this time dependence explicitly.

(ii) The functional is linear in the difference history. Therefore $\delta\mathcal{H}$ can be written as a linear combination of terms involving various components of $\mathbf{E}_{(t)}(s)$ with respect to the basis $\{\mathbf{e}^i\}$.

From (i) and (ii) we can write

$$\delta\mathcal{H} = \mathbf{e}^i \otimes \mathbf{e}^j \sum_{k,l} \delta \overset{\infty}{\underset{s=0}{\mathcal{H}}}_{ijkl} [\kappa, E_{(t)kl}(s)] \qquad (9.46)$$

Here $\delta \overset{\infty}{\underset{s=0}{\mathcal{H}}}_{ijkl}$ is a scalar-valued functional of the single component of the history $E_{(t)kl}(s)$ and a function of the shear rate κ.

Several additional restrictions can be added

(iii) $\mathbf{E}_{(t)}$ is symmetric, so $E_{(t)kl} = E_{(t)lk}$.
(iv) $\mathbf{T}$ is symmetric, so $\delta\mathcal{H}_{ijkl}{}^\dagger = \delta\mathcal{H}_{jikl}$.
(v) The extra stress $\mathbf{T}$ is traceless. Hence $\delta\mathcal{H}_{iijk} = 0$.
(vi) We have already noted that the influence of $\bar{\mathbf{C}}_{(t)}$ on $\delta\mathcal{H}$ is expressed through the variables $\mathbf{e}^i$ and κ. Because these quantities are related to $\bar{\mathbf{C}}_{(t)}$ through (7.24)–(7.26) their appearance in (9.46) is subject to certain symmetry requirements. For example, it is clear from (7.24) and (7.25) that a reflection of the 3-axis in the 1,2-plane can have no effect on $\bar{\mathbf{C}}_{(t)}$ and hence on $\delta\mathcal{H}_{ijkl}$. Recalling that the components $E_{(t)ij}$ are also written with respect to the basis $\{\mathbf{e}^i\}$, one concludes that, since $\delta\mathcal{H}_{ijkl}$ must be invariant to reflections of the 3-axis, any functional $\delta\mathcal{H}_{ijkl}$ in which the subscript 3 appears an odd number of times must vanish. Thus,

$$\delta\mathcal{H}_{3\alpha\beta\gamma} = \delta\mathcal{H}_{\alpha3\beta\gamma} = \delta\mathcal{H}_{\alpha\beta3\gamma} = \delta\mathcal{H}_{\alpha\beta\gamma3}$$
$$= \delta\mathcal{H}_{\alpha333} = \delta\mathcal{H}_{3\alpha33} = \delta\mathcal{H}_{33\alpha3} = \delta\mathcal{H}_{333\alpha} = 0 \qquad (9.47)$$

where, following the notation of Pipkin and Owen, we restrict the range of Greek indices to 1, 2.

(vii) For the same reason as that given in (vi), $\delta\mathcal{H}_{ijkl}$ must be invariant to simultaneous reflection of $\mathbf{e}^1$ and change of sign in κ, or to reflection of $\mathbf{e}^2$ with simultaneous sign change in κ. Thus functionals $\delta\mathcal{H}_{ijkl}$ with an odd number of subscripts equal to unity (or to 2) must be odd functions of κ. Functionals with an even number of subscripts equal to unity (or to 2) must likewise be even functions of κ.

Now let us incorporate these restrictions into the form of components $\delta\mathcal{H}_{ij}$ of $\delta\mathcal{H}$. Because of the special role of the 3-direction we consider components $\mathcal{H}_{i3}$ separately. Since, from (9.47) we know $\delta\mathcal{H}_{\alpha\beta3\gamma} = \delta\mathcal{H}_{\alpha\beta\gamma3} = 0$, we have

$$\delta\mathcal{H}_{\alpha\beta} = \delta\mathcal{H}_{\beta\alpha} = \delta\mathcal{H}_{\alpha\beta\gamma\delta}[\kappa, E_{(t)\gamma\delta}(s)] + \delta\mathcal{H}_{\alpha\beta33}[\kappa, E_{(t)33}(s)] \qquad (9.48)$$

and the nonzero components $\delta\mathcal{H}_{i3}$ are

$$\left.\begin{array}{l}
\delta\mathcal{H}_{\alpha3} = \delta\mathcal{H}_{3\alpha} = 2\delta\mathcal{H}_{\alpha3\beta3}[\kappa, E_{(t)\beta3}(s)] \\
\delta\mathcal{H}_{33} = \delta\mathcal{H}_{33\alpha\beta}[\kappa, E_{(t)\alpha\beta}(s)] + \delta\mathcal{H}_{3333}[\kappa, E_{(t)33}]
\end{array}\right\} \qquad (9.49)$$

Recall the additional requirement (v) that $\delta\mathcal{H}_{ii} = 0$. Counting up the independent scalar-valued functionals from (9.48) and (9.49) one then obtains a total of 19. However, we have not yet used the fact that $\mathbf{E}_{(t)}$ is constrained to motions which are isochoric. This of course requires (Section 5.6)

---

† We have suppressed explicit notation for the range of s between 0 and ∞.

$$\left.\begin{array}{l} \det[\mathbf{C}_t(s)] = \det[\overline{\mathbf{C}}_{(t)}(s) + \mathbf{E}_{(t)}(s)] = 1 \\ \det[\overline{\mathbf{C}}_{(t)}(s)] = 1 \end{array}\right\} \tag{9.50}$$

Since $\mathbf{E}_{(t)}(s)$ is a small perturbation on $\overline{\mathbf{C}}_{(t)}(s)$ it is useful to expand (9.50a), noting first the expression for the determinant of a sum of two tensors.

$$\det[\overline{\mathbf{C}}_{(t)} + \mathbf{E}_{(t)}] = \det[\overline{\mathbf{C}}_{(t)}]\det[\mathbf{I} + \overline{\mathbf{C}}_{(t)}^{-1}\mathbf{E}_{(t)}] = 1 \tag{9.51}$$

Equation (9.51) follows from the fact that $(\det \mathbf{A})(\det \mathbf{B}) = \det(\mathbf{AB})$ [14, p. 143]. But

$$\det[\mathbf{I} + \overline{\mathbf{C}}_{(t)}^{-1}\mathbf{E}_{(t)}] = 1 + \mathrm{tr}[\overline{\mathbf{C}}_{(t)}^{-1}\mathbf{E}_{(t)}] + \mathbf{O}[||\mathbf{E}||^2] \tag{9.52}$$

where $\mathbf{O}[||\mathbf{E}||^2]$ is a measure of the size of quadratic functions of $\mathbf{E}$. Equations (9.51) and (9.52) require that

$$\mathrm{tr}[\overline{\mathbf{C}}_{(t)}^{-1}\mathbf{E}_{(t)}] = \mathbf{O}[||\mathbf{E}||^2]$$

Since we are neglecting terms which contain powers of $\mathbf{E}_{(t)}$ greater than one, it is consistent to write

$$\mathrm{tr}[\overline{\mathbf{C}}_{(t)}^{-1}\mathbf{E}_{(t)}] = 0 \tag{9.53}$$

Equation (9.53) can be utilized in deducing the form of an expansion for $\delta\mathscr{H}$. For example we can interpret (9.53) to mean that only those difference histories $\mathbf{E}_{(t)}$ are allowed which are orthogonal to $\overline{\mathbf{C}}_{(t)}^{-1}$ (orthogonality being defined by (9.53)). Thus if one considers the (nonexistent) history $\mathbf{E}_{(t)}(s) = f(s)\overline{\mathbf{C}}_{(t)}^{-1}(s)$, we are free to write, for example,

$$\delta\mathscr{H}_{ijkl}[\kappa, f(s)\bar{c}_{(t)kl}^{-1}(s)] = 0 \tag{9.54}$$

But $\overline{\mathbf{C}}_{(t)}^{-1}$ is readily written in terms of κ and s from (7.24)–(7.26). In component form one finds

$$\bar{c}_{(t)ij}^{-1} = \delta_{ij} + \kappa s(\delta_{i1}\delta_{j2} + \delta_{i2}\delta_{j1}) + (\kappa s)^2\delta_{i2}\delta_{j2}$$

In view of the linear dependence of $\delta\mathscr{H}$ on $\overline{\mathbf{C}}_{(t)}^{-1}$ in (9.54) we can write

$$\delta\mathscr{H}_{ijkk}[\kappa, f(s)] + 2\delta\mathscr{H}_{ij12}[\kappa, \kappa sf(s)] + \delta\mathscr{H}_{ij22}[\kappa, (\kappa s)^2f(s)] = 0 \tag{9.55}$$

Now let us use (9.55) to reduce further the number of functionals required. For example, Pipkin and Owen chose to eliminate all $\delta\mathscr{H}_{ijkl}$ in which 3 appears as a subscript except for the functionals $\delta\mathscr{H}_{\alpha3\beta3}$. From (9.55)

$$-\delta\mathscr{H}_{\alpha\beta33}[\kappa, f(s)] = \delta\mathscr{H}_{\alpha\beta\gamma\gamma}[\kappa, f(s)] + 2\delta\mathscr{H}_{\alpha\beta12}[\kappa, \kappa sf(s)] + \delta\mathscr{H}_{\alpha\beta22}[\kappa, (\kappa s)^2f(s)] \tag{9.56}$$

From requirement (v)

$$\delta\mathscr{H}_{33\alpha\beta} = -\delta\mathscr{H}_{\gamma\gamma\alpha\beta} \tag{9.57}$$

while from (9.55) and (9.57),

$$\delta\mathscr{H}_{3333}[\kappa, f(s)] = \delta\mathscr{H}_{\alpha\alpha\beta\beta}[\kappa, f(s)] + 2\delta\mathscr{H}_{\alpha\alpha12}[\kappa, \kappa sf(s)] + \delta\mathscr{H}_{\alpha\alpha22}[\kappa, (\kappa s)^2f(s)] \tag{9.58}$$

Combining (9.56)–(9.58) with (9.48) and (9.49) we are left with *13* material functionals.

Of course, the 13 represent the maximum number of functionals which are necessary to describe a nearly viscometric flow in the sense that we have defined it. Without further inputs into the theory it is not possible to determine if all 13 will be equally important. Also, we have as yet said nothing about the three cornerstones $\tau(\kappa)$, $\mathscr{N}_1(\kappa)$, and $\mathscr{N}_2(\kappa)$ of Chapters 7 and 8. We can proceed by developing some connections between those functions and the $\delta\mathscr{H}_{ijkl}$ given here. For example, one can consider a viscometric history $\overline{\mathbf{C}}_{(t)}(s)$ and a second viscometric history $\overline{\overline{\mathbf{C}}}_{(t)}(s)$ which is close enough to $\overline{\mathbf{C}}_{(t)}(s)$ so that $\overline{\overline{\mathbf{C}}}_{(t)}(s)$

may be considered a small perturbation on the former. Thus $\mathbf{C}_{(t)}(s)$ is a "nearly" viscometric flow with respect to $\overline{\mathbf{C}}_{(t)}(s)$, and the expansion (9.44) (with $\mathbf{E}_{(t)} = \overline{\overline{\mathbf{C}}}_{(t)} - \overline{\mathbf{C}}_{(t)}$) can be applied. However, $\overline{\overline{\mathbf{C}}}_{(t)}(s)$ can also be treated as a viscometric flow in its own right. The matching of these two interpretations for $\overline{\overline{\mathbf{C}}}_{(t)}(s)$ is not difficult but is tedious and we shall not give the details here. The result [12] is a set of ten equations relating τ, $\mathcal{N}_1$, and $\mathcal{N}_2$, along with their derivatives, to the various $\delta\mathcal{H}_{ijkl}$.

9.6. CONCLUDING REMARKS

In the preceding section a framework has been developed for description of flows which are slightly removed from viscometric behavior. The chief result is an indication that these flows *may* require evaluation of a finite but large (any experimenter would call 13 large) number of material functionals. To proceed with application of this formalism one needs something more specific than the definition of a simple fluid in terms of some unspecified functional (7.11) of the history $\mathbf{C}_{(t)}(s)$. In addition to application of the ideas of Pipkin and Owen to specific constitutive equations, the approximation of $\overset{\infty}{\underset{s=0}{\mathcal{H}}}[\mathbf{C}_{(t)}(s)]$ by a series of integral expansions over the argument s provides a convenient starting point. Then one can reduce the functionals to known integral *functions* and can also use the connection between some of the integral expansions and the viscometric functions. By making approximations to both the flow *and* the fluid, considerable simplification is achieved. This technique has proved helpful in several instances. However, before discussing these applications, more must be said about constitutive equations, and that is the purpose of the next two chapters.

As a final comment we note that many of the ideas put forward by Pipkin and Owen were developed independently by Goddard and Miller [15]. They, however, were seeking means to attack a particular problem in hydrodynamic stability. Consequently, their method is less suitable for general exposition.

APPENDIX

9A1. THE CAYLEY–HAMILTON THEOREM

The Cayley–Hamilton theorem is a statement that a tensor $\mathbf{A}$ satisfies its own characteristic equation (3.93). Thus

$$-\mathbf{A}^3 + I_1\mathbf{A}^2 - I_2\mathbf{A} + I_3\mathbf{I} = 0 \qquad (9A1.1)$$

where I_1, I_2, and I_3 are the three principal invariants of $\mathbf{A}$ and are defined by (3.94). Proofs of (9A1.1) are available in most books on matrix analysis and linear algebra. We follow the development given by Hoffman and Kunze [14, p. 166] and Leigh [16, p. 67].

It is convenient to work with the matrix formed from the components of $\mathbf{A}$ with respect to an orthonormal basis $\{\mathbf{e}_i\}$. We shall need to use the matrix of the adjoint of a tensor $\mathbf{A}$, defined by

$$[\text{adj } \mathbf{A}] = \begin{bmatrix} \alpha_{11} & \alpha_{21} & \alpha_{31} \\ \alpha_{12} & \alpha_{22} & \alpha_{32} \\ \alpha_{13} & \alpha_{23} & \alpha_{33} \end{bmatrix}$$

where α_{ij} is the *cofactor* of a_{ij}. Recall that the cofactor of a_{ij} is defined as the determinant which is formed from the matrix of $\mathbf{A}$ when the ith row and jth column are removed and the result is multiplied by $(-1)^{i+j}$ [17, p. 5]. By performing the indicated operations one can verify that

$$[(\text{adj } \mathbf{A})_{\alpha i}][a_{j\alpha}] = (\det \mathbf{A})[\delta_{ij}] \tag{9A1.2}$$

where brackets have been used to emphasize that the indicated quantity is a matrix.

Given a tensor $\mathbf{A}$ one can readily show that the vector $\mathbf{Ae}_i$ can be written

$$\mathbf{Ae}_i = \sum_j [a_{ji}]\mathbf{e}_j \tag{9A1.3}$$

where the $\mathbf{e}_i$ are drawn from the orthonormal basis $\{\mathbf{e}_i\}$ and components a_{ji} of $\mathbf{A}$ are expressed with respect to that basis. An alternate form for (9A1.3) is

$$\sum_j \left\{ [\mathbf{A}[\delta_{ji}] - \mathbf{I}[a_{ji}]]\mathbf{e}_j \right\} = \mathbf{0} \tag{9A1.4}$$

Define

$$[B_{ji}] = [\mathbf{A}[\delta_{ji}] - \mathbf{I}[a_{ji}]] \tag{9A1.5}$$

so that $[\mathbf{B}]$ is a matrix with *tensor* elements. Then we can multiply (9A1.4) by $[(\text{adj } B)_{ik}]$ and apply (9A1.2) to obtain

$$\sum_j (\det \mathbf{B})[\delta_{kj}]\mathbf{e}_j = (\det \mathbf{B})\mathbf{e}_k = \mathbf{0} \tag{9A1.6}$$

But the value k of $\mathbf{e}_k$ is arbitrary. Hence det $\mathbf{B}$ must be zero. Computation of det $\mathbf{B}$ generates (9A1.1), and the theorem is proved.

A powerful consequence of the Cayley–Hamilton theorem concerns the form of a tensor-valued function $\mathbf{F}(\mathbf{A})$. If $\mathbf{F}$ can be expanded in a power series in $\mathbf{A}$, then one can use (9A1.1) to show that the tensor character of $\mathbf{F}(\mathbf{A})$ can always be expressed as a quadratic function of $\mathbf{A}$, with scalar coefficients which are functions of the three principal invariants of $\mathbf{A}$.[†]

9A2. PROOF OF EQUATION (9.13)

We wish to show that if a tensor $\mathbf{N}$ has the properties

$$|\mathbf{N}| = 1; \quad \mathbf{N}^2 = \mathbf{0} \tag{9A2.1}[‡]$$

then there is an orthogonal basis with respect to which physical components can be written

$$[n(ij)] = \begin{bmatrix} 0 & 1 & 0 \\ 0 & 0 & 0 \\ 0 & 0 & 0 \end{bmatrix} \tag{9A2.2}$$

Consider the symmetric tensor $\mathbf{M} = \mathbf{N} + \mathbf{N}^T$. In view of (9A2.1) we can write

$$\mathbf{M}^2 = \mathbf{N}\mathbf{N}^T + \mathbf{N}^T\mathbf{N}$$

[†] If $\mathbf{F}(\mathbf{A})$ is an isotropic tensor function of $\mathbf{A}$ (i.e., $\mathbf{QF}(\mathbf{A})\mathbf{Q}^T = \mathbf{F}(\mathbf{QAQ}^T)$), then the statement can be proved without resorting to a power series expansion [1, p. 32].

[‡] Recall that $|\mathbf{A}| = [\text{tr}(\mathbf{AA}^T)]^{1/2}$.

where $\mathbf{M}^2$ is a symmetric tensor and $\mathrm{tr}(\mathbf{M}^2) = 2$. Also, since $\mathbf{N}^2 = \mathbf{0}$,

$$\det \mathbf{N} = [\det(\mathbf{N}) \det(\mathbf{N})]^{1/2} = [\det(\mathbf{N}^2)]^{1/2} = 0 \qquad (9A2.3)$$

and since $\det(\mathbf{N}) = \det(\mathbf{N}^T)$, we may conclude

$$\det(\mathbf{M}) = 0 \qquad (9A2.4)$$

Now since $\mathbf{M}$ is symmetric, we can find an orthogonal basis with respect to which the matrix of $\mathbf{M}$ is diagonalized (Section 3.17). We can write this as

$$[m(ij)] = \begin{bmatrix} a & 0 & 0 \\ 0 & b & 0 \\ 0 & 0 & 0 \end{bmatrix} \qquad (9A2.5)$$

since $\det(\mathbf{M}) = 0$. We immediately deduce also that

$$\mathrm{tr}(\mathbf{M}^2) = a^2 + b^2$$

and therefore

$$a^2 + b^2 = 2 \qquad (9A2.6)$$

A second equation which governs a and b can be obtained from the Cayley–Hamilton theorem applied to $\mathbf{N}$. Since $\mathbf{N}^2 = \mathbf{0}$ we have from (9A1.1)

$$I_2 \mathbf{N} - I_3 \mathbf{I} = \mathbf{0} \qquad (9A2.7)$$

where $I_2 = \dfrac{1}{2}[(\mathrm{tr}\mathbf{N})^2 - \mathrm{tr}(\mathbf{N}^2)]$

$$I_3 = \det \mathbf{N}$$

From (9A2.1) and (9A2.3) it is evident that (9A2.7) reduces to

$$\tfrac{1}{2}(\mathrm{tr}\,\mathbf{N})^2 \mathbf{N} = \mathbf{0}$$

so that

$$\mathrm{tr}\,\mathbf{N} = 0 \qquad (9A2.8)$$

Consequently

$$\mathrm{tr}\,\mathbf{M} = 0$$

and

$$a + b = 0 \qquad (9A2.9)$$

Combining (9A2.6) and (9A2.9),

$$a = \pm 1; \quad b = \mp 1 \qquad (9A2.10)$$

One can readily demonstrate that a 45° rotation of base vectors 1 and 2 about the third base vector will yield a new orthogonal basis with respect to which the matrix of (9A2.5) will become

$$[m(ij)] = \begin{bmatrix} 0 & 1 & 0 \\ 1 & 0 & 0 \\ 0 & 0 & 0 \end{bmatrix} \qquad (9A2.11)$$

and that this can be decomposed into

$$[n(ij)] = \begin{bmatrix} 0 & 1 & 0 \\ 0 & 0 & 0 \\ 0 & 0 & 0 \end{bmatrix} \qquad (9A2.12)$$

where $\mathbf{M} = \mathbf{N} + \mathbf{N}^T$ and $\mathbf{N}$ satisfies the requirements of (9A2.1).

REFERENCES

1. Truesdell, C., and Noll, W., The non-linear field theories of mechanics, in *Handbuch der Physik*, Band III/3 (S. Flügge, ed.) (Berlin: Springer-Verlag, 1965).
2. Coleman, B. D., *Arch. Rational Mech. Anal.* **9**, 273 (1962).
3. Coleman, B. D., *Trans. Soc. Rheol.* **6**, 293 (1962).
4. Noll, W., *Arch. Rational Mech. Anal.* **11**, 97 (1962).
5. Huilgol, R. R., On the construction of motions with constant stretch history: I, Superposable viscometric flows, US Army Mathematics Research Center, The University of Wisconsin, Madison, Wisconsin, Report 954, 1968.
6. Coleman, B. D., and Noll. W., *Phys. Fluids* **5**, 840 (1962).
7. Coleman, B. D., *Proc. Roy. Soc.* (*London*), A **306**, 449 (1968).
8. Huilgol, R. R., On the construction of motions with constant stretch history: II, Motions superposable on simple extension and various simplified constitutive equations for constant stretch histories, US Army Mathematics Research Center, The University of Wisconsin, Madison, Wisconsin, Report 975, 1969.
9. Lodge, A. S., *Elastic Liquids* (New York: Academic Press, 1964).
10. Joseph, D. D., and Fosdick, R. L., *Arch. Rat. Mech. Anal.* **49**, 321 (1973).
11. Pipkin A. C., Approximate constitutive equations, in *Modern Developments in the Mechanics of Continua* (S. Eskinazi ed.), (New York: Academic Press, 1966).
12. Pipkin, A. C., and Owen, J. R., *Phys. Fluids* **10**, 836 (1967).
13. Volterra, V., *Theory of Functionals and of Integral and Integro-Differential Equations* (New York: Dover Publications, 1959).
14. Hoffman, K., and Kunze, R., *Linear Algebra* (Englewood Cliffs, New Jersey: Prentice-Hall, 1961).
15. Goddard, J. D., and Miller, C., A study of the Taylor–Couette stability of viscoelastic fluids, ORA Project 06673, University of Michigan, Ann Arbor, Michigan, 1967.
16. Leigh, D. C., *Nonlinear Continuum Mechanics* (New York: McGraw-Hill, 1968).
17. Amundson, N. R., *Mathematical Methods in Chemical Engineering* (Englewood Cliffs, New Jersey: Prentice-Hall, 1966).

PROBLEMS

9.1. Derive (9.12).

9.2. Following (9.12), it is stated that $|\mathbf{N_0}|=1$ implies $|\mathbf{N}|=1$. Prove this result.

9.3. It has been shown that a sufficient condition for a constant stretch history flow to be viscometric is $\mathbf{N} \neq \mathbf{0}$, but $\mathbf{N}^2 = \mathbf{0}$. Show that this condition is also necessary.

9.4. Show that for biaxial extension of a sheet

$$s_x - s_z = 3l(\lambda_1 - l\lambda_2) \tag{9.42}$$

and hence that the extensional viscosity for a Newtonian fluid is 6μ, where μ is the viscosity of the fluid.

9.5. Show how biaxial extension (case (iii) of Section 9.4) can be thought of as uniaxial *compression* (case (i) with a change of sign).

9.6. Verify (9.52).

CHAPTER 10

CONSTITUTIVE EQUATIONS: ELEMENTARY MODELS

10.1. INTRODUCTION

In contrast to the preceding chapters, which have for the most part been quite general and often rather formal, the discussion here will concentrate on very specific equations for constitutive behavior. Little of the material presented here fails to qualify as a special case of a class of materials already described or to be presented in Chapter 11 in more generality. However, so much of the motivation for research on constitutive equations has roots in some of the early specialized results, that a separate description is warranted.

The literature of rheology is filled with claims and counterclaims about the value of various approaches to the subject. This has been particularly true in the discussions surrounding a continuum vs. a molecular description of constitutive behavior. These controversies have frequently served to brighten technical literature, but they have also, it appears, overemphasized the limitations of each approach. The subject now seems to be achieving a degree of unity, and in recent years there has been a major effort to draw simultaneously upon the two descriptions. In both this chapter and the next the reader will be aware of the extent to which a molecular view can provide clues regarding the form and interpretation of constitutive equations. We have already seen the substantial advantages in breadth and analytical power which are afforded by continuum mechanics.

The purpose of this chapter is to introduce the reader to a few of the earliest and most widely used forms of constitutive equations. We first draw upon some far-reaching results of continuum mechanics for viscous fluids. Then a number of specific constitutive equations for viscous fluids are presented and discussed. The next portion of the chapter is devoted to ideas of classical linear viscoelasticity and, finally, some specific linear viscoelastic models which derive from a molecular interpretation of material behavior are considered.

10.2. THE CONSTITUTIVE EQUATION FOR A CLASS OF PURELY VISCOUS FLUIDS

We define a "purely viscous" fluid as one for which the stress at any given material point and time is a function of the velocity gradient evaluated at the point and time of interest. Since fluid response is characterized solely by motion at the present time, the

134

fluid is not capable of "memory". Then the functional equation (7.10) for an incompressible simple fluid reduces to

$$\mathbf{S} = \mathbf{G}(\mathbf{L}) \tag{10.1}$$

to within an isotropic pressure. Here, as in Chapter 5, $\mathbf{L} = \nabla\mathbf{v}$. We use the symbol $\mathbf{G}$ to mean a symmetric tensor-valued function of $\mathbf{L}$. The requirement, discussed in Section 9.4, that $\mathbf{G}$ be an isotropic function of $\mathbf{L}$ can be used to show that $\mathbf{G}$ can contain only the symmetric part $\mathbf{D} = 1/2[\nabla\mathbf{v} + (\nabla\mathbf{v})^T]$ of $\mathbf{L}$. To show this one can apply the principle of material objectivity, which was developed in Section 6.4. Recalling (5.52) we can write (10.1) in terms of its symmetric and antisymmetric parts

$$\mathbf{S} = \mathbf{G}(\mathbf{D} + \mathbf{W})$$

Now consider the change of frame (6.1) with $\tau' = \tau$. From (5.46), (5.52), and (6.2), one can show that at time t

$$\mathbf{W}' = \mathbf{Q}\mathbf{W}\mathbf{Q}^T + \dot{\mathbf{Q}}\mathbf{Q}^T$$

and, using (5.53), that

$$\mathbf{D}' = \mathbf{Q}\mathbf{D}\mathbf{Q}^T$$

Thus if

$$\mathbf{S}' = \mathbf{G}(\mathbf{D}' + \mathbf{W}')$$

where

$$\mathbf{S}' = \mathbf{Q}\mathbf{S}\mathbf{Q}^T$$

we have

$$\mathbf{S}' = \mathbf{G}(\mathbf{Q}\mathbf{D}\mathbf{Q}^T + \mathbf{Q}\mathbf{W}\mathbf{Q}^T + \dot{\mathbf{Q}}\mathbf{Q}^T)$$

But this violates the principle of material objectivity, which includes the requirement, discussed in Section 9.4, that $\mathbf{G}$ be an isotropic function of $\mathbf{L}$. Hence we must have

$$\mathbf{S}' = \mathbf{G}(\mathbf{Q}\mathbf{L}\mathbf{Q}^T) = \mathbf{G}(\mathbf{Q}\mathbf{D}\mathbf{Q}^T + \mathbf{Q}\mathbf{W}\mathbf{Q}^T)$$

and we are led to the conclusion that

$$\mathbf{S} = \mathbf{G}(\mathbf{D}) \tag{10.2}$$

where $\mathbf{G}$ now refers to any symmetric tensor-valued function of $\mathbf{D}$.

Note that (10.2) is identical to the tensor form of (9.28) except that we are no longer restricting the flow to the very special category discussed in Section 9.4. However, the same arguments used there may be applied again to (10.2) to derive the most general form of that equation. We follow, essentially, the argument of Truesdell and Noll [1, p. 32], which rests upon the requirements of an isotropic tensor function. Recall, from the definition of an isotropic tensor function that $\mathbf{Q}\mathbf{S}\mathbf{Q}^T = \mathbf{G}(\mathbf{Q}\mathbf{D}\mathbf{Q}^T)$. Also, since $\mathbf{D}$ is symmetric one can find three orthogonal principal directions for $\mathbf{D}$ (Section 3.17). Consider an orthonormal basis $\{\mathbf{e}_i\}$ for $\mathbf{D}$ where the $\mathbf{e}_i$ are in these principal directions. Call the corresponding principal values d_i. Now if $\mathbf{Q}$ is chosen to achieve a reflection of one of the principal directions, say $\mathbf{e}_k$, we have

$$\mathbf{Q}\mathbf{e}_k = -\mathbf{e}_k$$

But then

$$\mathbf{Q}\mathbf{D}\mathbf{Q}^T = \mathbf{D}$$

which also requires

$$\mathbf{Q}\mathbf{S}\mathbf{Q}^T = \mathbf{S}$$

so that

$$\mathbf{Q}\mathbf{S} = \mathbf{S}\mathbf{Q}$$

Then

$$\mathbf{Q}(\mathbf{S}\mathbf{e}_k) = \mathbf{S}(\mathbf{Q}\mathbf{e}_k) = -\mathbf{S}\mathbf{e}_k$$

so $\mathbf{Q}$ also reflects $\mathbf{S}\mathbf{e}_k$. Hence $\mathbf{S}\mathbf{e}_k$ must be a vector parallel to $\mathbf{e}_k$, which implies $\mathbf{S}\mathbf{e}_k = s_k \mathbf{e}_k$ (no sum). That is, $\mathbf{e}_k$ must also be a principal direction for $\mathbf{S}$, with principal value s_k. Thus $\mathbf{S}$ and $\mathbf{D}$ have the same principal directions. In terms of a matrix representation of components, we conclude that whenever the strain rate $\mathbf{D}$ is diagonalized, the stress $\mathbf{S}$ is also diagonalized.

Let us examine an interesting consequence of this result. From the component form of (10.2), written with respect to the basis in the principal directions $\{\mathbf{e}_i\}$, one can say

$$s_i = g_i(d_1, d_2, d_3)$$

If the principal values are distinct, we can write a set of three simultaneous equations

$$s_i = \phi_0 + \phi_1 d_i + \phi_2 d_i^2 \quad (i = 1, 2, 3, \text{ no sum}) \tag{10.3}$$

with a unique solution for the unknown functions $\phi_l(d_1, d_2, d_3)$. But we can regard the three equations (10.3) as a special representation of (10.2) when the base vectors are along the principal axes for $\mathbf{S}$ and $\mathbf{D}$. Since the ϕ_l are the same functions of d_1, d_2, and d_3 regardless of the value of i in (10.3) we conclude that (10.3) is a special form of a properly invariant tensor equation which, in basis-free notation, can be written

$$\mathbf{S} = \phi_0 \mathbf{I} + \phi_1 \mathbf{D} + \phi_2 \mathbf{D}^2 \tag{10.4}$$

where the ϕ_i can be written as scalar invariants of $\mathbf{D}$. We learned in Section 3.18 that the ϕ_i can then be expressed as functions of the three *principal* invariants of $\mathbf{D}$. For incompressible fluids (or isochoric motions) $I_1 = 0$ (or is of no consequence). Furthermore, we can absorb $\phi_0 \mathbf{I}$ into the undetermined isotropic pressure and write, finally, for the total stress[†]

$$\mathbf{S} = \mathbf{T} - p\mathbf{I} = -p\mathbf{I} + \phi_1(I_2, I_3)\mathbf{D} + \phi_2(I_2, I_3)\mathbf{D}^2 \tag{10.5}$$

Note that this treatment ensures unique solutions for the ϕ_i from (10.3) only if there are distinct principal values d_i.

The result (10.5) can also be obtained by assuming a power series expansion for $\mathbf{G}(\mathbf{D})$ and applying the Cayley–Hamilton theorem exactly as we did in Section 9.4.

Equation (10.2) and its reduction to (10.5) have an interesting history which is recounted in some detail by Truesdell and Noll [1, pp. 475 *et seq.*]. Equation (10.2) is a hypothesis used by Stokes in 1845. By adding the stipulation that $\mathbf{G}$ be *linear* in $\mathbf{D}$ he arrived at the general form of the constitutive equation for what we call a Newtonian fluid. Extension to (10.5) was reported in important papers by Reiner [2] and Rivlin [3], and as a consequence (10.5) is often referred to as the equation for a *Reiner–Rivlin fluid*.[‡]

Experimental evidence does not support the actual existence of a Reiner–Rivlin fluid for which $\phi_2 \neq 0$ (i.e., a Reiner–Rivlin fluid which exhibits normal stress effects).

Nonzero normal stresses seem characteristic of fluids with *elasticity* in the sense that fluid memory of past behavior is important in determining rheological response. Equation (10.5) has no provision for effects of memory. Furthermore, in laminar shearing flow (Section 7.4) it follows from (7.18), (7.33), and (10.5) that

---

[†] In reading the literature on this subject one must be cautious about the definitions of $\mathbf{T}$ and p. When discussing flow of simple fluids, $\mathbf{T}$ is almost always defined as the extra stress, in the sense of Section 7.2, so that $\mathrm{tr}\,\mathbf{T} = 0$. However, that practice is not generally followed in treatments of (10.5) or its equivalent.

[‡] Some authors (see, for example, [4, p. 68]) also call (10.5) the constitutive equation of a *Stokesian* fluid. This practice has been criticized by Truesdell and Noll [1, p. 486].

$$\tau(\kappa) = \frac{1}{2}\,\phi_1(\kappa)\kappa; \quad \mathcal{N}_1(\kappa) = 0; \quad \mathcal{N}_2(\kappa) = \frac{1}{4}\,\phi_2(\kappa)\kappa^2 \tag{10.6}$$

However, most data for polymer solutions characteristically show $|\mathcal{N}_1| > |\mathcal{N}_2|$ (cf. Section 8.9). Data on normal stresses of other kinds of rheologically complex fluids are less complete, but certainly do not offer strong support to (10.6).

In view of these highly undesirable features, one asks why the Reiner–Rivlin fluid deserves any notice at all. Several reasons can be given. The weight which one ascribes to these responses depends in good measure upon the attributes which one seeks in a constitutive equation. A practicing engineer and a specialist in macromolecular chemical research may have rather different criteria for a "good" constitutive equation. Positive features of the Reiner–Rivlin equation include

(i) Mathematical simplicity. It is the least-complicated model which allows, in a properly invariant way, introduction of nonzero normal stress differences.

(ii) Some, albeit uncertain, value as an indicator of the qualitative behavior of fluids with elasticity. Though the model allows only inelastic response, in the sense that elasticity is associated with memory, the fact that some nonzero normal stress differences are permitted can be of value in fixing ideas about behavior of viscoelastic liquids.

(iii) Most models of inelastic fluid response are contained in the special case where $\phi_2 = 0$ in (10.5).

(iv) In most cases the user of a constitutive equation is not seeking one which will model *exactly* the constitutive behavior of a real material. Rather, he wishes to find a relatively *simple* equation which will model to a reasonable *approximation* those *features* of constitutive behavior which are important to the user over a *limited range* of flow conditions.

10.3 SOME SPECIFIC EXAMPLES FOR REINER–RIVLIN FLUIDS

In this section we comment briefly about certain special cases of Reiner–Rivlin fluids. When evaluating various constitutive equations, it is important that the evaluator retain comment (iv) above as a guide.

(i) IDEAL (INCOMPRESSIBLE) INVISCID FLUID

$$\mathbf{S} = -p\mathbf{I}; \quad \phi_1 = \phi_2 = 0$$

This is of course the constitutive assumption which leads to Euler's equation of motion for inviscid incompressible fluids [5, p. 135].

(ii) INCOMPRESSIBLE NEWTONIAN FLUID

This most serviceable of all fluid constitutive models is obtained by setting

$$\phi_1 = \text{const} = \mu; \quad \phi_2 = 0.$$

(iii) $\phi_1 \neq$ **CONST**; $\phi_2 = 0$

This category, sometimes referred to as "generalized Newtonian fluids" [6], includes most of the constitutive models which have been in use for many years. Some are still useful for engineering design of apparatus in which non-Newtonian fluids are used.

At this point we digress to record the behavior typically found with polymer melts, polymer solutions, or solid-in-liquid suspensions when an approximation to laminar shearing flow is generated in a viscometer and the shear stress $\tau_{12} = \tau$ is measured. Some typical rheograms are shown in Fig. 10.1. A Newtonian fluid is of course characterized by

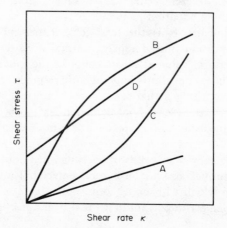

Fig. 10.1. Representative rheograms. A, Newtonian fluid; B, pseudoplastic fluid; C, dilatant fluid; D, Bingham plastic.

a straight line through the origin. The slope of the line is the viscosity μ. Polymer melts and solutions usually exhibit the behavior shown by curve B. At low shear rates κ, rheogram B approaches a straight line and hence Newtonian behavior. At sufficiently high values of κ straight-line behavior is again approached. Some materials, for example certain concentrated suspensions, show an opposite behavior in the sense that the slope of the τ vs. κ curve *increases* with increasing κ (curve C). In the former case one has a decrease in the effective viscosity (defined either as the instantaneous slope of the curve, or as the slope τ/κ of a straight line drawn from the origin to the point of interest). Such *shear-thinning* materials are referred to as *pseudoplastic*. The latter, shear-thickening, behavior is often called *dilatant* [7]. If time dependent effects are present (i.e., the curves depend upon the length of time the material has been sheared at a given value of κ) the two types of behavior are said to exhibit, respectively, *thixotropy* or *rheopexy*. It is an understatement to say that these as well as numerous other terms in rheology do not have a universally accepted interpretation [8, 9].

We consider now some properties of a few specific forms for ϕ_1:

(a) *The Power-law or Ostwald-de Waele Model and its Modifications*. A two-parameter function which has been useful in fitting rheological data for a large variety of pseudoplastic and dilatant flows is

$$\tau(\kappa) = K|\kappa|^{n-1}\kappa \tag{10.7}$$

where K and n are determined empirically, and the quantity $\mu_{\text{eff}} = K|\kappa|^{n-1}$ is often called the *effective viscosity*. Equation (10.7) is probably the most criticized, most maligned, and most widely used equation in all of rheology. The appeal of the power law is evident. When n, the power-law index, is unity (10.7) reduces to the description of a Newtonian fluid with viscosity K. For $0 < n < 1$ (10.7) describes a rheogram with the shape of a pseudoplastic fluid such as Curve B of Fig. 10.1. For $n > 1$ a curve characteristic of dilatant behavior is found. Hence three of the four shapes of curves shown in Fig. 10.1 can be described by (10.7). Furthermore, from values of n one can readily obtain a qualitative picture of the degree of shear-rate dependence shown by a given material. The breadth of application of (10.7) is illustrated in Table 10.1, where a list of power-law constants has been compiled.

TABLE 10.1. Representative power-law constants for equation (10.7)
(numbers apply at room temperature unless otherwise stated)

Material	Approximate range of κ (s^{-1})	K (dyne cm^{-2} sec^n)	n	Reference
54.3% cement rock in water	10–200	25.1	0.153	[7]
23.3% Illinois yellow clay in water	1800–6000	55.5	0.229	[7]
Polystyrene at 300°F	0.03–3	1.6×10^6	0.4	[10]
1.5% carboxymethylcellulose (CMC) in water	10^2–10^4	97	0.4	[11, p. 17]
0.7% CMC in water	2×10^3–3×10^4	15	0.5	[12]
3% polyisobutylene (Enjay Vistanex L-100) in decalin	25–200	9.4	0.77	[13]

Proper use of the power law depends upon recognition of its strengths and weaknesses First, it is not a law at all, but merely an attempt at empirical curve fitting with maximum simplicity. Even though (10.7) may fail to fit the total range of experimental data for some material, the expression can be very useful for a two-constant fit of rheological data over a restricted range of shear rate.

If the linear behavior of the pseudoplastic and dilatant curves shown in Fig. 10.1 at very high and low shear rates is universal, then it is clear that the power law must fail in the limits $\kappa \to 0$ and $\kappa \to \infty$. Neither a lower nor an upper limiting viscosity is predicted by (10.7). By paying the price of an additional empirical constant some of these shortcomings are readily overcome. For example, Bird and his associates have worked extensively with an "Ellis model" [6], in one form of which μ_{eff}, defined by

$$\tau(\kappa) = \mu_{\text{eff}}\,\kappa \tag{10.8}$$

is given by

$$\frac{1}{\mu_{\text{eff}}} = \frac{1}{\mu_0}\left[1 + \left(\frac{\tau}{\tau_{1/2}}\right)^{(1-n)/n}\right] \quad (0 < n < 1) \tag{10.9}$$

where $\tau_{1/2}$ is the shear stress at which $\mu_{\text{eff}} = \frac{1}{2}\mu_0$. It is apparent that at very low values of shear stress (and hence shear rate) Newtonian behavior with viscosity μ_0 is approached. As τ becomes large with respect to $\tau_{1/2}$, power-law behavior is approached. Thus the Ellis model avoids a problem as $\kappa \to 0$. However, an "upper Newtonian regime" is not predicted with the restrictions placed on n in (10.9). An example of Ellis constants is presented in Table 10.2.

TABLE 10.2. Ellis model constants for 6.1 % polyisobutylene in cetane at room temperature (see (10.9)) (Data of Hoffman [14])

$\mu_0 = 32.5$ poise	$\tau_{1/2} = 831$ dynes/cm²
$n = 0.424$	

Somewhat more subtle problems which are common to these one-dimensional empirical forms for ϕ_1 become apparent when one considers extension of (10.7) to more complex flows. Recall that, according to the Reiner–Rivlin equation (10.5), we have $K|\kappa|^{n-1} = \phi_1(I_2, I_3)$ where, by (3.94),

$$I_2 = -\tfrac{1}{2}d_{\alpha\beta}d^{\alpha\beta}; \quad I_3 = \det(\mathbf{D})$$

Suppose that we consider first a case of two-dimensional flow, the example being flow in a two-dimensional boundary layer over a flat plate or flow in a converging channel formed by two large and nonparallel planes. Since the flow is two dimensional, $I_3 = 0$ and $\phi_1 = \phi_1(I_2)$. Then to form a properly invariant form for ϕ_1 which is consistent with (10.7) we need merely write for the laminar shear flow

$$I_2 = -\frac{\kappa^2}{4}$$

and

$$t_{ij} = 2K(-4I_2)^{(n-1)/2}d_{ij} \tag{10.10}$$

Equation (10.10) has been very useful as a constitutive equation which can be used to handle problems in flows of engineering interest. We shall return to it repeatedly in a later chapter.

Note that connection of (10.10) with (10.7) was made possible by the restriction $I_3 = 0$. In a flow for which $I_3 \neq 0$ it is impossible to make any *a priori* statement about the form of $\phi_1(I_2, I_3)$ from measurement of power-law constants in a laminar shearing experiment for which $I_3 = 0$. Periodically, there have been conjectures over the importance of I_3 in $\phi_1(I_2, I_3)$, and the question is still not conclusively settled [15]. In any event, there are no data suggesting a strong dependence of ϕ_1 on I_3, and the approximation

$$\phi_1(I_2, I_3) \cong \phi_1(I_2) \tag{10.11}$$

seems consistent with the level of other approximations used in solving problems to which (10.11) is applied.

(b) *The Bingham Plastic.* This model incorporates the idea of a finite yield stress, which is that stress to be overcome before the material will sustain a nonzero rate of strain. The behavior is illustrated by curve D in Fig. 10.1 and is described by

$$\begin{aligned} \tau - \tau_0 &= \mu_p \kappa & |\tau| \geqslant \tau_0 \\ \kappa &= 0 & |\tau| \leqslant \tau_0 \end{aligned} \tag{10.12}$$

where τ_0 is the yield stress and μ_p is often called the plastic viscosity. The Bingham model has enjoyed considerable popularity as a means for characterization of rheological properties of drilling muds used in the petroleum industry [16]. If one is willing to ignore any possible effect of I_3, (10.12) is readily rewritten in proper tensor form. Thus

$$t_{ij} = \left[\mu_p - \frac{\tau_0}{\pm \sqrt{-4I_2}} \right] 2d_{ij} \quad \text{for } \frac{1}{2} t_{\alpha\beta} t^{\alpha\beta} \geqslant \tau_0{}^2$$

$$\tag{10.13}$$

$$d_{ij} = 0 \qquad\qquad \text{for } \frac{1}{2} t_{\alpha\beta} t^{\alpha\beta} \leqslant \tau_0{}^2$$

Other special forms for $\phi_1(I_2)$ are discussed in several textbooks [4, ch. 4; 11, pp. 99 et seq.].

10.4. FORMS OF THE RIVLIN–ERICKSEN FLUID

All of the constitutive equations discussed in the previous section were predicated upon a unique relation between the stress and the instantaneous value of the velocity gradient. We now consider the consequences of a slight relaxation of that restriction to include higher time derivatives of the deformation gradient, the first being the velocity gradient. Note, however, that these derivatives are still to be evaluated at the material point and time for which the stress is desired.

In Section 9.5 we noted that a functional could be expanded to connect a nonviscometric history to a related viscometric flow. We now consider a rather different expansion of the functional (7.11) for a simple fluid. Suppose that the combination of fluid properties and nature of the flow are such that only values of $\mathbf{C}_{(t)}(s)$ for $s \cong 0$ contribute to the functional $\underset{s=0}{\overset{\infty}{\mathscr{H}}}(\mathbf{C}_{(t)}(s))$. Then one can imagine an approximation to $\mathscr{H}$ that is a Taylor-type expansion of $\mathbf{C}_{(t)}(s)$ around $s=0$. One form of the result is

$$\mathbf{T}(t) = \mathbf{S} + p\mathbf{I} = \mathbf{F}(\mathbf{A}_1, \mathbf{A}_2, \ldots, \mathbf{A}_n) \tag{10.14}$$

where the $\mathbf{A}_i$ are the Rivlin–Ericksen tensors defined in (5.65). This result, though published by Rivlin and Ericksen in 1955 [17], is implicit in a form of the metric tensor used earlier by Oldroyd [18], and about which we shall say more in Chapter 11.

A special case of (10.14) is the restricted form $\mathbf{T} = \mathbf{F}(\mathbf{A}_1, \mathbf{A}_2)$. If $\mathbf{F}$ is well behaved (can be expanded as a polynomial of $\mathbf{A}_1$ and $\mathbf{A}_2$) then it can be shown to reduce to [1, p. 481, eqn. (13.7)]

$$\mathbf{T} = \mathbf{S} + p\mathbf{I} = \alpha_0 \mathbf{A}_1 + \alpha_1 \mathbf{A}_2 + \alpha_2 \mathbf{A}_1^2 + \alpha_3 \mathbf{A}_2^2 + \alpha_4 (\mathbf{A}_1 \mathbf{A}_2 + \mathbf{A}_2 \mathbf{A}_1) + \text{higher terms} \tag{10.15}$$

where the α_i are scalar functions of the joint invariants which can be formed from $\mathbf{A}_1$ and $\mathbf{A}_2$.

The Rivlin–Ericksen formulation has seldom been used as a starting point for development of constitutive equations, and that is why we have omitted the details leading to the results stated above. However, various forms of (10.14) often occur as limiting results of more general theories. A result of particular interest occurs when one considers viscometric flows. It is immediately clear from (7.27) that viscometric flow data will never allow a distinction between (10.14), (10.15), or a simple fluid since $\mathbf{A}_n = \mathbf{0}$ for $n > 2$ in all viscometric flows. It is perhaps worth noting here the contrasting situation for steady extension, in which case $\mathbf{A}_n \neq \mathbf{0}$ for all values of n. This is consistent with the distinctions noted in Section 9.2 where flows with constant stretch history were studied, and we see here again the fundamental differences between viscometric and extensional flows.

A special case of (10.15) is the equation of a "second-order fluid"

$$\mathbf{T} = \alpha_0 \mathbf{A}_1 + \alpha_1 \mathbf{A}_2 + \alpha_2 \mathbf{A}_1^2 \tag{10.16}$$

where here the α_i are constants. This equation is highly restrictive, not even allowing for a shear-dependent viscosity α_0. Also, it cannot be applied indiscriminately to time-dependent problems, since nonphysical results can occur [19]. However, (10.16) arises as the result of an orderly truncation of the general theory of simple fluids, and hence the equation is of interest as a limiting case for simple fluids [20].

10.5. LINEAR VISCOELASTICITY

Certainly the most striking feature connected with deformation of a polymeric material (solid or liquid) is its simultaneous exhibition of properties normally associated with elastic solids and viscous liquids. Experimenters were quick to notice this dual behavior, and it is not surprising that early quantitative descriptions were built around the notion of a linear combination of viscous and elastic properties. An authoritative treatment of this subject has been provided by Ferry [19]. Mathematical aspects are presented concisely in the monograph by Gross [22]. Most of the early workers in this field were polymer chemists who wished to describe the deformation properties of solid polymers. These properties could be determined rather nicely through experiments that were time-dependent in nature. In particular, the following experiments have been (and still are) physically convenient and mathematically simple:

(a) strain change upon cessation or initiation of stress (creep);
(b) stress change with a step-change of strain (stress buildup or stress relaxation);
(c) amplitude and phase relations for a sinusoidal oscillation of stress and strain.

In all three cases it is assumed that stresses and strains are "small" in the sense that one can reasonably expect a linear relation between the two, and that the stress and strain are homogeneous throughout the sample. Typically, creep and relaxation experiments are conducted with a rod which is stretched in a tensile-test machine or, alternatively, in a shearing experiment in which a small block of material is held between parallel plates undergoing relative motion. Oscillatory experiments can be similarly conducted.

In these exceedingly simple situations the complexities that abound in any multi-dimensional finite-strain description of material response (Chapters 5 and 6) disappear. Hence in this section we can dispense with tensor notation and merely refer to a stress τ and strain e. For the shearing experiment described above, strain is defined as distance traveled by one of the parallel plates relative to the other, divided by thickness of the sample between the parallel plates.

Response in the nonoscillatory experiments described above is often characterized by a modulus G which is the ratio τ/e. In dealing with oscillatory experiments it is convenient to describe the stress and strain by

$$\tau = \mathbf{R}[\tau_0 e^{i\omega t}]$$
$$e = \mathbf{R}[e_0 e^{i\omega t}] \tag{10.17}$$

where τ_0 and e_0 are in general complex and $\mathbf{R}$ refers to the real part of the terms following. A complex modulus is defined by

$$G^* = G' + iG'' = \tau_0/e_0 \tag{10.18}$$

G' is referred to as the *storage* modulus and G'' as the *loss* modulus, G' and G'' both being real numbers. It is readily shown that the dissipation associated with an oscillatory experiment is governed by G'', while G' represents elastic response. Thus if the stress and strain are in phase ($G''=0$) the behavior is that which one obtains from an ideal elastic (Hookean) material. If $G'=0$ but $G'' \neq 0$ the stress and strain are 90° out of phase, and one finds that all of the energy introduced into the system is dissipated into heat. This is of course the *viscous* portion of the material response.

Alternatively, but equivalently, one can describe material response in terms of the complex *compliance* $\mathcal{J}^*$ defined by

$$\mathcal{J}^* = \frac{1}{G^*} = \mathcal{J}' - i\mathcal{J}'' \tag{10.19}$$

Relations between G', G'' and $\mathcal{J}'$, $\mathcal{J}''$ are easily established [21, p. 15].

In contrast to the rest of this book, we have dealt here with materials as if they are primarily "solid-like" in their behavior—though to be sure one of the functions of the subject of rheology is to show how imprecise our notions of the words "solid" and "liquid" are. If one wishes to incorporate a modulus more characteristic of liquids, that is readily done by defining a *complex viscosity*

$$\mu^* = \mu' - i\mu'' = \tau/\dot{e} \tag{10.20}$$

where $\dot{e}=de/dt$. Since the strain is considered to be infinitesimal we are not bothered by the problems, discussed elsewhere (Chapters 5 and 11), associated with time derivatives. The connections between G^* and μ^* are readily established for an oscillatory experiment described by (10.17):

$$\mu^* = G^*/i\omega; \quad \mu' = G''/\omega; \quad \mu'' = G'/\omega \tag{10.21}$$

μ' is known as the *dynamic* viscosity and, of course, corresponds with the dissipative character of the material.

An obvious characterization of the relative importance of viscous and elastic components of a material is afforded by the *loss tangent*, tan δ, defined by

$$\tan \delta = G''/G'$$

Relations between the various quantities described above are conveniently illustrated on polar diagrams such as the one shown in Fig. 10.2. One notes the similarity to descriptions of simple alternating-current electrical circuits. This analogy has been used extensively in the treatment of linear viscoelasticity by Gross [22].

One would of course hope that the general features of curves of, say, G' and G'' plotted against ω could be interpreted in terms of molecular properties of the material. That in fact has been the object of much work by polymer chemists, and the reader interested in details is again referred to Ferry [21]. To give an idea of the range of responses provided by different polymeric materials, a few examples are presented in Fig. 10.3. As expected, several general classifications of behavior can be associated with polymer species and architecture. Interpretation of linear viscoelastic response in terms of molecular parameters continues to be an active area for research.

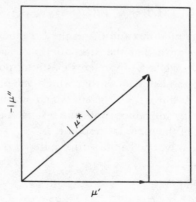

Fig. 10.2. Polar diagram for the complex viscosity: $|\mu^*| = [(\mu')^2 + (\mu'')^2]^{1/2}$.

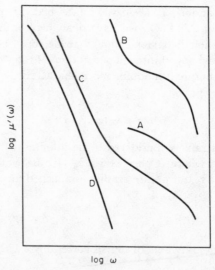

Fig. 10.3. Schematic drawing of effect of molecular properties on real part of the dynamic viscosity in oscillatory shear or extension [21, p. 46]. Curves are intended to show changes in shape, not relative numerical values. A, amorphous polymer with low molecular weight (10,500); B, amorphous polymer with high molecular weight (180,000); C, lightly cross-linked amorphous polymer; D, highly crystalline polymer.

10.6. SOME LINEAR VISCOELASTIC MODELS

(i) THE MAXWELL MODEL

In view of what has been said above about description of material response by linear combination of viscous and elastic contributions, one is not surprised to learn that a cornerstone of linear viscoelasticity is the *Maxwell model*, which is synthesized by a series combination of viscous and elastic behavior. The common mental visualization (for experiments in tension *or* shear!) of this behavior is a Hookean spring, with spring constant K linked to a "viscous dashpot" with viscosity μ (Fig. 10.4). The total strain is the sum of both viscous and elastic parts, but the stress in each element must be the same.

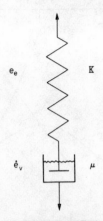

Fig. 10.4. The Maxwell element.

Thus

$$\tau = K e_e = \mu \dot{e}_v$$

The strain has been divided into elastic and viscous portions

$$e = e_e + e_v$$

so that

$$\dot{e} = \frac{\dot{\tau}}{K} + \frac{\tau}{\mu}$$

and

$$\tau + \lambda \dot{\tau} = \mu \dot{e} \qquad (10.22)$$

where $\lambda = \mu/K$.

This is the constitutive equation, or equation of state, of a *Maxwell* fluid. The feature which distinguishes (10.22) from a Newtonian fluid is of course the term reflecting an influence of the rate of change of stress. Its importance is weighed by a relaxation time λ. For example, let us consider the response in a stress relaxation experiment. The material, in a state of stress τ_0 at time $t=0$, is maintained at constant strain ($\dot{e}=0$) for all $t>0$. Then

$$\tau = \tau_0 e^{-t/\lambda} \qquad (10.23)$$

and we see that the relaxation process is controlled by λ.

The definitions (10.21) and (10.22), and the description (10.17) of an oscillatory experiment may be used to relate the complex modulus and complex viscosity to the relaxation time of a Maxwell fluid. Thus

$$\left. \begin{aligned} G'(\omega) &= \frac{K(\omega\lambda)^2}{1+(\omega\lambda)^2}; \quad \mu' = \frac{K\lambda}{1+(\omega\lambda)^2} \\[2mm] G''(\omega) &= \frac{K\omega\lambda}{1+(\omega\lambda)^2}; \quad \mu'' = \frac{K\omega\lambda^2}{1+(\omega\lambda)^2} \end{aligned} \right\} \qquad (10.24)$$

The effect of frequency on relative importance of recoverable and lost energy is seen from (10.24) as well as the curve for the loss modulus and $\tan \delta$, shown in Fig. 10.5. Though a real

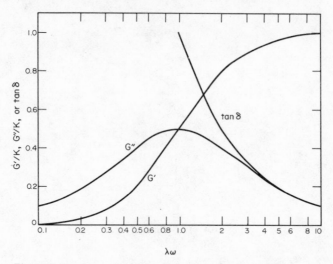

Fig. 10.5. Response of a Maxwell fluid as a function of frequency.

material surely will not obey the response (10.22) of a Maxwell fluid in any quantitative sense, the qualitative idea of relaxation and its characterization with a relaxation time are central to physical interpretations of rheological behavior. In subsequent chapters we shall draw frequently upon (10.22) or a related equation.

(ii) THE VOIGT MODEL

One of the features of the Maxwell model is its predominant "fluid-like" response. For example, according to (10.22) the material will continue to flow ($\dot{e} \neq 0$) as long as it is subjected to a constant nonzero stress, i.e., the material has infinite capacity for creep. A more solid-like response is exhibited by the *Voigt model* for a viscoelastic material. The mechanical description of this model is a parallel connection of spring and dashpot, shown in Fig. 10.6. In this case

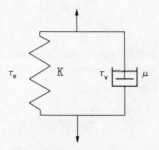

Fig. 10.6. The Voigt element.

$$\left. \begin{array}{l} \tau_e = Ke; \quad \tau_v = \mu\dot{e} \\ \tau = \tau_v + \tau_e = Ke + \mu\dot{e} \end{array} \right\} \tag{10.25}$$

Here we can define a retardation time $\lambda' = \mu/K$. This governs the change of strain in a creep experiment, since if at $t=0$ τ is suddenly set to zero, the strain changes as $e^{-t/\lambda'}$ for $t>0$. Relations similar to those found for the Maxwell fluid are readily obtained for the Voigt model.

(iii) SUPERPOSITION OF MODELS

One of the highly desirable features of the linear theories is that they can be conveniently superposed. Thus combined properties of Maxwell and Voigt models can be obtained through a series combination of the two [23]. The constitutive equation

$$\tau + \lambda\dot{\tau} = \mu(\dot{e} + \lambda'\ddot{e}) \tag{10.26}$$

accounts for some of the properties of each model. Its derivation, however, does not follow from straightforward superposition of (10.22) and (10.25) [24, p. 146]. One specific physical system from which (10.26) follows is a dilute suspension of elastic spheres in a Newtonian liquid [25].

10.7. A SPECTRUM OF RELAXATION TIMES

One can readily imagine the increased flexibility which becomes available through superposition of an ever-increasing number of Maxwell and Voigt elements. Of course the constants which must be evaluated increase correspondingly. For example, if one considers extension of N Maxwell elements (Fig. 10.7) connected in parallel, the stress is

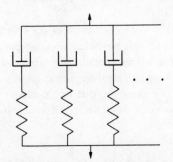

Fig. 10.7. Superposition of Maxwell elements.

the sum of contributions from each of the N elements. Strain is of course the same in each element. For discussion of stress relaxation when the material is at some constant strain, say $e = e_0$ at $t=0$, we can define a modulus

$$\lambda = \frac{\mu}{q}$$

$$G(t) = \tau(t)/e_0 = \sum_{i=1}^{N} G_{0_i} e^{-t/\lambda_i} \tag{10.27}$$

G_{0_i} being the contribution of the ith element to $G(0)$. It is natural to think of the limit as $N \to \infty$. Then one can define a continuous spectrum of relaxation times, for example,

$$G(t) = G_e + \int_{-\infty}^{\infty} H(\lambda)e^{-t/\lambda}\, d\ln \lambda \tag{10.28}$$

Thus $H(\lambda)$ (or $H(\lambda)/\lambda$ if one wishes to use an arithmetic scale) is a distribution function for relaxation times. $H(\lambda)$ weights the importance of the infinity of relaxation times available to the material. G_e is an equilibrium modulus which allows for extension of the model to materials capable of supporting a nonzero stress at $t \to \infty$.

The example above was drawn from stress relaxation for a sample under an initial tensile stress τ_0. Clearly, we can develop the same formalism for relaxation of shear stress.

A typical curve for $H(\lambda)$ against λ is shown in Fig. 10.8. These curves constitute an

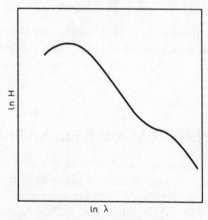

Fig. 10.8. Schematic drawing of the relaxation spectrum of an amorphous polymer with high molecular weight (example B of Fig. 10.3) [21, p. 64].

important "fingerprint" for linear viscoelastic behavior. In particular, if one can relate properties of the $H(\lambda)$ curve to fundamental molecular parameters, then there is hope of associating viscoelastic response with molecular architecture.

A similar approach can be used to analyze the response of a superposition of Maxwell elements in oscillatory flow. For example, one can consider the expressions for $\mu'(\omega)$ and $\mu''(\omega)$ in (10.24). Summing over N elements and passing to the limit as $N \to \infty$, a relaxation modulus $M(\lambda)$ is defined by

$$\left.\begin{aligned}
\mu'(\omega) &= \int_{0}^{\infty} \frac{1}{1+(\omega\lambda)^2}\, M(\lambda)d\lambda \\[2ex]
\mu''(\omega) &= \int_{0}^{\infty} \frac{\omega\lambda}{1+(\omega\lambda)^2}\, M(\lambda)d\lambda
\end{aligned}\right\} \tag{10.29}$$

We shall see in Section 10.9 that $M(\lambda) \equiv H(\lambda)$.

10.8. HIGHER TIME DERIVATIVES

A mathematical generalization of the physical superposition of elements discussed above is readily obtained. We write a general linear connection between the stress $\mathbf{T}$ and its time derivatives and the strain $\mathbf{C}$ and its time derivatives, up to an unspecified isotropic part, by

$$\mathscr{P}\mathbf{T}(t) = 2\mathscr{Q}\mathbf{C}(t) \qquad (10.30)$$

The linear operators $\mathscr{P}$ and $\mathscr{Q}$ are defined by

$$\mathscr{P} = 1 + \lambda_0 \frac{\partial}{\partial t} + \lambda_1 \frac{\partial^2}{\partial t^2} + \cdots$$

$$\mathscr{Q} = \eta \left(\lambda'_{-1} + \lambda'_0 \frac{\partial}{\partial t} + \lambda'_1 \frac{\partial^2}{\partial t^2} + \cdots \right)$$

where λ_i, λ'_i, and η are constants. We have reverted to a tensor expression for the stress and used the right Cauchy–Green tensor. Since we are restricting ourselves to infinitesimal strains the choice of strain tensor is not important. The right-hand side of the equation can be obtained from the Rivlin–Ericksen equation (10.14) in the limit of linearization with respect to strain.

10.9. INTEGRAL MODELS: THEIR RELATION TO DIFFERENTIAL MODELS

We now consider superposition from a somewhat different standpoint than in Section 10.8. We can, for example, imagine the stress to be due to a summation of a number of small partial stresses, each corresponding to a partial strain, and each stress relaxing according to some relaxation law which is a function of the difference between present time and the time at which the partial strain was imposed. Thus at time t, for material point X,

$$\left. \begin{aligned} \mathbf{T}(X, t) &= \sum_{i=1}^{N} \Delta\mathbf{T}_i \\[2mm] \Delta\mathbf{T}_i(X, t) &= 2\psi(t-t')\Delta\mathbf{C}_i(X, t') \end{aligned} \right\} \qquad (10.31)$$

or, taking the limit as $N \to \infty$,

$$\mathbf{T}(X, t) = 2 \int_{(\mathbf{C})_{t'=-\infty}}^{(\mathbf{C})_{t'=t}} \psi(t-t')d\mathbf{C} = 2 \int_{-\infty}^{t} \psi(t-t')\mathbf{D}(t')dt' \qquad (10.32)$$

since

$$\dot{\mathbf{C}}(t') = \frac{d\mathbf{C}(t')}{dt'} = \mathbf{D}$$

for infinitesimal strains. Equation (10.32) has an immediate physical appeal because the notion of a memory function $\psi(t-t')$ is made explicit. The stress at time t is seen to depend upon the strain rate at the present time *and* at all past times. One of course expects $\psi(\tau) \to 0$ as $\tau \to \infty$. The speed with which ψ decreases as its argument increases is a measure of the capacity for memory in the fluid.

Clearly, one could alternately choose to weight those increments in stress over past time which contribute to present strain, and the result would be of the form

$$\mathbf{C}(t) = \frac{1}{2} \int_{-\infty}^{t} \phi(t-t')\dot{\mathbf{T}}(t')dt' \tag{10.33}$$

If one agrees to set a reference time $t=0$ such that $\mathbf{T}=\mathbf{C}=\mathbf{0}$ for all $t \leqslant 0$, (10.32) and (10.33) can be related by taking Laplace transforms and applying the convolution theorem [26, pp. 35 *et seq.*]

$$\int_{0}^{t} \phi(t-t')\psi(t')dt' = \int_{0}^{t} \phi(t')\psi(t-t')dt' = t$$

The approach which leads to (10.32) and (10.33) is less different from the differential formulation of (10.30) than may be apparent. Suppose we set $\underline{\lambda}'_{-1}=0$ and $\underline{\lambda}'_{0}=1$ in the operator $\mathscr{2}$. If we again assume that there is some time $t=0$, say, before which the material was in an unstrained and unstressed state, then we can take the Laplace transformation of (10.30) and obtain

$$\overline{\mathbf{T}}(X, s) = 2\eta \left[\frac{1 + s\underline{\lambda}'_1 + \cdots}{1 + s\underline{\lambda}_0 + s^2\underline{\lambda}_1 + \cdots} \right] \overline{\mathbf{D}}(X, s) \tag{10.34}$$

where s now refers to the Laplace transform variable. But we can also express the Laplace transform of (10.32) by [26, pp. 35 *et seq.*]

$$\overline{\mathbf{T}}(X, s) = 2\overline{\psi}(s)\overline{\mathbf{D}}(X, s) \tag{10.35}$$

and the two expressions can be linked if we set

$$\left.\begin{aligned} [\overline{\psi}(s)] &= \eta \left[\frac{1 + s\underline{\lambda}'_1 + \cdots}{1 + s\underline{\lambda}_0 + s^2\underline{\lambda}_1 + \cdots} \right] \\ \psi(t) &= \eta \, \mathscr{L}^{-1} \left[\frac{1 + s\underline{\lambda}'_1 + \cdots}{1 + s\underline{\lambda}_0 + s^2\underline{\lambda}_1 + \cdots} \right] \end{aligned}\right\} \tag{10.36}$$

From this result several of the specific models discussed earlier can be generated by introducing truncated forms of the differential operators of (10.30). Thus for a Maxwell fluid we can write, from (10.22) and (10.36),

$$\psi(t) = \mu\mathscr{L}^{-1} \left\{ \frac{1}{1+s\lambda} \right\} \tag{10.37}$$

The inversion is readily accomplished, giving

$$\psi(t) = \frac{\mu}{\lambda} e^{-t/\lambda} \tag{10.38}$$

Hence an integral equivalent to the Maxwell model (10.22) is

$$\mathbf{T}(t) = 2 \, (\mu/\lambda) \int_{0}^{t} e^{-(t-t')/\lambda}\mathbf{D}(t')dt' \tag{10.39}$$

and, not surprisingly, one obtains a memory function which decays exponentially—probably the first form one would attempt to use in any sort of empirical integral relationship. Further connections between integral and differential models are cited by Fredrickson [4, pp. 125 *et seq.*].

We are now in a position to understand the equality of $M(\lambda)$ in (10.29) and $H(\lambda)$ in (10.28) for N parallel Maxwell models in the limit as $N \to \infty$. Recall that (10.28) describes the relaxation modulus of such a model. One can associate that specific result with the general idea of relaxation of partial stresses $\Delta \mathbf{T}_i$ which result from partial strains $\Delta \mathbf{C}_i$ (see (10.31)). If, for simplicity, we eliminate the possibility of an equilibrium stress, we can write, for (10.31),

$$\Delta \mathbf{T}_i = G(t - t') \Delta \mathbf{C}_i$$

so that (10.32) becomes

$$\mathbf{T}(X, t) = \int_{-\infty}^{t} G(t - t') \mathbf{D}(t') dt'$$

Then it is easy to show, by combination of the above equation with an oscillatory stress and strain of the form (10.17), that the expression $M(\lambda)$ in (10.29) can be replaced by $H(\lambda)$.

10.10. CONNECTIONS BETWEEN EXPERIMENTS

In preceding sections the response predicted from several rheological models was noted for a number of different experiments. One would expect that the material functions which describe one type of experiment (e.g., stress relaxation) should be related, through basic molecular parameters, to the material functions appropriate for another kind of experiment (e.g., oscillatory shear). A number of these connections have been found, by both empirical and fundamental studies. Ferry [21, chs. 3 and 4] takes up several of these, and Middleman [11, pp. 187 *et seq.*] presents a concise summary. To enter into a thorough discussion of these interrelations is beyond the scope of the present work. However, one connection is so fundamental to experimental rheology that we discuss it briefly.

The reader may have noted that steady laminar shearing flow, the central feature of Chapter 8, has been conspicuously absent from our analysis of linear viscoelastic response. One reason for this is that, when dealing with extremely viscous materials, it is difficult if not impossible to conduct steady shear experiments under controlled laboratory conditions and at shear rates high enough to correspond to flow situations of practical importance. For this reason there has been great interest in using the results of oscillatory shear experiments, which can be executed at small amplitudes but extremely high frequency, to infer behavior in steady shear flow. A particularly intriguing suggestion has been made by Cox and Merz [27]; namely, that one can associate oscillatory data relating $\mu^*(\omega)$ to ω with steady shear results by merely making the transformations

$$\omega \to \kappa; \quad |\mu^*(\omega)| \to \mu(\kappa) \tag{10.40}^\dagger$$

† The correlation actually presented in reference [27] is between $\mu^*(\omega)$ and the apparent viscosity in a capillary tube, defined as the ratio of wall shear stress to average velocity gradient across the viscometer tube. However, the authors note that the general shape of a curve of true viscosity (wall shear stress divided by wall shear rate) vs. shear rate "remains virtually unchanged".

κ being the shear rate $(\kappa = \dot{e})$. This analogy between steady and oscillatory flow has been a subject of considerable controversy. It involves fundamental questions such as the applicability of a linear analysis to flows in which the strain becomes arbitrarily large. We shall see later that a number of rather fundamental constitutive models suggest that though an analogy may exist, it is less straightforward than the relation (10.40).

10.11. MOLECULAR THEORIES OF VISCOELASTICITY: THE DUMBBELL MODEL

For the rest of this chapter we shall present a brief survey of some ideas for constitutive equations which are derived from models of *molecular* behavior. The term "molecular theories" is used rather loosely in rheology. In most cases it would be more accurate to call them "discrete" or "two-phase" theories since, ultimately, the calculation involves use of some hydrodynamic law for flow of a continuum past a solid body. To be sure, statistical considerations and Brownian effects may be imposed. However, there is no clear dividing line between the subjects of *suspension rheology* and *molecular theories of visco-elasticity*.

As an example we can consider the dumbbell model of viscoelasticity. Though the name may appear unambiguous, a search of the literature will reveal an almost endless variety of conceptions which are considered to be dumbbell models. These range from the very literal interpretation, in which a dumbbell model is a suspension of macroscopic objects, each consisting of two masses (spherical or otherwise) connected by a rigid or flexible connector [28], to a dumbbell which is meant to represent some statistical mean of the extension or orientation of a macromolecular chain with a constantly changing shape and position [29].

The use of rigid dumbbells as a model for macromolecular behavior has been described in a thorough review by Bird and coworkers [30]. They have shown, in fact, that many properties of polymer solutions are modeled more satisfactorily by rigid dumbbells than by masses joined with a Hookean spring. Unless the latter model is adjusted to provide some nonzero rest length for the connector, it approaches zero length when all stresses are removed. Hence the Hookean spring model does not include rigid dumbbells as a limit. This situation can of course be rectified by introduction of a spring force law which accounts for separation between the two dumbbell elements under conditions of total relaxation. The idea of a molecule modeled by a series of connected dumbbells has proved quite fruitful, and we consider one such model in the next section.

10.12. THE BEAD–SPRING MODELS OF ROUSE AND ZIMM

An extension of the modeling of a molecule by a dumbbell is a molecular model of $N+1$ "beads" connected by N massless springs. There have been many variants of this basic idea; in fact, the literature on the subject is so extensive that even reviews of various aspects of bead–spring formulations are numerous and rather highly specialized (see, for example, [31]). Bead–spring approaches combine an interesting mixture of continuum mechanics and statistical theory. By describing one particular treatment of a bead–spring model we shall be able to demonstrate the primary mathematical and physical features of this side of rheology, and yet avoid the considerable space that would be necessary to include a comprehensive description of the subject. The model proposed by Zimm [32] has been chosen as an example because it exhibits essential features of the bead–spring approach and it embraces, as limiting cases, some of the important earlier work. Furthermore, Zimm-like expressions result as special cases of some of the more complicated models which will be considered in Chapter 11.

We choose here to follow the derivation given by Lodge and Wu [33] rather than Zimm's original formulation. Lodge and Wu have clarified and broadened several features of Zimm's development.

Formulation of molecular theories requires use of several important and, in some cases, highly specialized and technical ideas of macromolecular kinetic theory. The description below is limited to an exposition of the physical concepts associated with ingredients for Zimm's theory. An appendix is provided in which more details are available for the interested reader.

The essential physical content of Zimm's theory (and bead–spring models generally) is the following:

(i) One wishes to model a polymer solution which is sufficiently dilute so that no consideration need be given to interactions between solute molecules.

(ii) Each polymer molecule is a long chain composed of monomer units. The molecule is not modeled by an analysis of the motion of each monomer unit. Rather, a model is conceived to approximate the average properties of groups of monomer units, each of these groups being called a submolecule. The submolecules are joined to form the full polymer molecule.

(iii) To each submolecule are assigned the properties of a massless spring joining two beads which have negligible inertia. Then the full molecule consists of $N+1$ identical beads joined by N springs. The springs are joined to the beads by freely rotating linkages.

We wish to write down an equation of motion appropriate to each bead of the model. To do this it is necessary to know the forces acting upon each bead. These are taken to be:

(a) force of the surrounding solvent, which is treated as a continuum;
(b) force which is transmitted to the bead in question by adjacent springs;
(c) force due to Brownian motion: the force results from bombardment of the macro-molecule by molecules from the "continuum". This is a correction to the continuum hypothesis of (a).

Let us consider quantitative estimation of these forces:

(a_1) Bead-continuum force. Initially we approximate the solvent motion as if it takes place unhindered by the presence of any beads. Then it makes physical sense to formulate the continuum force of the solvent upon a bead situated at position $\mathbf{r}_p$ with respect to a fixed origin by

$$\mathbf{F}_{s,p} = -f_0(\dot{\mathbf{r}}_p - \mathbf{v}'(\mathbf{r}_p)) \tag{10.41}$$

The constant f_0 is a bead "friction coefficient" and $\dot{\mathbf{r}}_p$ is the bead velocity. As a zeroth approximation, subject to correction in the next paragraph, $\mathbf{v}'(\mathbf{r}_p)$ is taken to be the undisturbed velocity $\mathbf{v}'_s(\mathbf{r}_p)$ of solvent at position $\mathbf{r}_p$.

(a_2) In general the force of (a_1) will have to be modified because of the perturbation of the flow at $\mathbf{r}_p$ from its undisturbed value, the perturbation being caused of course by the presence of other "beads" in the solvent. Thus we consider that the introduction of beads at $\mathbf{r}_0, \mathbf{r}_1, \mathbf{r}_2, \ldots \mathbf{r}_q, \ldots, \mathbf{r}_N$, $(q \neq p)$ will perturb the velocity at $\mathbf{r}_p$ from its undisturbed value. $\mathbf{F}_{s,p}$ of (10.41) is then computed for a bead p placed at $\mathbf{r}_p$ after the solvent velocity $\mathbf{v}'(\mathbf{r}_p)$ has been corrected for the presence of the other beads. This correction, often called *hydrodynamic interaction*, is given by (see Appendix, Section 10A3)

$$\sum_{\substack{q=0 \\ (q \neq p)}}^{N} <\mathbf{T}_{pq}> \mathbf{F}_q \qquad (10.42)$$

where $\mathbf{F}_q$ is the resultant force on bead q from spring and Brownian forces, and

$$<\mathbf{T}_{pq}> = [6\pi^3 |p-q|]^{-1/2}(\mu_s b)^{-1}\mathbf{I} \quad (p \neq q) \qquad (10.43)$$

μ_s is the solvent viscosity and b is explained below, following (10.44).

(b) *Spring force.* Very special springs are used for this model. They are Hookean, but have a force constant obtained from a statistical analysis of the properties of a submolecule. The simplest treatment of submolecule statistics leads to what is commonly referred to as a "Gaussian spring" (see Appendix, Section 10A1), defined as

$$\mathbf{F}_{S,p} = \frac{3kT}{b^2}(\mathbf{r}_p - \mathbf{r}_{p-1}) \qquad (10.44)$$

where $\mathbf{F}_{S,p}$ is the force exerted on the $(p-1)$th bead by the pth spring. The Boltzmann constant is given by k, T is the absolute temperature, and b^2 is the mean-square end-to-end distance of two adjacent beads for a system at macroscopic equilibrium, i.e., in a fluid which, from a macroscopic standpoint, has been at rest for an arbitrarily long time.

(c) *Brownian force.* Although the solute molecule is large relative to the solvent it is generally not so large that Brownian motion can be neglected. This is essentially a correction to the continuum hypothesis which has been applied to the solvent. The Brownian force on the pth bead is (see Appendix, Section 10A2)

$$\mathbf{F}_{B,p} = -kT\nabla_{\mathbf{r}_p} \ln \Psi \qquad (10.45)$$

At this point we have recognized the statistical nature of the problem by introduction of the distribution function for orientation of the bead–spring model. Here $\Psi(\mathbf{r}_0, \mathbf{r}_1, \ldots, \mathbf{r}_N, t)d\mathbf{r}_0 d\mathbf{r}_1 \ldots d\mathbf{r}_N{}^{\dagger}$ is the probability that at time t a given polymer molecule will be situated so that its bead–spring equivalent will have bead number 0 located in a differential volume element $d\mathbf{r}_0$, etc.

When the solution is in macroscopic equilibrium there is a Gaussian distribution of submolecule chain lengths, so that

$$\Psi_o \sim \exp\left[-\frac{3}{2b^2}\sum_{p=1}^{N}(\mathbf{r}_p - \mathbf{r}_{p-1})^2 \right] \qquad (10.46)$$

The proportionality constant is fixed by normalizing Ψ so that its integral over the entire $3(N+1)$ space is unity.

These ingredients permit one to write a system of coupled equations of motion for the beads.

The next task is to solve the resulting equations of motion. Then one has an expression for the motion of the beads in the bead–spring model when the "undisturbed" solvent

---

† We use the convention that $d\mathbf{r}_i$ symbolizes a differential element of volume at $\mathbf{r}_i$.

velocity is $\mathbf{v}'_s(\mathbf{r}_p)$. Usual practice in these formulations is to associate $\mathbf{v}'_s(\mathbf{r}_p)$ with the macroscopic continuum velocity $\mathbf{v}(\mathbf{r})$ which one would measure in the flow at position $\mathbf{r}$. Batchelor [34] has shown that this procedure is valid for the special case of flow between two parallel places in steady relative shearing motion (plane Couette flow). However, rigorous justification for general flow fields does not exist. One would not expect the error to be important in dilute solutions.

Solutions to the bead motion equations permit evaluation of $(\mathbf{r}_p - \mathbf{r}_{p-1})$ and hence the tension (10.44) in the Gaussian springs. This tension is taken to be the contribution of the macromolecule to the stress-supporting capacity of the polymer solution. Then the total contribution of the polymer to the stress tensor becomes (see Appendix, Section 10A4).

$$\mathbf{T}_M = \Sigma \mathbf{T}_{M_p} = c \int_0 \dots \int_N \sum_{p=1}^{N} \Psi \mathbf{F}_{S,p} \otimes (\mathbf{r}_p - \mathbf{r}_{p-1}) d\mathbf{r}_0 \dots d\mathbf{r}_N - Nck T \mathbf{I} \tag{10.47}$$

where c is the number density of polymer molecules in solution. The total stress, to within an isotropic pressure, is then taken to be the sum of the contribution of solute plus Newtonian solvent. Thus

$$\mathbf{T} = 2\mu_s \mathbf{D} - \mathbf{T}_M \tag{10.48}$$

The fundamental ideas employed by Zimm were not new at the time of his work (see Zimm's paper [32] for a brief review). However, he was the first person to indicate how a transformation of the equations of motion of the beads to a set of $3(N+1)$ normal coordinates could be effected when one allowed for hydrodynamic interaction between beads of a molecule. Earlier workers (see Rouse [35]) had found the transformation for the special case of negligible hydrodynamic interaction.

Zimm's original paper is restricted to application of his model to specific flow fields such as small amplitude oscillatory shearing. The derivation of Lodge and Wu, however, is applicable to a general rate of deformation, as long as its magnitude is small. Each bead–spring submolecule contributes an amount $\mathbf{T}_{M_p}$ to the polymer-solvent constitutive equation, where $\mathbf{T}_{M_p}$ obeys the equation (see Appendix, Section 10A5)

$$\left(\tau_p \frac{\partial}{\partial t} + 1\right)\mathbf{T}_{M_p} = ck T \, \tau_p (\mathbf{L} + \mathbf{L}^T) \tag{10.49}$$

and there are N equations. Here $\tau_p = f_0 b^2 / (6\pi k T \lambda_p)$ and $\mathbf{L} = \nabla \mathbf{v}$. The new quantity λ_p cannot be found without specification of chain length N and the degree of hydrodynamic interaction.

Equation (10.49) displays some striking features. First of all, the left-hand side is reminiscent of the Maxwell model (10.22). For example if one considers stress relaxation for the pth submolecule the stress behavior, to within a constant factor, matches that of the Maxwell model. The relaxation time constant in this case is τ_p. In that context one can interpret the Zimm model as a system of subunits, each responding to the motion as a linear viscoelastic element (Maxwell element) with submolecule time constant τ_p. Though the form of (10.49) is such that one may be tempted to imagine each submolecule acting independently of the others, interdependence of submolecules through parameters (N, degree of hydrodynamic interaction) must not be forgotten. Naturally, an expression for the total polymer molecule stress requires a solution of (10.49) so that the partial stresses can be added. One obtains

$$\mathbf{T}_M = \sum_{p=1}^{N} \mathbf{T}_{M_p} = \int_{-\infty}^{t} \mu(t-t')2\mathbf{D}(t')dt' \qquad (10.50)$$

where

$$\mu(\tau) = ckT \sum_{p=1}^{N} e^{-\tau/\tau_p}$$

$$\mathbf{D} = \frac{1}{2}(\mathbf{L}+\mathbf{L}^T)$$

To give some idea of the form of results from the Zimm model, we consider a few special cases. In an oscillatory shear experiment, for example, one can obtain the form of the complex viscosity following the procedure of Section 10.6. Assuming that the solvent viscosity is μ_s and that solvent and polymer contributions are additive,

$$\mu^* = \mu_s + ckT \sum_{p=1}^{N} \frac{\tau_p}{1+i\omega\tau_p} \qquad (10.51)$$

Computation of μ^* or other rheological moduli requires knowledge of τ_p, which in turn means that the principal values $\lambda_p \left(\tau_p = \dfrac{f_o b^2}{6kT\lambda_p}\right)$ of (10A5.21) must be computed. Lodge and Wu [33] have shown that the λ_p used here are identical to those originally employed by Zimm [32].

We note in passing that considerable simplification is possible in limiting cases. In the limit of vanishing hydrodynamic interaction Zimm has found [32]

$$\lambda_p = \pi^2 p^2/N^2 \qquad (10.52)$$

which is the case treated by Rouse [35]. The Rouse relaxation times (10.52) are often related to the *intrinsic viscosity* $[\mu]$ of a polymer solution, defined by

$$[\mu] = \lim_{c \to 0} \left[\frac{1}{\mu_s m} \frac{d\mu}{dc} \right] \qquad (10.53)$$

where m is the mass of a polymer molecule. If we take the limit of (10.51) as $\omega \to 0$ to represent the viscosity in steady shear flow, then the corresponding intrinsic viscosity $[\mu]_0$ is readily found from (10.51), (10.52), and (10.53).

$$[\mu]_0 = \frac{f_o b^2 N^2}{36 m \mu_s} \qquad (10.54)$$

where the approximation

$$\sum_{p=1}^{N} \frac{1}{p^2} \cong \frac{\pi^2}{6} \quad (N \gg 1)$$

has been used. Since the bead–spring model results in a viscosity which is linear with concentration, it follows from (10.54), (10.52), and the definitions of τ_p that in steady shear flow

$$\tau_p = \frac{6m(\mu - \mu_s)}{\pi^2 p^2 ck T} \tag{10.55}$$

In contrast to (10.54) Zimm [32] has reported that in the limit of dominant hydrodynamic interaction the intrinsic viscosity is given, for large N, by

$$[\mu]_0 = \frac{0.472 \, N^{3/2} \, b^3}{m} \tag{10.56}$$

It is clear that the additional flexibility afforded by the hydrodynamic interaction effect of the Zimm model is a useful asset when one wishes to correlate data. Many such attempts have been made, a notable case being the work of Osaki *et al.* [36]. They used values of λ_i computed by Lodge and Wu [37], who did not employ the approximate technique of Zimm but evaluated the principal values of the Zimm matrix without making the approximation $N \to \infty$. For various values of N and the hydrodynamic interaction parameter, reasonable agreement with experiment was obtained.

The chief asset of the Rouse model is its combination of simplicity and high yield of qualitatively useful information. For example, the Rouse or Zimm theory provides a link between the idea of a relaxation time derived from continuum (or mechanical) models and relatively more fundamental molecular parameters.

A feature of the Rouse–Zimm theory (and necessarily of any linear viscoelastic theory) is the independence of shear viscosity and shear rate. It is clear from earlier sections that shear dependent viscosity is one of the primary features of many rheologically interesting materials. Lack of this provision is a major failure of the Rouse–Zimm and other linear viscoelastic models. In Chapter 11 we shall describe some of the attempts which have been made to remedy this shortcoming.

10.13. NETWORK THEORIES: THE RUBBERLIKE LIQUID

There is another approach to description of flow behavior which is motivated from physical ideas very different from those of the preceding section but which leads to surprisingly similar results. Because the concept of a polymer melt or solution as a network has been the basis for so many constitutive theories, we briefly discuss the physical ideas here.

Attempts to model the deformation characteristics of rubber by a network of elastic members subject to laws of statistical mechanics have proved quite successful and form the basis for quantitative treatments of rubber elasticity [38]. These ideas have been adapted to polymer solutions by adding the notion that junctions are continually being created and destroyed, and that the terms in a "balance" equation for network junctions are affected by the macroscopic properties of the flow field. In this approach one derives an expression for the Helmholtz free energy of the system in terms of the length of network strands at a given time. These lengths are weighted over all possible values with a distribution function for strand lengths. However, the distribution function at time t depends upon its value over past times since the function is obtained from solution of a time-dependent differential balance equation. In this way the concept of memory is introduced into the expression for the free energy. Finally, the free energy is related to the stress by considering the work required to deform an element of the polymer solution, and hence to alter its free energy. By noting that the work performed is a double contraction between stress and strain tensors, and by assuming that the network deforms as if it were embedded in the fluid, one can identify the network stress and strain with the continuum stress and strain of the polymer solution. The final expression for stress (neglecting solvent) can be written, to within an arbitrary isotropic stress [33, 39],

$$\mathbf{T} = kT \int_0^\infty \mathcal{N}(s)\mathbf{C}_{(t)}^{-1}(s)\,ds \tag{10.57}$$

where $s = t - t'$ and $\mathbf{C}_{(t)}^{-1}$ is the inverse of the right relative Cauchy–Green strain tensor (5.35). $\mathcal{N}(s)$ is a relaxation function which has the form

$$\mathcal{N}(t - t') = \sum_i C_i e^{-(t-t')/\tau_{n_i}} \tag{10.58}$$

C_i and τ_{n_i} being constants which characterize the creation rate and the average lifetime of the ith type of junction.

Equation (10.57) can be integrated by parts. If the deformation is infinitesimal one can use

$$\frac{d\mathbf{C}_{(t)}^{-1}}{dt'} \cong -\frac{d\mathbf{C}_{(t)}}{dt'} \cong -2\mathbf{D}(t')$$

This approximation leads to the result that the constitutive equation (10.57), derived for solutions so concentrated that they can be modeled by a network of entangled polymer molecules, is identical in form to that obtained with the bead–spring model, in which the solution is assumed to be so dilute that each molecule acts entirely independently of all others. Although we have shown an equivalence for small deformations, the similarity also holds when one generalizes the treatment to finite deformations. This result, first pointed out decisively by Lodge and Wu [33], accounts for the success of each approach in describing phenomena under conditions for which the basic model would seem to be a very poor approximation.

Constitutive behavior of a rubber-like model, which is described by an equation of the form (10.57) and hence also represents the bead–spring model, has been discussed at length by Lodge [40, ch. 6]. Although, as noted in the previous section, these models fail to include a shear dependent viscosity, the qualitative features of a number of time-dependent flows are adequately described by (10.57) or its equivalent. The response in oscillatory shear is shown in Fig. 10.9.

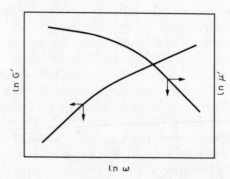

Fig. 10.9. Qualitative behavior of bead–spring and network models in small-amplitude oscillatory shearing.

10.14. CONCLUDING REMARKS

By now the reader should be aware of the striking contrasts between the approach to constitutive equations taken in Chapter 7 and that employed in the last sections of this chapter. In the present instance we have been very explicit about the type of fluid model which is envisioned. As a result one hopes, though not always with justification, that the constitutive parameters will have a direct physical interpretation. Furthermore, by being explicit in the description of the fluid, one would expect applicability to wider classes of flows than was found possible with the simple fluid model. Thus far in our exposition that hope has not been fully realized because of constraints under which the models were formulated, the most notable limitation being a restriction to small strains. The purpose of Chapter 11 is to discuss the limitations of small deformations in some detail and to explore possibilities for a more general formulation.

APPENDIX

10A1. THE GAUSSIAN SPRING

Recall that in the bead–spring model each submolecule or bead–spring combination represents many monomer units. Hence one can envision the submolecule as a chain built up from many freely rotating links, the alignment of each link being independent of the direction of neighboring links. We are interested in the end-to-end distance of the submolecule, i.e., the distance between two neighboring beads. If an origin is fixed in one bead, the configuration at some instant of time may appear as shown in Fig. 10A1.

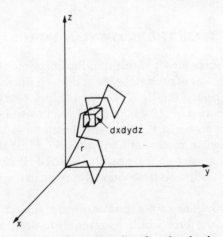

Fig. 10.A1. Representation of a submolecule.

We desire the form of a probability distribution function $P(\mathbf{r})$, where $P(\mathbf{r})dxdydz$ is the probability of one end (bead) of the submolecule residing in a volume element $dxdydz$ about position $\mathbf{r}$ with respect to the other end (bead). The reader will recognize this as a three-dimensional random walk problem, familiar from many areas of physics. One expects all directions in space to be equally probable, so that the result should depend only upon $|\mathbf{r}|$. The Gaussian probability distribution function is readily derived (see, for example, Tobolsky [41, pp. 297, *et seq.*] for a brief exposition relevant to the present application).

$$P(r)\,dxdydz = \left(\frac{3}{2\pi b^2}\right)^{3/2} \exp\left(\frac{-3r^2}{2b^2}\right) dxdydz \qquad (10\text{A}1.1)$$

where b^2 is the mean-square separation distance and $r \equiv |\mathbf{r}|$.

To obtain an expression for the spring tension in a bead–spring model of the situation shown in Fig. 10A1, we consider that the work required to change the separation from r to $r+dr$ is given by the change in Helmholtz free energy A between the two states [38, p. 61]. Then the magnitude of the force F for separation r is

$$F = \frac{dA}{dr} = - T \frac{dS}{dr}$$

where S is the entropy and T the absolute temperature. The entropy is connected to configuration probability through Boltzmann's equation [42, pp. 368 *et seq.*]

$$S = k \ln P + \text{const}$$

so that

$$F = \frac{3kT}{b^2} r \qquad (10\text{A}1.2)$$

This is the force law of a "Gaussian spring" connecting two beads with a distance r between centers. Equation (10.44) is the vector generalization of (10A1.2). It is clear that in (10.44) the force $F_{S,p}$ is assumed to be independent of the presence of other submolecules in the system.

10A2. THE BROWNIAN FORCE

The primary modern source for a treatment of Brownian motion is the work of Chandrasekhar [43], from which most of the development below is drawn. More recent extensions and generalizations have been given by Kirkwood [44, pp. 1–29].

To show the origin of equation (10.45) in detail would require a substantial diversion from the main subject of this book. We limit the presentation here to some of the basic ideas which apply to Brownian motion of a free particle. From the results, extension to a coupled system such as the bead–spring model is inferred. Because of the complexity of the subject, several aspects of Brownian motion are presented without the rigor which is available elsewhere [43].

Consider a particle moving through a fluid and hence subject to a drag force from the fluid. Over and above a drag force which is amenable to description by methods of continuum mechanics, we wish to account for the effect of erratic bombardment of the particle by molecular collision with fluid molecules. This additional effect gives rise to the "Brownian force" of (10.45).

Particle motion is governed by the Langevin equation [43]

$$\frac{d\mathbf{u}}{dt} = - \beta\mathbf{u} + \mathbf{A}(t) \qquad (10\text{A}2.1)$$

where **u** is the particle velocity and β is a coefficient which accounts for drag exerted on the particle by the surrounding continuum.

The function $\mathbf{A}(t)$ represents the Brownian contribution. It is a rapidly fluctuating quantity and is uncorrelated with itself in time increments Δt over which **u** changes by a perceptible amount. $\mathbf{A}(t)$ is only defined in a statistical sense. Consequently, we must describe **u** in terms of a probability function. It is instructive to note that, in view of the interpretation of $\mathbf{A}(t)$, we can write

$$\Delta\mathbf{u} = -\beta\mathbf{u}\Delta t + \mathbf{B}(\Delta t) \tag{10A2.2}$$

where

$$\mathbf{B}(\Delta t) = \int_{t}^{t+\Delta t} \mathbf{A}(\xi)\,d\xi$$

is the change in velocity over Δt because of Brownian effects.

We must next make some inferences about the nature of the probability distribution function which describes **u**. It seems reasonable to expect that at times sufficiently far removed from any initial conditions placed upon solutions for **u** from (10A2.1), the distribution function for **u** should be Gaussian. Furthermore, we would expect the average kinetic energy of the particle to correspond to the temperature of the fluid. Hence the distribution of velocities should tend to the Maxwellian form

$$\left(\frac{m}{2\pi kT}\right)^{3/2} \exp\left(\frac{-mu^2}{2kT}\right)$$

where m is the particle mass, k the Boltzmann constant, and T the absolute temperature.

Given this asymptotic form for the distribution function governing **u** one might expect, in view of (10A2.2), that a similar form of distribution function should apply to $\mathbf{B}(\Delta t)$. Chandrasekhar [43] shows such an expectation to be true. One finds that the probability that a particle will undergo a change in velocity $\mathbf{B}(\Delta t)$ over time interval t is given by the transition probability distribution function

$$\phi(\mathbf{B}(\Delta t)) = (4\pi q\Delta t)^{-3/2} \exp\left[\frac{-[\mathbf{B}(\Delta t)]^2}{4q\Delta t}\right] \tag{10A2.3}$$

where $q = \beta kT/m$. Using (9A2.2) we rewrite (9A2.3)

$$\phi(\mathbf{u};\Delta\mathbf{u}) = (4\pi q\Delta t)^{-3/2} \exp\left[\frac{-[\Delta\mathbf{u}+\beta\mathbf{u}\Delta t]^2}{4q\Delta t}\right] \tag{10A2.4}$$

Our next task is to use $\phi(\mathbf{u};\Delta\mathbf{u})$ to obtain a differential equation for $W(\mathbf{r}, \mathbf{u}, t)$, where

$$W(\mathbf{r}, \mathbf{u}, t)\,dx\,dy\,dz\,du_x\,du_y\,du_z$$

is the probability that at time t a particle will be found in some element of volume $dx\,dy\,dz$ about **r** and will have a velocity within a volume $du_x\,du_y\,du_z$ in velocity space about **u**. From the meaning of $\phi(\mathbf{u};\Delta\mathbf{u})$ we can write

$$W(\mathbf{r}+\mathbf{u}\Delta t, \mathbf{u}, t+\Delta t) = \int W(\mathbf{r}, \mathbf{u}-\Delta\mathbf{u}, t)\,\phi(\mathbf{u}-\Delta\mathbf{u};\Delta\mathbf{u})\,d(\Delta\mathbf{u}_x)\,d(\Delta\mathbf{u}_y)\,d(\Delta\mathbf{u}_z) \tag{10A2.5}$$

Expanding W and ϕ in Taylor series and considering the limit as $\Delta t \to 0$, one eventually obtains [43], with obvious notation,

$$\frac{\partial W}{\partial t} + \nabla_r \cdot (W\mathbf{u}) = \beta \nabla_u \cdot (W\mathbf{u}) + q\nabla_u^2 W \tag{10A2.6}$$

Next we transform $W(\mathbf{r}, \mathbf{u}, t)$ into a *spatial* distribution function by integrating over velocity space. Thus, define

$$w(\mathbf{r}, t) = \int W(\mathbf{r}, \mathbf{u}, t)\, du_x du_y du_z \tag{10A2.7}$$

where

$$\int w\, dx dy dz = 1$$

Equation (10A2.6) is transformed to a differential equation for w by integrating (10A2.6) over velocity space and requiring that $|W\mathbf{u}| \to 0$ as $|\mathbf{u}| \to \infty$. The result is readily seen to be

$$\frac{\partial w(\mathbf{r}, t)}{\partial t} + \nabla_r \cdot (w <\mathbf{u}>) = 0 \tag{10A2.8}$$

where

$$w <\mathbf{u}> \ = \int \mathbf{u} W\, du_x du_y du_z$$

A second differential equation for w can be derived. One can imagine obtaining an expression for particle *displacement* by integration of (10A2.1) or its incremental form (10A2.2). Then, by using arguments analogous to those leading to (10A2.3), one finds the transition probability for incremental displacement to be

$$\phi(\Delta r) = (4\pi D \Delta t)^{-3/2} \exp\left[\frac{(\Delta \mathbf{r})^2}{4D\Delta t}\right]$$

for $\Delta t \gg \beta^{-1}$, where $D = kT/(m\beta)$. Next, the analogy to (10A2.5) is formed.

$$w(\mathbf{r}, t + \Delta t) = \int w(\mathbf{r} - \Delta \mathbf{r}, t)\, \phi(\Delta r)\, d(\Delta x) d(\Delta y) d(\Delta z)$$

By expansion and consideration of the limit as $\Delta t \to 0$ there results

$$\frac{\partial w}{\partial t} = D\nabla_r^2 w \tag{10A2.9}$$

We see that D is aptly called the Brownian diffusion coefficient. Combination of (10A2.8) and (10A2.9) yields

$$\nabla \cdot (w <\mathbf{u}>) = -D\nabla^2 w = -D\nabla \cdot [(\nabla \ln w)w]$$

where the subscript has been dropped on ∇_r. It follows that

$$<\mathbf{u}> \ = -D\nabla \ln w \tag{10A2.10}$$

Now consider integration of (10A2.1) over velocity space. If the result is multiplied by the mass m, the result is an equation of average motion for the particle. Neglecting inertia we find

$$m\beta <\mathbf{u}> = m<\mathbf{A}(t)>$$

or

$$m<\mathbf{A}(t)> = -kT\nabla \ln w \qquad (10A2.11)$$

and $m<\mathbf{A}(t)>$ can be interpreted as a smoothly varying force acting on the particle because of Brownian effects. When the particle is part of a larger system, for example a bead in the bead–spring system of Section 10.12, a distribution function for the full system must be used. The result is (10.45).

10A3. HYDRODYNAMIC INTERACTION

Oseen's tensor, a formalism with which one can account for hydrodynamic interaction, is often quoted and used but is seldom explained. Rather than presuming a knowledge of Green's functions, we develop (10.42) and (10.43) in an *ad hoc* manner.

Consider an incompressible Newtonian fluid in creeping motion, i.e., inertial effects are negligible relative to viscous effects. At various points in the fluid there are disturbances to a simple flow field (for example a velocity which is everywhere constant). Suppose that these disturbances can be equated to action of body forces located in the flow field in specified regions of arbitrarily small size. What perturbations do these forces cause to the simple flow field?

The starting point is the equation of motion (4.26) written for a Newtonian incompressible fluid

$$\mathbf{S} = -p\mathbf{I} + 2\mu\mathbf{D}$$

Since inertial effects are negligible, we neglect acceleration in (4.26). Let $\mathbf{B}$ denote the body force per unit mass at a point in the field. Neglecting gravitational effects and using Cartesian coordinates, we find

$$\mu\nabla^2 u_j = \frac{\partial p}{\partial x_j} - \rho B_j; \quad \frac{\partial u_i}{\partial x_i} = 0 \qquad (10A3.1)$$

We assume that the force B_j is conservative so that it can be related to a potential ϕ. Let

$$B_j = -\partial\phi/\partial x_j$$

and

$$P = p + \rho\phi$$

Then the equations to be solved become

$$\mu\nabla^2 u_j = \frac{\partial P}{\partial x_j}; \quad \frac{\partial u_i}{\partial x_i} = 0 \qquad (10A3.2)$$

We wish to know the effect which a force $\mathbf{B}$ acting at point $P(x_1, x_2, x_3)$ will have upon the velocity at another point $P^{(0)}(x_1^{(0)}, x_2^{(0)}, x_3^{(0)})$. It is useful to define new functions

$$t_{jk} = \frac{\delta_{jk}}{r} + \frac{(x_j - x_j^{(0)})(x_k - x_k^{(0)})}{r^3}$$

$$P_k = 2\mu \frac{(x_k - x_k^{(0)})}{r^3}$$

(10A3.3)

where $r^2 = (x_i - x_i^{(0)})^2$, and it readily follows from these definitions that

$$\mu \nabla^2 t_{jk} - \frac{\partial P_k}{\partial x_j} = 0$$

(10A3.4)

$$\frac{\partial t_{jk}}{\partial x_j} = 0$$

The quantities u_j, P, t_{jk}, and P_j can be combined through an application of Green's theorem [45, p. 60], which states that if f and g are two continuous functions with continuous derivatives over a volume V enclosed by surface S, then

$$\int_V (f\nabla^2 g - g\nabla^2 f)\,dV = \int_S \left(f\frac{\partial g}{\partial n} - g\frac{\partial f}{\partial n} \right) dS$$

(10A3.5)

where n refers to the outer normal of surface S at the point in question. Let us identify f with u_j and g with t_{jk}. Volume V is a volume of arbitrarily large size surrounding $P^{(0)}$, but from which $P^{(0)}$ is excluded by a small sphere with radius ε (see Fig. 10A2). From (10A3.5), (10A3.2), and (10A3.4) we have

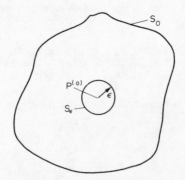

Fig. 10.A2. Application of Green's theorem.

$$\int_V \left(u_j \frac{\partial P_k}{\partial x_j} - t_{jk} \frac{\partial P}{\partial x_j} \right) dV = \mu \int_S \left(u_j \frac{\partial t_{jk}}{\partial n} - t_{jk} \frac{\partial u_j}{\partial n} \right) dS$$

(10A3.6)

But from (10A3.1) and (10A3.4) we know that

$$u_j \frac{\partial P_k}{\partial x_j} = \frac{\partial}{\partial x_j}(u_j P_k); \quad t_{jk}\frac{\partial P}{\partial x_j} = \frac{\partial}{\partial x_j}(t_{jk}P)$$

so that the divergence theorem (4.25) can be applied to (10A3.6) with the result

$$\int_S u_j \left[\mu\frac{\partial t_{jk}}{\partial n} - P_k n_j \right] dS = \int_S t_{jk}\left[\mu\frac{\partial u_j}{\partial n} - Pn_j\right] dS \qquad (10A3.7)$$

Assuming no singular behavior for u_j as $x_i \to x_i^{(0)}$, one readily sees from (10A3.3) that the right-hand side of (10A3.7) vanishes over the surface $r = \varepsilon$ as $\varepsilon \to 0$. Then, denoting the outer surface (Fig. 10A2) by S_0 we have

$$\lim_{\varepsilon \to 0} \int_{S_\varepsilon} u_j\left[\mu\frac{\partial t_{jk}}{\partial n} - P_k n_j\right] dS = \int_{S_0}\left\{ t_{jk}\left[\mu\frac{\partial u_j}{\partial n} - Pn_j\right] - u_j\left[\mu\frac{\partial t_{jk}}{\partial n} - P_k n_j\right]\right\} dS$$

$$(10A3.8)$$

The left-hand side is evaluated by expanding u_j in a Taylor series about $P^{(0)}$. The result is $8\pi\mu u_k^{(0)}$. Now recall that $P = p + \rho\phi$. Then (10A3.8) can be written

$$8\pi\mu u_k{}^{(0)} = \int_{S_0}\left\{ t_{jk}\left[\mu\frac{\partial u_j}{\partial n} - pn_j\right] - u_j\left[\mu\frac{\partial t_{jk}}{\partial n} - P_k n_j\right]\right\} dS - \int_{V_0 + V_\varepsilon}\frac{\partial}{\partial x_j}(\rho\phi t_{jk})dV$$

Let us take S_0 to be a surface with $r \to \infty$, and assume that the integrand approaches zero with increasing r faster than r^{-2}. This is reasonable in view of (10A3.3) and the known behavior of the velocity field far from a disturbance in a creeping flow. Furthermore, in view of (10A3.4b) and the definition of ϕ we can then write

$$u_k{}^{(0)} = \frac{1}{8\pi\mu}\int_{V_0 + V_\varepsilon} t_{jk}\rho B_j dV \qquad (10A3.9)$$

Suppose that the disturbance $u_k^{(0)}$ at $P^{(0)}$ is due to the presence of a force **F** with components F_j acting in an arbitrarily small neighborhood around P. Then

$$F_j = \int_{V_0 + V_\varepsilon} \rho B_j dV$$

and (10A3.9) becomes

$$u_k{}^{(0)} = \frac{F_j}{8\pi\mu}t_{jk} = \frac{F_j}{8\pi\mu}\left[\frac{\delta_{jk}}{r} + \frac{(x_j - x_j^{(0)})(x_k - x_k^{(0)})}{r^3}\right] \qquad (10A3.10)$$

or, in vector form,

$$\mathbf{u} = \frac{1}{8\pi\mu r}\left[\mathbf{I} + \frac{\mathbf{rr}}{r^2}\right]\cdot\mathbf{F} \qquad (10A3.11)$$

where dyadic notation has been used to make clear that a contraction is indicated. In (10A3.11) **u** is the disturbance at the origin due to a point force **F** acting at **r**. Equation (10A3.11) is often written in the form

$$\mathbf{u} = \mathbf{T} \cdot \mathbf{F} \tag{10A3.12}$$

where $\mathbf{T} = \dfrac{1}{8\pi\mu r}\left[\mathbf{I} + \dfrac{\mathbf{rr}}{r^2}\right]$ is Oseen's tensor.

Since we are dealing with linear equations, the disturbance at some position in the flow field due to multiple point forces $\mathbf{F}_j$ can be found by simple summation of the individual contributions (10A3.12).

Now we wish to apply (10A3.12) to the bead–spring system of Section 10.12. Specifically, we wish to estimate the velocity perturbation $\mathbf{u}_i$ at the point occupied by bead i when the perturbation is the linear combination of all of the perturbations due to the presence of other beads j in the bead–spring system. The presence of bead j is modeled by a point force $\mathbf{F}_j$ at the location of bead j. Taking bead i as the origin we assume a normal distribution of beads j ($j \neq i$) about the origin and hence adapt the probability distribution function (10A1.1), except that b^2 becomes $|i-j|b^2$, $|i-j|$ being the number of "springs" separating beads i and j. Then, following Kirkwood and Riseman [46], we write

$$\mathbf{u}_i = \sum_{\substack{j \\ (j \neq i)}} < \mathbf{T}_{ij} > \cdot \mathbf{F}_j \tag{10A3.13}$$

where

$$\mathbf{T}_{ij} = \left(\frac{3}{2\pi|i-j|b^2}\right)^{3/2} \frac{1}{8\pi\mu} \int \left[\frac{1}{r}\mathbf{I} + \frac{\mathbf{rr}}{r^3}\right] \exp\left[\frac{-3r^2}{2b^2|i-j|}\right] dxdydz \tag{10A3.14}$$

and r is the distance from bead i to bead j. The integration is readily carried out, the result being

$$<\mathbf{T}_{ij}> = \frac{\mathbf{I}}{\mu b}(6\pi^3|i-j|)^{-1/2} \tag{10A3.15}$$

This is the explanation of (10.43).

10A4. CONTRIBUTION OF THE BEAD–SPRING SYSTEM TO THE STRESS

To derive (10.47) we use a combination of the derivation of Lodge and Wu [33] and of Bird, Warner, and Evans [30], the latter having been developed for a dumbbell model of polymer behavior. We have already noted in the text that the stress is assumed to be a summation of the contribution of pure solvent and the added effect of the bead–spring model, the latter identified by $\mathbf{T}_M$. To ascertain the origin of $\mathbf{T}_M$ consider an arbitrary plane with unit vector $\mathbf{n}$ (positive direction arbitrary). The plane is located in the fluid at the point under consideration and is moving with the bulk fluid velocity. Allow this plane to divide a cube into equal parts as shown in Fig. 10A3. The cube is assumed to be small with respect to distances over which macroscopic quantities (such as bulk fluid velocity) change appreciably. If c is the number density of polymer molecules in the solution the cube is taken to have a volume $1/c$, so that on average it contains just one polymer molecule. In the figure we wish, for the moment, to consider only a portion of the stress vector acting on the plane with normal $\mathbf{n}$; namely, the portion $\mathbf{t}_{(\mathbf{n})}^s$ which is due to the intersection through the plane of the "spring" between beads $p-1$ and p. To do

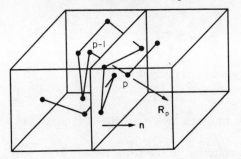

Fig. 10.A3. Contribution of polymer to the bulk stress.

this we need to use some forms of probability distribution functions. First of all, the polymer is assumed to be uniformly distributed throughout the solvent. Then the distribution function for the beads of a bead–spring system depends only on the distances between beads and not on their actual positions. This can be stated quantitatively by

$$\Psi'(r_0, r_1, \ldots, r_N, t) = \phi(\mathbf{r}_c, t)\psi(\mathbf{R}_1, \mathbf{R}_2, \ldots, \mathbf{R}_N, t) \tag{10A4.1}$$

where $\phi(r_c, t) = \text{const}$ is the distribution function for position of the centroid of a bead–spring system and $\mathbf{R}_p = \mathbf{r}_p - \mathbf{r}_{p-1}$. If we normalize these distribution functions so that

$$\int \phi d\mathbf{r}_c = 1$$

$$\int \psi(\mathbf{R}_1, \ldots, \mathbf{R}_N, t) d\mathbf{R}_1, \ldots, d\mathbf{R}_N = 1 \tag{10A4.2}$$

it is clear that for the present case $\phi = 1/V$, the reciprocal of the volume of the system under consideration.

Now consider the probability that a bead–spring system in the volume shown in Fig. 10A3 will have a spring connecting beads $p-1$ and p with direction and magnitude $\mathbf{R}_p$. This is given by the probability distribution function $\psi_{\mathbf{R}_p} d\mathbf{R}_p$, where

$$\Psi'_{\mathbf{R}_p} = \int \psi d\mathbf{R}_1 \ldots d\mathbf{R}_{p-1} d\mathbf{R}_{p+1} \ldots d\mathbf{R}_N \tag{10A4.3}$$

The probability that such a configuration for $\mathbf{R}_p$ will not only exist for $\mathbf{R}_p$ but that $\mathbf{R}_p$ will also intersect the plane of Fig. 10A3 is

$$\frac{(1/c)^{2/3}\mathbf{R}_p \cdot \mathbf{n}}{1/c} (\psi_{\mathbf{R}_p} d\mathbf{R}_p)$$

Then the average force exerted across the plane from left to right (recalling that $\mathbf{F}_{S,p}$ is the force exerted on bead $(p-1)$ by the spring connecting $(p-1)$ and p) is, for all orientations of $\mathbf{R}_p$,

$$\int \mathbf{F}_{S,p}(\mathbf{R}_p \cdot \mathbf{n})c^{1/3}\psi_{\mathbf{R}_p} d\mathbf{R}_p$$

The contribution to the stress vector is obtained by dividing by the planar surface area $(1/c)^{2/3}$. The next step of course is to sum over all values of p to get the total polymer contribution to the stress vector, $\mathbf{t}^s_{(n)M}$

$$\mathbf{t}^s_{(n)M} = c \int \sum_{p=1}^{N} \psi_{\mathbf{R}_p} \mathbf{F}_{S,p} (\mathbf{R}_p \cdot \mathbf{n}) d\mathbf{R}_p \tag{10A4.4}$$

Recalling that the stress tensor is defined by

$$\mathbf{t}^s_{(n)M} = \mathbf{T}^s_M \mathbf{n}$$

and rewriting (10A4.4) in terms of Ψ (see (10A4.1) and (10A4.3)) one obtains the first term of (10.47).

There is another contribution to the stress tensor. This contribution is due to the flux of momentum carried by beads which cross from one side of the plane to the other. Let us consider a pth bead which at time t has velocity $\dot{\mathbf{r}}_p$. We first ask how many pth beads from the c bead–spring systems per unit volume will cross the area of Fig. 10A3 with surface normal $\mathbf{n}$ (and, say, surface area S) in a small time Δt. Following the familiar procedure of Hirschfelder, Curtiss, and Bird [47, p. 455] we construct a volume on S which is swept out in Δt and we multiply that volume by c. Thus

$$c[\Delta(\text{volume})] = c|\dot{\mathbf{r}}_p| \Delta t \, S \left(\mathbf{n} \cdot \frac{\dot{\mathbf{r}}_p}{|\dot{\mathbf{r}}_p|} \right)$$

Note that we have considered the surface S to be stationary. The momentum flux at S due to p-type beads is then

$$m\dot{\mathbf{r}}_p c (\mathbf{n} \cdot \dot{\mathbf{r}}_p)$$

The next step is to sum the contribution from p-beads which have velocities $\dot{\mathbf{r}}_p$ ranging over all values. To do this we have to say something about a distribution of velocities for the beads. We now *assume* that this distribution is Gaussian. Specifically, we assume the Maxwellian distribution function

$$G(\dot{\mathbf{r}}_p) d\dot{\mathbf{r}}_p = (a/\pi)^{3/2} \exp(-a\dot{\mathbf{r}}_p^2) d\dot{\mathbf{r}}_p \tag{10A4.5}$$

where $a = \dfrac{m}{2kT}$ and

$$\int G(\dot{\mathbf{r}}_p) d\dot{\mathbf{r}}_p = 1$$

Thus the momentum flux from pth beads which have velocities distributed according to a Maxwellian distribution is

$$(a/\pi)^{3/2} cm \mathbf{n} \cdot \int \dot{\mathbf{r}}_p \dot{\mathbf{r}}_p \exp(-a\dot{\mathbf{r}}_p^2) d\dot{\mathbf{r}}_p \tag{10A4.6}$$

which integrates to

$$ckT(\mathbf{n} \cdot \mathbf{I})$$

Since each bead behaves, on average, identically, we find the total contribution by simply multiplying by the number $(N+1)$ of beads to obtain

$$(N+1)ckT(\mathbf{n} \cdot \mathbf{I})$$

Finally, we correct for the fact that the surface S was taken to be stationary. It is assumed that the center of mass of each macromolecule moves at the local bulk fluid velocity (the "no-drift" hypothesis [30]). To correct the above expression one merely subtracts the contribution of momentum flux from the motion of the center of mass (which in this case is identical with the centroid $\mathbf{r}_c$).

$$\mathbf{r}_c = \frac{1}{\mathcal{N}+1} \sum_{p=0}^{N} \mathbf{r}_p$$

Thus we have

$$c \int (\mathcal{N}+1)m \frac{1}{(\mathcal{N}+1)^2} (\mathbf{n} \cdot \dot{\mathbf{r}}_c)\dot{\mathbf{r}}_c G(\dot{\mathbf{r}}_0) \ldots G(\dot{\mathbf{r}}_N)d\dot{\mathbf{r}}_0 \ldots d\dot{\mathbf{r}}_N \qquad (10A4.7)$$

since $(\mathcal{N}+1)m$ is the mass associated with $\mathbf{r}_c$. Because of the independence of the distribution functions associated with the individual beads, one obtains

$$ckT(\mathbf{n} \cdot \mathbf{I})$$

so that the contributions $\mathbf{t}^m{}_{(n)M}$ and $\mathbf{T}_M^m$ to the stress vector and stress tensor, respectively, which are due to polymer momentum flux are

$$\mathbf{t}^m{}_{(n)M} = -\mathcal{N}ckT(\mathbf{n} \cdot \mathbf{I})$$

and

$$\mathbf{T}_M^m = -\mathcal{N}ckT\mathbf{I} \qquad (10A4.8)$$

which is the second term of (10.47). Note that, since we are using the convention that tensions are positive, we must subtract the momentum flux.

10A5. CONSTITUTIVE BEHAVIOR OF THE ZIMM MODEL

We follow the treatment of Lodge and Wu [33]. Recall that each bead is subjected to three kinds of forces. These are:

(a) spring force $\mathbf{F}_{S,p} = \dfrac{3kT}{b^2} (\mathbf{r}_p - \mathbf{r}_{p-1})$ $\qquad (10.44)$

(b) Brownian force $\mathbf{F}_{B,p} = -kT\nabla_{\mathbf{r}_p} \ln \Psi'$ $\qquad (10.45)$

(c) solvent force $\mathbf{F}_{s,p} = -f_0(\dot{\mathbf{r}}_p - \mathbf{v}'(\mathbf{r}_p))$ $\qquad (10.41)$

Since inertia is being neglected,

$$\mathbf{F}_{S,p} + \mathbf{F}_{B,p} + \mathbf{F}_{s,p} = 0 \qquad (10A5.1)$$

The spring forces can be conveniently described in terms of a Gaussian distribution function

$$\Psi'_0 \sim \exp\left\{ \sum_{p=1}^{N} \left[\frac{-3}{2b^2} (\mathbf{r}_p - \mathbf{r}_{p-1})^2 \right] \right\} \qquad (10A5.2)$$

This is true because

$$\mathbf{F}_{S,0} = \frac{3kT}{b^2} (\mathbf{r}_1 - \mathbf{r}_0)$$

$$\mathbf{F}_{S,p} = \frac{3kT}{b^2} (\mathbf{r}_{p+1} - \mathbf{r}_p) - \frac{3kT}{b^2} (\mathbf{r}_p - \mathbf{r}_{p-1}) \Bigg\} (p \neq 0, N)$$

$$\mathbf{F}_{S,N} = - \frac{3kT}{b^2} (\mathbf{r}_N - \mathbf{r}_{N-1})$$

from which it readily follows that

$$\mathbf{F}_p = \mathbf{F}_{B,p} + \mathbf{F}_{S,p} = - kT\nabla_{\mathbf{r}_p} \ln(\Psi/\Psi_0) \tag{10A5.3}$$

To incorporate the solvent force we need an expression for $\mathbf{v}'$. The undistorted velocity at the position of bead p is given by

$$\mathbf{v}_0 + \mathbf{L}\mathbf{r}_p$$

where $\mathbf{v}_0$ is the velocity at the origin (fixed in space) and $\mathbf{L}$ is the velocity gradient tensor ($\nabla\mathbf{v}$) in the undisturbed flow. Although such an expression implies that $\mathbf{L}$ is spatially constant, one can argue that this constraint is not essential. For example, if the origin is taken close to the macromolecule in question, $\mathbf{L}$ may be taken as constant on a scale important for the polymer molecule even though it varies on a macroscopic scale. Then, using (10.42)

$$\mathbf{v}'(\mathbf{r}_p) = \mathbf{v}_0 + \mathbf{L}\mathbf{r}_p + \sum_{\substack{q=0 \\ (q \neq p)}}^{N} <\mathbf{T}_{pq}> \mathbf{F}_q \tag{10A5.4}$$

Recalling (10.43), it is convenient to define

$$T_{pq} = \frac{1}{[6\pi^3|p-q|]^{1/2}\mu_s b} \quad (p \neq q)$$

$$T_{pq} = 0 \quad (p = q)$$

Then (10A5.1) can be written as

$$\dot{\mathbf{r}}_p - \mathbf{v}_0 - \mathbf{L}\mathbf{r}_p = - \frac{kT}{f_0} \sum_{q=0}^{N} H_{pq}\nabla_{\mathbf{r}_p} \ln(\Psi/\Psi_0) \tag{10A5.5}$$

where

$$H_{pp} = 1 \quad \text{(no sum)}$$

$$H_{pq} = f_0 T_{pq} \quad (p \neq q)$$

and (10A5.5) is the equation of motion of the pth bead.

In principle we can use (10A5.5) along with a statement of conservation of mass to solve for Ψ. The conservation statement is

$$\frac{d}{dt} \int \Psi d\mathbf{r}_0 \dots d\mathbf{r}_N = 0$$

Using the arguments applied in Chapter 4 to derive the continuity equation, one obtains

$$\frac{\partial \Psi}{\partial t} + \sum_{p=0}^{N} \nabla_{\mathbf{r}_p} \cdot (\Psi \dot{\mathbf{r}}_p) = 0 \tag{10A5.6}$$

Once Ψ is known, it can be substituted into (10.47) to obtain an expression for the stress. One can, however, derive the desired constitutive equation without ever solving for Ψ explicitly. This is done by making one or more sufficiently clever coordinate transformations which enable one to express results in terms of Ψ-averaged quantities which are directly related to macroscopically measurable properties.

Lodge and Wu employed two transformations. The first is similar to that used in Section 10A4 and results in a distribution function the arguments of which are distances *between* adjacent beads. We make the transformation

$$\mathbf{r}_0, \mathbf{r}_1, \ldots, \mathbf{r}_N \rightarrow \mathbf{R}_0, \mathbf{R}_1, \ldots, \mathbf{R}_N$$

where

$$\mathbf{R}_0 = \sum_{p=0}^{N} l_p \mathbf{r}_p$$

$$\mathbf{R}_p = \mathbf{r}_p - \mathbf{r}_{p-1} \quad (p \neq 0)$$

and l_p is subject to the constraint

$$\sum_{p=0}^{N} l_p = 1 \tag{10A5.7}$$

We can also write

$$\mathbf{R}_p = \sum_{q=0}^{N} G_{pq} \mathbf{r}_q \tag{10A5.8}$$

where

$$G_{0q} = l_q \qquad\qquad (q=0, 1, \ldots, N)$$
$$G_{pq} = \delta_{pq} - \delta_{p-1,q} \quad \begin{array}{l}(p=1, 2, \ldots, N)\\(q=0, 1, \ldots, N)\end{array}$$

By applying elementary operations to the matrix of G_{pq} [48, ch. 2] one can show that

$$\det(G_{pq}) = \sum_{p=0}^{N} l_p = 1 \tag{10A5.9}$$

Inspection of the matrix also reveals that

$$\sum_{q=0}^{N} G_{pq} = 0 \quad (p=1, 2, \ldots, N) \tag{10A5.10}$$

Now multiply (10A5.5) by G_{kp} and sum over p. When the independent variables are converted to $\mathbf{R}_p$ according to (10A5.8) and the properties of G_{pq} given above are used, one obtains after straightforward but tedious manipulation

$$\left.\begin{array}{l}\dot{\mathbf{R}}_0 - \mathbf{v}_o - \mathbf{L}\mathbf{R}_0 = -\dfrac{kT}{f_o}\left\{S_0 \nabla_{\mathbf{R}_0} + \sum_{q=1}^{N} Z_q \nabla_{\mathbf{R}_q}\right\} \ln(\Psi/\Psi_o)\\[4mm]\dot{\mathbf{R}}_k - \mathbf{L}\mathbf{R}_k = -\dfrac{kT}{f_o}\left\{Z_k \nabla_{\mathbf{R}_0} + \sum_{q=1}^{N} B_{kq} \nabla_{\mathbf{R}_q}\right\} \ln(\Psi/\Psi_o)\end{array}\right\} \tag{10A5.11}$$

where Ψ' and Ψ'_0 are now written in terms of the new set of independent variables $\mathbf{R}_p$. The following new quantities have been used:

$$S_0 = \sum_{p,q=0}^{N} G_{0p}H_{pq}G_{0q}$$

$$Z_k = \sum_{p=0}^{N} G_{0p}H_{pq}G_{kq} = \sum_{p=0}^{N} l_p(H_{pk} - H_{p,k-1})$$

$$B_{kl} = B_{lk} = \sum_{p,q=0}^{N} G_{kp}H_{pq}G_{lq} = H_{kl} - H_{k-1,l} - H_{k,l-1} + H_{k-1,l-1}$$

Recall that, except for the constraint (10A5.9), the l_p are not yet defined. They are now fixed by the requirement

$$Z_k = 0 \quad (k=1, 2, \ldots, N) \tag{10A5.12}$$

Lodge and Wu call the vector $\mathbf{R}_0$ the center of resistance of the macromolecule. In the special case of no hydrodynamic interaction $(T_{pq}=0)$ we have $H_{pq}=0$ $(p \neq q)$ and

$$l_p = 1/(N+1) \quad (p=0, 1, \ldots, N) \tag{10A5.13}$$

from which it follows that $\mathbf{R}_0$ is the position vector of the centroid of the bead–spring model of a macromolecule.

We are now in a position to see how the equation of motion for the center of resistance can be separated from those for the individual beads. Writing $\Psi' = \Psi'(\mathbf{R}_0, \mathbf{R}_1, \ldots, \mathbf{R}_N)$, simple application of the chain rule transforms (10A5.6) into

$$\frac{\partial \Psi'}{\partial t} + \sum_{p=0}^{N} \nabla_{\mathbf{R}_p} \cdot (\Psi' \dot{\mathbf{R}}_p) = 0 \tag{10A5.14}$$

In view also of (10A5.11) and (10A5.12) we are led to the fact that Ψ' can be written in separable form (analogous to (10A4.1))

$$\Psi'(\mathbf{R}_0 \ldots \mathbf{R}_N, t) = \phi(\mathbf{R}_0, t)\psi(\mathbf{R}_1, \ldots, \mathbf{R}_N, t) \tag{10A5.15}$$

Then the governing equations become

$$\left. \begin{aligned} &\frac{\partial \phi}{\partial t} + \nabla_{\mathbf{R}_0} \cdot (\phi \dot{\mathbf{R}}_0) = 0 \\[2ex] &\dot{\mathbf{R}}_0 - \mathbf{v}_0 - \mathbf{L}\mathbf{R}_0 = -\frac{kT}{f_0} S_0 \nabla_{\mathbf{R}_0} \ln(\phi/\phi_0) \end{aligned} \right\} \tag{10A5.16}$$

$$\left. \begin{aligned} &\frac{\partial \psi}{\partial t} + \sum_{p=1}^{N} \nabla_{\mathbf{R}_p} \cdot (\psi \dot{\mathbf{R}}_p) = 0 \\[2ex] &\dot{\mathbf{R}}_p - \mathbf{L}\mathbf{R}_p = -\frac{kT}{f_0} \sum_{q=1}^{N} B_{pq} \nabla_{\mathbf{R}_q} \ln(\psi/\psi_0) \quad (p=1 \ldots N) \end{aligned} \right\} \tag{10A5.17}$$

In applying the separation (10A5.15) to (10A5.14) an arbitrary function of time has been set to zero to be consistent with a constant total number of macromolecules in the system [33]. In addition, the functions ϕ and ψ are separately normalized so that

$$\int \phi(\mathbf{R}_0, t)d\mathbf{R}_0 = 1$$

$$\int \psi(\mathbf{R}_1 \ldots \mathbf{R}_N, t)d\mathbf{R}_1 \ldots d\mathbf{R}_N = 1$$

Then the significance of $\phi d\mathbf{R}_0$ and $\psi d\mathbf{R}_1 \ldots d\mathbf{R}_N$ is obvious. Furthermore, since our attention is confined to solutions which, on average, contain no concentration gradients, ϕ takes on the trivial form $\phi(\mathbf{R}_0, t) = \text{const} = 1/V$, where V is the volume of the system under consideration. Note that a consequence of this is the prediction, from the second of (10A5.16), that the center of resistance moves with the free stream velocity $\mathbf{v}_0 + \mathbf{LR}_0$.

In view of our new decomposition of Ψ and the definitions (10A4.2) and (10A4.3) we can write, from (10A4.4),

$$\mathbf{T}_M^s = c < \sum_{p=1}^{N} \mathbf{F}_{S,p} \otimes \mathbf{R}_p > \qquad (10A5.18)$$

as the basis for our constitutive equation. Here the brackets refer to an average over ψ, i.e.,

$$< \sum_{p=1}^{N} \mathbf{F}_{S,p} \otimes \mathbf{R}_p > = \sum_{p=1}^{N} \int \mathbf{F}_{S,p} \otimes \mathbf{R}_p \psi d\mathbf{R}_1 \ldots d\mathbf{R}_N$$

To utilize the first of (10A5.17) we take the partial time derivative of (10A5.18) after substituting (10.44) for $\mathbf{F}_{S,p}$. Some manipulation leads to

$$\frac{\partial \mathbf{T}_M^{s,}}{\partial t} = -cK_0 \int \sum_{p,q=1}^{N} \nabla_{\mathbf{R}_p} \cdot [\psi \dot{\mathbf{R}}_p \otimes \mathbf{R}_q \otimes \mathbf{R}_q]d\mathbf{R}_1 \ldots d\mathbf{R}_N$$

$$+ cK_0 \int \sum_{p=1}^{N} \psi[\dot{\mathbf{R}}_p \otimes \mathbf{R}_p + (\dot{\mathbf{R}}_p \otimes \mathbf{R}_p)^T]d\mathbf{R}_1 \ldots d\mathbf{R}_N$$

where $K_0 = 3kT/b^2$. Now the first integral above can be transformed by Gauss' theorem (4.25) into an integration over the multidimensional surface $\mathbf{R}_p \to \infty$ for all values of p between 1 and N. We assume that $\psi \to 0$ sufficiently rapidly with increasing $\mathbf{R}_p$ so that the first integral is zero. Furthermore, we eliminate $\dot{\mathbf{R}}_p$ from the second integral by using the second of (10A5.17).

$$\frac{\partial \mathbf{T}_M^s}{\partial t} = cK_0 \int \sum_{p=1}^{N} \psi \left\{ \left[\mathbf{LR}_p - \frac{kT}{f_0} \sum_{q=1}^{N} B_{pq} \nabla_{\mathbf{R}_q} \ln(\psi/\psi_0) \right] \otimes \mathbf{R}_p \right.$$

$$\left. + \mathbf{R}_p \otimes \left[\mathbf{LR}_p - \frac{kT}{f_0} \sum_{q=1}^{N} B_{pq} \nabla_{\mathbf{R}_q} \ln(\psi/\psi_0) \right] \right\} d\mathbf{R}_1 \ldots d\mathbf{R}_N \qquad (10A5.19)$$

The terms containing $\mathbf{L}$ are readily seen to reduce to $\mathbf{LT}_M^s + \mathbf{T}_M^s \mathbf{L}^T$. Reduction of the remaining terms is straightforward but lengthy. One utilizes the Gaussian form of ψ_0 to obtain a contribution from the terms containing ψ_0. Remaining terms are cast into a form which again permits use of Gauss' theorem to convert from a volume to a surface integral. The result is

$$\frac{\partial \mathbf{T}_M^s}{\partial t} = \mathbf{L}\mathbf{T}_M^s + \mathbf{T}_M^s\mathbf{L}^T + \frac{2kTcK_0}{f_0}\mathbf{I}\sum_{p=1}^{N}B_{pp} - \frac{cK_0^2}{f_0}\sum_{p,q=1}^{N}B_{pq} < \mathbf{R}_p \otimes \mathbf{R}_q + \mathbf{R}_q \otimes \mathbf{R}_p >$$

$$(10A5.20)$$

Note that, except for the final term, we have succeeded in our attempt to eliminate the need for knowledge of the distribution function ψ. If, however, the term can be diagonalized, then we should be able to express the averages in terms of $\mathbf{T}_M$.

From the definition of B_{pq} following (10A5.11) we see that the B_{pq} form a symmetric matrix. If one imagines a tensor formed from an N-dimensional orthonormal basis and the coefficients B_{pq}, then it is a trivial extension of Section 3.17 to show that there exists an orthogonal transformation $\mathbf{Q}$ which will diagonalize the tensor coefficients. We write in component form

$$(q^T)_{i\alpha}B_{\alpha\beta}q_{\beta j} = \Lambda_{ij}$$

where

$$[\Lambda] = \begin{bmatrix} \lambda_1 & 0 & \cdots & & & 0 \\ 0 & \lambda_2 & 0 & \cdots & & 0 \\ 0 & & \lambda_3 & 0 & \cdots & \\ \cdot & \cdot & & \cdot & & \cdots \\ \cdot & & & & \cdot & \\ \cdot & & & & & \cdot \\ \cdot & & & & & \cdot \\ 0 & \cdot & & \cdot & \cdots & \lambda_N \end{bmatrix}$$

$$(10A5.21)$$

The λ_i are of course the principal values of the tensor $\mathbf{B}$. Next we transform the bead-to-bead distance $\mathbf{R}_p$ by

$$\mathbf{u}_i = \sum_{j=1}^{N} q_{ij}\mathbf{R}_j \qquad (10A5.22)$$

and define a distribution function in terms of the new variables

$$\Theta(\mathbf{u}_1 \ldots \mathbf{u}_N, t) = \psi(\mathbf{R}_1 \ldots \mathbf{R}_N, t)$$

Then

$$\mathbf{T}_M^s = cK_0 \int \Theta \sum_{i=1}^{N} \mathbf{u}_i \otimes \mathbf{u}_i du_1 \ldots du_N \qquad (10A5.23)$$

Application of the chain rule to (10A5.17) yields

$$\left.\begin{array}{c} \dfrac{\partial \Theta}{\partial t} + \sum_{i=1}^{N} \nabla \mathbf{u}_i \cdot (\Theta \dot{\mathbf{u}}_i) = 0 \\[2em] \Theta \dot{\mathbf{u}}_i = \mathbf{M}_i \Theta \quad (i=1, 2, \ldots, N) \end{array}\right\} \qquad (10A5.24)$$

where

$$\mathbf{M}_i = \mathbf{L}\mathbf{u}_i - \frac{kT\lambda_i}{f_0}\left(\nabla \mathbf{u}_i + \frac{3}{b^2}\mathbf{u}_i\right)$$

The second equation is obtained by using the Gaussian form for Θ_0.

It is clear from (10A5.24) that Θ can be expressed as a product of its separate spatial variables

$$\Theta = \prod_{i=1}^{N} \theta_i(\mathbf{u}_i, t)$$

where each θ_i satisfies

$$\frac{\partial \theta_i}{\partial t} + \nabla \mathbf{u}_i \cdot (\mathbf{M}_i \theta_i) = C_i(t)\theta_i \tag{10A5.25}$$

$C_i(t)$ is an arbitrary function of time and is a result of the separation of Θ into product functions. All of the desired properties of the distribution functions θ_i are possible with the $C_i(t)$ set equal to zero, which we do [33]. Each θ_i is separately normalized so that

$$\int \theta_i d\mathbf{u}_i = 1$$

We can write the contribution of the model springs to the stress by transforming (10A5.18) to the new basis. Then

$$\mathbf{T}_{M_i}^s = cK_0 \int \theta_i \mathbf{u}_i \otimes \mathbf{u}_i d\mathbf{u}_i \tag{10A5.26}$$

and

$$\mathbf{T}_M^s = \sum_{i=1}^{N} \mathbf{T}_{M_i}^s \tag{10A5.27}$$

Now one can proceed exactly as we did previously in going from (10A5.18) to (10A5.20). However, in the present case only diagonal elements of B_{ij} will contribute and one obtains

$$\frac{\partial \mathbf{T}_{M_i}^s}{\partial t} = \mathbf{L}\mathbf{T}_{M_i}^s + \mathbf{T}_{M_i}^s \mathbf{L}^T + \frac{2cK_0\lambda_i}{f_0}\left[kT\mathbf{I} - K_0 \int \theta_i \mathbf{u}_i \otimes \mathbf{u}_i d\mathbf{u}_i\right] \tag{10A5.28}$$

and the last term can be combined with (10A5.26).

Finally, we introduce the momentum contribution which, from (10A5.8), we can write as

$$\mathbf{T}_{M_i}^m = ckT\mathbf{I}$$

so that

$$\mathbf{T}_{M_i} = \mathbf{T}_{M_i}^s + \mathbf{T}_{M_i}^m$$

Then (10A5.28) becomes

$$\frac{\partial \mathbf{T}_{M_i}}{\partial t} - \mathbf{L}\mathbf{T}_{M_i} - \mathbf{T}_{M_i}\mathbf{L}^T = (\mathbf{L}+\mathbf{L}^T)ckT - \frac{2K_0}{f_0}\lambda_i\mathbf{T}_{M_i}$$

Recalling that $K_0 = 3kT/b^2$,

$$\left(\tau_i\frac{\partial}{\partial t} + 1\right)\mathbf{T}_{M_i} - \tau_i(\mathbf{L}\mathbf{T}_{M_i}+\mathbf{T}_{M_i}\mathbf{L}^T) = ckT\tau_i(\mathbf{L}+\mathbf{L}^T) \tag{10A5.29}$$

where $\tau_i = f_0 b^2/(6kT\lambda_i)$. Now if the velocity gradient is small, the stress $\mathbf{T}_{M_i}$ is also expected to be small. Under those conditions (10A5.29) can be reduced, approximately, to a linear relation between stress and twice the bulk deformation rate $(\mathbf{L}+\mathbf{L}^T)$

$$\left(\tau_i \frac{\partial}{\partial t} + 1\right) \mathbf{T}_{M_i} = ckT\tau_i(\mathbf{L}+\mathbf{L}^T) \tag{10A5.30}$$

which is the same as (10.49).

REFERENCES

1. Truesdell, C., and Noll, W., The non-linear field theories of mechanics, in *Handbuch der Physik*, Band III/3 (S. Flügge, ed.), (Berlin: Springer-Verlag, 1965).
2. Reiner, M., *Am. J. Math.* **67**, 350 (1945).
3. Rivlin, R. S., *Proc. Roy. Soc.* (*London*) A, **193**, 260 (1948).
4. Fredrickson, A. G., *Principles and Applications of Rheology* (Englewood Cliffs, New Jersey: Prentice-Hall, 1964).
5. Serrin, J., Mathematical principles of classical fluid mechanics, in *Handbuch der Physik*, Band VIII/1 (S. Flügge, ed.), (Berlin: Springer-Verlag, 1959).
6. Bird, R. B., *Can. J. Chem. Eng.* **43**, 161 (1965).
7. Metzner, A. B., Non-Newtonian technology: fluid mechanics, mixing, and heat transfer, in *Advances in Chemical Engineering*, vol. 1, pp. 79 *et seq.* (T. B. Drew and J. W. Hoopes, Jr., eds.), (New York: Academic Press, 1956).
8. Reiner, M., and Scott Blair, G. W., Rheology terminology, in *Rheology*, vol. 4, pp. 461 *et seq.* (F. R. Eirich, ed.), (New York: Academic Press, 1967).
9. Pearson, J. R. A., Review of *Rheology*, vol. 4, in *J. Fluid Mech.* **44**, 396 (1970).
10. Ballman, R. L., *Rheol. Acta* **4**, 137 (1965).
11. Middleman, S., *The Flow of High Polymers* (New York: Interscience, 1968).
12. Metzner, A. B., Houghton, W. T., Sailor, R. A., and White, J. L., *Trans. Soc. Rheol.* **5**, 133 (1961).
13. Dierckes, A. C., Jr., and Schowalter, W. R., *Ind. Eng. Chem. Fund.* **5**, 263 (1966).
14. Hoffman, R. L., PhD thesis, Princeton University, 1968.
15. Tanner, R. I., *Ind. Eng. Chem. Fund.* **5**, 55 (1966).
16. Gatlin, C., *Petroleum Engineering, Drilling and Well Completions*, ch. 7 (Englewood Cliffs, New Jersey: Prentice-Hall, 1960).
17. Rivlin, R. S., and Ericksen, J. L., *J. Rational Mech. Anal.* **4**, 323 (1955).
18. Oldroyd, J. G., *Proc. Roy. Soc.* (*London*) A, **200**, 523 (1950).
19. Etter, I., and Schowalter, W. R., *Trans. Soc. Rheol.* **9**, 351 (1965).
20. Coleman, B. D., and Noll, W., *Arch. Rational Mech. Anal.* **6**, 355 (1960).
21. Ferry, J. D., *Viscoelastic Properties of Polymers*, 2nd edn. (New York: Wiley, 1970).
22. Gross, B., *Mathematical Structure of the Theories of Elasticity* (Paris: Hermann, 1953).
23. Leaderman, H., in *Rheology*, vol. 2, p. 1. (F. R. Eirich, ed.), (New York: Academic Press, 1958).
24. Reiner, M., *Deformation, Strain and Flow* (London: H. K. Lewis, 1960).
25. Fröhlich, H., and Sack, R., *Proc. Roy. Soc.* (*London*), A, **185**, 415 (1946).
26. Churchill, R. V., *Operational Mathematics*, 2nd edn. (New York: McGraw-Hill, 1958).
27. Cox, W. P., and Merz, E. H., *J. Pol. Sci.* **28**, 619 (1958).
28. Giesekus, H., *Rheol. Acta* **2**, 112 (1962).
29. Peterlin, A., *Pure Applied Chem.* **12**, 563 (1966).
30. Bird, R. B., Warner, H. R., and Evans, D. C., *Advances in Polymer Science*, vol. 8, pp. 1–90 (Berlin: Springer-Verlag, 1971).
31. Jerrard, H. G., *Chem. Revs.* **59**, 345 (1959).
32. Zimm, B. H., *J. Chem. Phys.* **24**, 269 (1956).
33. Lodge, A. S., and Wu, Y.-j. *Rheol. Acta* **10**, 539 (1971).
34. Batchelor, G. K., *J. Fluid Mech.* **41**, 545 (1970).
35. Rouse, P. E., Jr., *J. Chem. Phys.* **21**, 1272 (1953).
36. Osaki, K., Schrag, J. L., and Ferry, J. D., *Macromolecules* **5**, 144 (1972).
37. Lodge, A. S., and Wu, Y.-j. Exact relaxation times and dynamic functions for dilute polymer solutions from the bead/spring model of Rouse and Zimm, The University of Wisconsin, Madison, Mathematics Research Center Report 1250, 1972.
38. Treloar, L. R. G., *The Physics of Rubber Elasticity*, 2nd edn. (London: Oxford University Press, 1958).
39. Lodge, A. S., *Rheol. Acta* **7**, 379 (1968).
40. Lodge, A. S., *Elastic Liquids* (New York: Academic Press, 1964).
41. Tobolsky, A. V., *Properties and Structure of Polymers* (New York: Wiley, 1960).
42. Kennard, E. H., *Kinetic Theory of Gases* (New York: McGraw-Hill, 1938).
43. Chandrasekhar, S., *Rev. Mod. Phys.* **15**, 1 (1943).

44. Kirkwood, J. G., *Selected Topics in Statistical Mechanics* (R. W. Zwanzig, ed.), (New York: Gordon & Breach, 1967).
45. Aris, R., *Vectors, Tensors, and the Basic Equations of Fluid Mechanics* (Englewood Cliffs, New Jersey: Prentice-Hall, 1962).
46. Kirkwood, J. G., and Riseman, J., *J. Chem. Phys.* **16**, 565 (1948).
47. Hirschfelder, J. O., Curtiss, C. F., and Bird, R. B., *Molecular Theory of Gases and Liquids* (New York: Wiley, 1954).
48. Amundson, N. R., *Mathematical Methods in Chemical Engineering* (Englewood Cliffs, New Jersey: Prentice-Hall, 1966).

PROBLEMS

10.1. Verify the relations (10.24).

10.2. The relaxation modulus $G(t)$ is related to the storage and loss moduli by Fourier transforms

$$G'(\omega) = G_e + \omega \int_0^\infty [G(t) - G_e] \sin \omega t \, dt$$

$$G''(\omega) = \omega \int_0^\infty [G(t) - G_e] \cos \omega t \, dt$$

Derive these expressions.

Hint: Use (10.21), (10.28), (10.29), the equivalence $M(\lambda) \equiv H(\lambda)$, and $\exp(i\omega t) = \cos \omega t + i \sin \omega t$.

10.3. A fluid is at rest between two parallel plates in a Couette viscometer. At time $t=0$, a stress τ is applied to the movable plate and is held constant for all $t>0$. Compare the ensuing motion if the fluid is described by (a) the Maxwell model (10.22) or (b) the model (10.26).

10.4. Suppose that a "cylinder" of Maxwell fluid, initially in a rest state, is subjected to a uniaxial extension (the extension of a circular cylinder described in Section 9.4) $\partial v_z / \partial z = d$ for all $t>0$. Find an expression for the stress $(t_{zz} - t_{rr})$ for $t>0$. Neglect inertia.

10.5. Show that, in small-amplitude oscillatory shear, a fluid represented by (10.50) has a complex viscosity given by (10.51).

CHAPTER 11

CONSTITUTIVE EQUATIONS: GENERALIZATIONS OF ELEMENTARY MODELS

11.1. INTRODUCTION

The material presented in the previous chapter was, to a considerable extent, independent of some of the more technical and abstract notions developed earlier. We have seen in particular that the subject of linear viscoelasticity is free from the vexing distinctions between different measures of strain and of time differentiation. However, the price to be paid for the simplicity of linear viscoelasticity is a considerable restriction in material response. In laminar shear flows, for example, the viscosity is independent of shear rate. This prediction is contrary to experimental results obtained with most polymer solutions and melts under conditions of interest.

In spite of these disadvantages linear viscoelastic theories can be useful for predicting certain time-dependent features of fluid behavior. Additionally, the physical content of linear theories is often less obscure than that of the general functional expressions employed in Chapter 7. As a result, there is motivation for connecting the previous chapter with more general formulations of constitutive equations. Indeed, some of the general expressions have evolved from physical ideas expressed in the linear treatments.

The purpose of this chapter is, first of all, to make some connections. We wish to show how linear viscoelasticity fits into a more general formulation of rheological behavior, including the functional statement of (7.10). That will link Chapter 10 to the earlier portions of the book in which a restriction to linear response was avoided. Second, we shall discuss a number of nonlinear but nevertheless specific constitutive equations which have been proposed and used to describe fluid behavior.

We are embarking here upon a difficult subject, not because large numbers of new concepts are required, but rather because the subject has developed from many origins and, seemingly, toward many goals. The frustrations awaiting the student of constitutive equations and the reasons for these frustrations have been well stated by Walters and his associates [1]. Much untidiness could be avoided by following a single formalism to the subject. However, important research literature is based on a multiplicity of approaches. This chapter has been organized to acquaint the reader with several points of entry, and also to emphasize those fundamental concepts which permeate the various "systems".

11.2. AN INTEGRAL EXPANSION FOR SIMPLE FLUIDS

Reference was made in Section 7.1 to the work of Rivlin and coworkers [2–4] as an alternate to the functional approach of Noll [5]. We state here a connection between the two. However, in order to make the connection one must ascribe certain smoothness properties to the functions $\mathscr{H}$ of equation (7.10). Fréchet showed (cf. Volterra [6, p. 21]) that every functional $\mathscr{H}\limits_{t=a}^{t=b} [y(t)]$ which is continuous over the field of continuous functions $y(t)$ can be represented by

$$k_0 + \int_a^b k_1(\xi)\, y(\xi)d\xi + \int_a^b \int_a^b k_2(\xi_1,\, \xi_2)\, y(\xi_1)\, y(\xi_2)d\xi_1 d\xi_2 + \cdots$$

$$+ \int_a^b \cdots \int_a^b k_n(\xi_1, \ldots \xi_n)\, y(\xi_1) \cdots y(\xi_n)d\xi_1 \cdots d\xi_n + \cdots.$$

The extension to tensor-valued functionals is apparent. It is convenient to convert to a deformation measure which is $\mathbf{0}$ for a fluid at the present time. Then for a simple fluid we write, recalling (7.11),

$$\mathbf{T}(t) = \mathscr{H}_{s=0}^{\infty}\left(\mathbf{C}(_{(t)}s)\right) = \mathscr{G}_{s=0}^{\infty}\left[\mathbf{G}_{(t)}(s)\right] \tag{11.1}$$

where $\mathbf{G}_{(t)}(s) = \mathbf{C}_{(t)}(s) - \mathbf{I}$. Assuming a deformation history for which an integral expansion is valid, one can write an expansion by direct analogy to the result given above for scalars. This can be represented in compact form by

$$\mathbf{T}(t) = \sum_{n=1}^{\infty} \int_0^{\infty} \cdots \int_0^{\infty} \mathbf{g}(s_1 \ldots s_n)[\mathbf{G}_{(t)}(s_1) \ldots \mathbf{G}_{(t)}(s_n)]ds_1 \ldots ds_n$$

One can imagine that the form of the integrand in this equation would be subject to certain invariance requirements by virtue of the necessity to satisfy the principle of material objectivity. In fact, the integrand must be a multilinear and isotropic function of the $\mathbf{G}_{(t)}(s_1) \ldots \mathbf{G}_{(t)}(s_n)$. It is shown, for example, in Truesdell and Noll [7, p. 99] that the invariance constraints lead to an expansion of the form

$$\mathbf{T}(t) = \int_0^{\infty} M_1(s)\mathbf{G}_{(t)}(s)ds + \int_0^{\infty}\int_0^{\infty} M_2(s_1, s_2)\mathbf{G}_{(t)}(s_1)\mathbf{G}_{(t)}(s_2)ds_1 ds_2$$

$$+ \int_0^{\infty}\int_0^{\infty}\int_0^{\infty} \{M_3(s_1, s_2, s_3)\mathbf{G}_{(t)}(s_1)\mathrm{tr}[\mathbf{G}_{(t)}(s_2)\mathbf{G}_{(t)}(s_3)]$$

$$+ M_4(s_1, s_2, s_3)\mathbf{G}_{(t)}(s_1)\mathbf{G}_{(t)}(s_2)\mathbf{G}_{(t)}(s_3)\}ds_1 ds_2 ds_3 + \cdots \tag{11.2}$$

Because of the symmetry of $\mathbf{T}$ it is clear that the M_i must possess certain symmetry properties. For example, $M_2(s_1, s_2) = M_2(s_2, s_1)$. The smoothness properties necessary for (11.2) to be equivalent to (7.11) invoke restrictions on the present formulation which do not apply to (7.11). However, these restrictions are not likely to be important in practical applications.

It is sometimes convenient to view the integral expansion in terms of strain *rate* (i.e., deformation rate) rather than strain. This can be done through integration of (11.2) by parts and use of (5.63). If we assume that one can write

$$M_1(s) = -\frac{1}{2}\frac{\partial \mathcal{N}_1(s)}{\partial s}; \quad M_2 = \frac{1}{4}\frac{\partial^2 \mathcal{N}_2(s_1, s_2)}{\partial s_1 \partial s_2}; \quad \text{etc.}$$

where the $\mathcal{N}_i$ decrease to zero sufficiently rapidly as any of their arguments approach ∞, then partial integration of (11.2) leads to

$$\mathbf{T}(t) = \int_0^\infty \mathcal{N}_1(s)\mathbf{F}_{(t)}^T(s)\mathbf{D}(s)\mathbf{F}_{(t)}(s)ds$$

$$+ \int_0^\infty \int_0^\infty \mathcal{N}_2(s_1, s_2)\mathbf{F}_{(t)}^T(s_1)\mathbf{D}(s_1)\mathbf{F}_{(t)}(s_1)\mathbf{F}_{(t)}^T(s_2)\mathbf{D}(s_2)\mathbf{F}_{(t)}(s_2)ds_1ds_2 + \ldots \quad (11.3)$$

11.3. REDUCTION TO INFINITESIMAL AND FINITE LINEAR VISCOELASTICITY

Let us now consider the truncation of (11.2) in the limit where the deformation is very small. To show the relation of (11.2) to theories of infinitesimal linear viscoelasticity, define the infinitesimal strain tensor [8, p. 21]

$$\mathbf{E}_{(t)}(s) = \frac{1}{2}\left(\mathbf{F}_{(t)}(s) + \mathbf{F}_{(t)}^T(s) - 2\mathbf{I}\right) \quad (11.4)$$

Now let us make the restriction

$$\{\text{tr}[(\mathbf{F}_{(t)}(s) - \mathbf{I})(\mathbf{F}_{(t)}(s) - \mathbf{I})^T]\}^{1/2} \leqslant \varepsilon \ll 1$$

for all values of s. Then we can write, since $\mathbf{G}_{(t)} = \mathbf{F}_{(t)}^T\mathbf{F}(t) - \mathbf{I}$

$$\mathbf{G}_{(t)} = 2\mathbf{E}_{(t)} + \mathbf{O}(\varepsilon^2) \cong 2\mathbf{E}_{(t)} \quad (11.5)$$

Thus by omitting terms $\mathbf{O}(\varepsilon^2)$ and higher, (11.2) reduces to

$$\mathbf{T}(t) = 2\int_0^\infty M_1(s)\mathbf{E}_{(t)}(s)ds \quad (11.6)$$

But this is just the equation of classical linear viscoelasticity. Equation (11.6) can be expressed in several alternate forms, given that $\dot{K}(s) = M_1(s)$ and that $K(s) \to 0$ sufficiently rapidly as $s \to \infty$. Then (11.6) can be integrated by parts to give

$$\mathbf{T}(t) = \int_0^\infty \dot{K}(s)\mathbf{E}_{(t)}(s)ds = -\int_0^\infty K(s)\mathbf{D}(s)ds \quad (11.7)$$

which corresponds in form to the expression (10.32) obtained from the idea of superposition in Section 10.9. Thus our present description of infinitesimal viscoelasticity corresponds to the linear viscoelastic models discussed in the last chapter.

It is worth noting that, because of the truncation step of (11.5), we have lost an important property of the original integral expansion (11.2): Equation (11.6) is not in accord with the principle of material objectivity, as is immediately apparent from (11.4) and a recollection from Section 6.4 that $\mathbf{F}$ does not transform objectively. This is an unattractive feature of the infinitesimal theory. It is also an inconsequential feature, however, since the departure from an objective transformation for these infinitesimal motions is $O(\varepsilon^2)$. We can easily recover an objectively proper equation by being less severe in the truncation (11.5). If, instead of that approximation, we maintain $\mathbf{G}_{(t)}(s)$ as the measure of strain and require the allowable motions to be such that all integrals in (11.2) beyond the first are negligible because of a sufficiently small (and appropriately defined) norm for $\mathbf{G}_{(t)}(s)$, then we have the equation

$$\mathbf{T}(t) = \int_0^\infty M_1(s)\mathbf{G}_{(t)}(s)\,ds \tag{11.8}$$

which is known as the constitutive equation for *finite linear viscoelasticity* The distinction, along with a description of attributes, of (11.8) as opposed to (11.6) has been carefully stated by Coleman and Noll [9] in an important paper which shows in detail how the theory of simple fluids can be truncated systematically to the finite linear and to the infinitesimal theories of viscoelasticity.

11.4. A DIFFERENTIAL EXPANSION: THE RELATION TO AN INTEGRAL EXPANSION

One may alternately choose to make slightly different assumptions about the properties of the functional $\mathscr{G}$ of (11.1). In Section 9.5 we spoke briefly of the Fréchet differential of a functional, and used it to infer the behavior of a flow slightly removed from a viscometric flow. If the functional $\mathscr{G}$ is n-times Fréchet differentiable, then $\mathscr{G}$ can be expanded as a sum of functionals each of which is a polynomial in $\mathbf{G}(s)$ of degree equal to or less than n. Furthermore, if the motion is sufficiently *slow*, Coleman and Noll have shown that the polynomial can then be written as a polynomial *function* of the Rivlin–Ericksen tensors $\mathbf{A}_1 \ldots \mathbf{A}_n$, *all evaluated at the present time t.* A demonstration of this reduction is not warranted here, and the interested reader is urged to consult more specialized sources [7, p. 108; 10]. An essential ingredient is that one can consider expansion of a functional in a manner similar to a Taylor series expansion of a function; that is, the argument of the stress functional can be expanded into a series of successively higher time derivatives of $\mathbf{G}_{(t)}(s)$, each evaluated at $s = 0$ and multiplied by s to some power. The functional then reduces to a *function* of the time derivatives $[\partial^n \mathbf{G}_{(t)}(s)/\partial s^n]_{s=0}$. But these are directly related to the Rivlin–Ericksen tensors (cf. (5.62)), so one obtains for (11.1)

$$\mathbf{T} \cong \sum_{i=1}^k \mathbf{H}_i(\mathbf{A}_1 \ldots \mathbf{A}_i)$$

To various orders of approximation the form of the functions $\mathbf{H}_i$ can be determined by objectivity and symmetry requirements. If the expansion process is truncated at successively higher order, there results the sequence of constitutive equations

$$\left.\begin{aligned}
\mathbf{T} &= \alpha_0\mathbf{A}_1 \\
\mathbf{T} &= \alpha_0\mathbf{A}_1 + \alpha_1\mathbf{A}_2 + \alpha_2\mathbf{A}_1\mathbf{A}_1 \\
\mathbf{T} &= \alpha_0\mathbf{A}_1 + \alpha_1\mathbf{A}_2 + \alpha_2\mathbf{A}_1\mathbf{A}_1 + \beta_1(\text{tr}(\mathbf{A}_1\mathbf{A}_1))\mathbf{A}_1 + \beta_2\mathbf{A}_3 + \alpha_4(\mathbf{A}_1\mathbf{A}_2 + \mathbf{A}_2\mathbf{A}_1)
\end{aligned}\right\} \quad (11.9)$$

where the α_i and β_i are constants. The utility of the approximations (11.9) rests upon validity of the assumption that the deformation is small against some norm $\|\mathbf{G}_{(t)}(s)\|$ for s small (time near the present). At times far from the present, the memory of the fluid is taken to be sufficiently short so that even if $\|\mathbf{G}_{(t)}(s)\|$ is not small as s becomes large, that is of no consequence to the present fluid response. Thus (11.9) is an expansion around the present time. It is often called a retarded motion expansion, and is seen to result in equations which are special forms of the Rivlin–Ericksen fluid (Section 10.4). Since (11.9) is obtained by an orderly truncation of simple fluids, the expansions are often referred to as models of first, second, and third-order fluids, respectively. This is the origin of the second-order fluid, discussed in Section 10.4.

The utility of expansions leading to (11.2) and (11.9) is of course highly dependent upon both the fluid and the flow under consideration. If requirements for both expansions are met, then one can proceed to relate the constants of (11.2) and (11.9) [11]. From a Taylor series expansion of $\mathbf{G}_{(t)}(s)$ in (11.2) around $s=0$ and application of the Cayley–Hamilton theorem (9A1.1) one readily obtains

$$\left.\begin{aligned}
\alpha_0 &= -\int_0^\infty M_1(s)s\,ds; \quad \alpha_1 = \frac{1}{2}\int_0^\infty M_1(s)s^2\,ds \\[2mm]
\alpha_2 &= \int_0^\infty\int_0^\infty M_2(s_1, s_2)s_1s_2\,ds_1ds_2 \\[2mm]
\alpha_4 &= -\frac{1}{2}\int_0^\infty\int_0^\infty M_2(s_1, s_2)(s_1s_2{}^2 + s_1{}^2s_2)\,ds_1ds_2 \\[2mm]
\beta_1 &= -\int_0^\infty\int_0^\infty\int_0^\infty \left[M_3(s_1, s_2, s_3) + \frac{1}{2}M_4(s_1, s_2, s_3)\right]s_1s_2s_3\,ds_1ds_2ds_3 \\[2mm]
\beta_2 &= -\frac{1}{6}\int_0^\infty M_1(s)s^3\,ds
\end{aligned}\right\} \quad (11.10)$$

11.5. RELATION TO EARLIER WORK

The results of the previous sections should be helpful in unifying the isolated examples of Chapter 10 into the broad framework of constitutive theory. For example, consider the expression for infinitesimal deformation of a bead–spring model (10.50). We now see that this falls into the class of infinitesimal linear viscoelasticity and can be readily extended to a finite linear theory by introduction of $\mathbf{C}_{(t)}(s)$ into the integral expression (10.50), which, except for a different choice of strain measure, corresponds also to the rubberlike liquid (10.57). Additionally, a connection has been made between integral and differential expressions, and the relation of both to a simple fluid has been noted.

Of course in any practical problem the continual dilemma is one of knowing which truncation can be usefully effected and how severe a truncation will still admit meaningful results. From the foregoing examples it is now clear that the answers depend upon both the fluid (how persistent the memory function is) and the flow (how large some suitable norm $\|\mathbf{G}_{(t)}(s)\|$ is).

11.6. THE OLDROYD APPROACH TO CONSTITUTIVE EQUATIONS

Our development of constitutive theory in Chapter 7 was quite general. We learned in fact that for sufficiently elementary flows, such as viscometric flows, it is unnecessary to be specific about the form of the functional $\mathscr{H}$. However, to deal intelligently with specific constitutive models, such as those hinted at in the previous chapter, it is necessary to have a systematic means for relating the response history of material points (i.e., a collection of particles) to the stress behavior over time at specified *places* in the flow configuration. Furthermore, the effect of the history must be explicitly stated in the constitutive model. Generally, this means that both differentiations and integrations over past time and for a given material point are relevant to constitutive equations. However, the flow boundary conditions and hence final equations of motion usually deal with regions *fixed* in space and through which material points are flowing. It was recognition of the need for explicit treatment of time dependent behavior in a material system and the need to express this time dependence in equations of motion which are to be solved relative to a space-fixed coordinate system, that prompted Oldroyd in 1950 to publish his fundamental discussion on formulation of rheological equations of state [12].

The difficulty perceived by Oldroyd as well as the motivation for wishing to overcome it is readily appreciated by citing an example to which he referred. In 1946 Fröhlich and Sack [13] offered an interesting extension of Einstein's treatment for the constitutive behavior of a dilute suspension of rigid spheres in a Newtonian fluid [14]. The model differed from Einstein's in that the spheres were taken to be perfectly *elastic* bodies capable of small deformations from sphericity. The final result of Fröhlich and Sack for the constitutive equation representative of the combination of viscous and elastic phases is

$$\left(1 + \lambda_1 \frac{\partial}{\partial t}\right) \mathbf{T}(\mathbf{x}, t) = 2\mu \left(1 + \lambda_2 \frac{\partial}{\partial t}\right) \mathbf{D}(\mathbf{x}, t) \tag{11.11}$$

where the constants can be related to material properties of the individual phases of the suspension. One immediately recognizes (11.11) as identical to one of the variants, (10.26), of a spring–dashpot model of the previous chapter.

Oldroyd recognized a problem with (11.11), and the problem arises from the terms with a time derivative $\partial/\partial t$. Equation (11.11) was derived from a consideration of tensor components in a coordinate system with origin fixed in the center of one of the particles. In use of (11.11), however, one would certainly apply it to a coordinate system fixed in the laboratory. The two become identical only in the limit as the deformation approaches zero, i.e., as one enters the regime of infinitesimal linear viscoelasticity. Oldroyd's objection to (11.11) was the failure of it to obey what we have called the principle of material objectivity. He realized very well the dilemma associated with preservation of invariance under a "scaleup" to arbitrary deformations of results obtained under restricted conditions of flow [12]:

"But, in general, simple relationships connecting the familiar stress, strain, and rate-of-strain tensors of the classical theories of elasticity and hydrodynamics will not suffice to describe the rheological properties of a material completely. In the most general case contemplated, the equations of state defining the properties of an element at any instant may involve all the kinematic and dynamic quantities which define the states of the same element during its previous history. On the other hand, experiments designed to measure the rheological properties of particular materials must, for easy interpretation, relate to simple types of motion, and the resulting simple equations of state will require generalization before they can be applied to more complicated types of motion. There may be many ways of generalizing the results of a single experiment, and the method of generalization so as to give a universally valid form of the equations of state is by no means obvious. The form of the completely general equations must be restricted by the requirement that the equations describe properties independent of the frame of reference."

Whereas Noll and his followers wished to present formalisms free from specification of ingredients of the functional $\mathscr{H}$, Oldroyd realized that a form of $\mathscr{H}$ is ultimately necessary to the person who wishes to solve flow problems or to pose specific constitutive equations. Oldroyd developed a recipe which insures correct invariance properties for differential and integral functions appearing in $\mathscr{H}$.

11.7. TIME DIFFERENTIATION OF TIME-DEPENDENT BASE VECTORS

We have acknowledged since Chapter 7 that the stress of a material point is determined by the history of prior motion of the material point. Hence if one is interested in *formulation* of specific theories of constitutive behavior it makes sense to develop these theories with respect to a coordinate system embedded in the body rather than with respect to a space-fixed coordinate system. Then, as a final step, one wishes to transform the tensor equation written for a body-fixed coordinate system to a space-fixed coordinate system.

One's first response is likely to be that the whole of tensor analysis is built upon the presumption of invariance to change of coordinate system, and that much of Chapter 3 was devoted to the rules which insure invariance under transformation of tensor components. That is true. However, it is important to note that the transformations illustrated in Chapter 3 were all for base vector systems which were *independent of time*. We now wish to explore the implications of relaxing that assumption.

Students of mechanics are well aware of the necessity for specifying the state of the observer whenever reference is made to time differentiation. In fact we have already noted in Section 5.7 the important distinction to be made between time differentiation with material points held constant and time differentiation with position in space held constant ("with respect to the fixed stars" [7, p. 41]). In this section we wish to discuss some new complications which arise when one considers transformations between reference frames which have a time-varying connection [15].

A reference frame consists of a measure of time, a measure of length, an origin, and a basis. Let us consider a frame, say frame 2 with basis $\{\mathbf{e}_i(\tau)\}$, which may vary continuously over time with respect to another frame, which we call frame 1. The time rate of change of $\{\mathbf{e}_i(\tau)\}$ as viewed by an observer employing length and time scales in frame 1 can be expressed by

$$\frac{d_1 \mathbf{e}_i(\tau)}{d\tau} = \mathbf{M}(\tau)\mathbf{e}_i(\tau) \tag{11.12}$$

where (11.12) is a defining expression for $\mathbf{M}(\tau)$. We shall find it convenient to decompose $\mathbf{M}$ into symmetric and antisymmetric parts

$$\left.\begin{aligned}\mathbf{D} &= \frac{1}{2}(\mathbf{M} + \mathbf{M}^T) \\[2.5em] \mathbf{W} &= \frac{1}{2}(\mathbf{M} - \mathbf{M}^T)\end{aligned}\right\} \tag{11.13}$$

Then

$$\frac{d_1 \mathbf{e}_i(\tau)}{d\tau} = (\mathbf{D} + \mathbf{W})\mathbf{e}_i(\tau) \tag{11.14}$$

A source of confusion over time differentiation in rheology becomes evident when we consider $d_1/d\tau$ of the basis *reciprocal* to $\mathbf{e}_i(\tau)$ in (11.12). Recall that $\{\mathbf{e}^i\}$, the basis reciprocal to $\{\mathbf{e}_j\}$, is defined by

$$\mathbf{e}^i(\tau) \cdot \mathbf{e}_j(\tau) = \delta^i_j \tag{2.35}$$

which, incidentally, can be shown [16, p. 121; Chapter 2, problem 2.1] to be equivalent to

$$\mathbf{e}^i(\tau) = \frac{\mathbf{e}_i(\tau) \times \mathbf{e}_j(\tau)}{\mathbf{e}_1(\tau) \cdot [\mathbf{e}_2(\tau) \times \mathbf{e}_3(\tau)]} \quad (i, j, k \text{ in cyclic order}) \tag{11.15}$$

Then

$$\left[\frac{d_1 \mathbf{e}^i}{d\tau} + \mathbf{M}^T \mathbf{e}^i\right] \cdot \mathbf{e}_j = 0 \tag{11.16}$$

for all i and j. This implies

$$\frac{d_1 \mathbf{e}^i}{d\tau} = -\mathbf{M}^T \mathbf{e}^i = -(\mathbf{D} - \mathbf{W})\mathbf{e}^i \tag{11.17}$$

and provides, from comparison with (11.12), the interesting result that an observer in frame 1 will record different time derivatives for the bases $\{\mathbf{e}_i\}$ and $\{\mathbf{e}^i\}$ *unless* $\mathbf{M} = -\mathbf{M}^T$.

We can prepare to apply these results by associating them with two specific reference frames. Consider two neighboring material points X and $X + dX$ (recall that these are labels, not coordinates). We form a reference frame 2 using a basis $\mathbf{e}_k(\tau)$ and coordinates $\{\zeta^k\}$ such that at any time τ the two points have coordinates $\{\zeta^i\}$ and $\{\zeta^i + d\zeta^i\}$, respectively, and positions $\mathbf{r}$ and $\mathbf{r} + d\mathbf{r}$ given, respectively, by

$$\left.\begin{aligned}\mathbf{r} &= \bar{\zeta}^i \mathbf{e}_i(\tau) \\ \mathbf{r} + d\mathbf{r} &= (\bar{\zeta}^i + d\bar{\zeta}^i)\mathbf{e}_i(\tau)\end{aligned}\right\} \tag{11.18}$$

The overbar signifies a *component* and does not necessarily coincide with the corresponding *coordinate.* Note that the coordinates $\{\zeta^i\}$ for any material point X do not change with time. They are sometimes called *convected* coordinates.

We can identify another frame (frame 1) in which coordinates of X and $X + dX$ are given, respectively, by $\{x^i\}$ and $\{x^i + dx^i\}$ with respect to a basis $\{\mathbf{e}_i(t)\}$. We shall require that at $\tau = t$

$$\zeta^i = x^i$$

Then, since the $\{\zeta^i\}$ are fixed for a given material point, we can think of $\{\zeta^i\}$ as coordinates "fixed" or "embedded" in the material being described, while the x^i are fixed in physical space. At the special time $\tau = t$ the two frames coincide. The relation between $\{\zeta^i\}$ and $\{x^i\}$ is shown for a particular example in Fig. 11.1. At any time we have for material point X

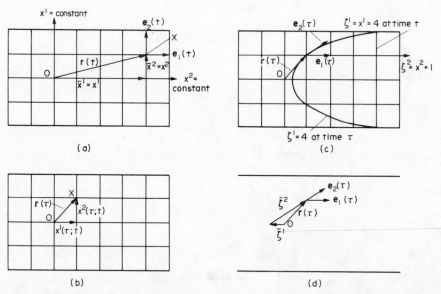

Fig. 11.1. Body-fixed (convected) coordinates and space-fixed coordinates. (a) Consider a two-dimensional slice of laminar flow (Section 7.5) between two parallel planes. At time t the location of material points is described by the coordinates x^1, x^2, the origin O, and basis $\mathbf{e}_1(t)$, $\mathbf{e}_2(t)$. A particular material point is singled out at position $\mathbf{r}(t)$. The point is labeled X and is located at $x^1 = 4$, $x^2 = 1$. (b) At some earlier time τ, material point X is at position $\mathbf{r}(\tau)$, which can be described by coordinates $x^1(\tau; t) = 1$, $x^2(\tau; t) = 1$, origin O and basis $\mathbf{e}_1(t)$, $\mathbf{e}_2(t)$. (c) At time τ point X can also be described by convected coordinates ζ^1, ζ^2. These are formed by considering the coordinates x^1 and x^2, shown in (a), to have been embedded in the material. (d) The convected coordinate system is neither orthogonal nor Cartesian, and the components $\bar{\zeta}^i$ do not coincide with the coordinates ζ^i.

$$\mathbf{r} = \bar{\zeta}^i \mathbf{e}_i(\tau) = \bar{x}^i(\tau; t)\mathbf{e}_i(t) \tag{11.19}$$

In applications we shall take t as the present instant of time for which some quantity, for example the stress, is desired. Then τ can be taken to represent previous times, and hence deformation over τ is the history of the deformation.

An observer using a distance scale in the $(\mathbf{x}, t)$ reference frame (physical space) can write for the distance ds between X and $X + dX$

$$(ds(\tau))^2 = \gamma_{ij}(\zeta^k, \tau)d\zeta^i d\zeta^j \tag{11.20}$$

The γ_{ij} are recognized as components of the metric tensor

$$\mathbf{\Gamma}(\zeta^k, \tau) = \gamma_{ij}\mathbf{e}^i(\tau) \otimes \mathbf{e}^j(\tau)$$

If we consider the change in $(ds)^2$ between two material points over a time interval from τ to $\tau + \delta\tau$ and let $\delta\tau \to 0$, we have

$$\frac{d}{d\tau}(ds(\tau))^2 = \frac{\partial\gamma_{ij}(\zeta^k, \tau)}{\partial\tau} d\zeta^i d\zeta^j \tag{11.21}$$

Since the material points are unchanged, the coordinates ζ^k are also unchanged over the time interval. The derivative $(\partial\gamma_{ij}/\partial\tau)$ is then a measure of the rate at which adjacent particles in the material are separated during a motion. Hence we might expect this quantity to be, as indeed it is, a useful measure of *strain rate*.

Finally, the notation of (11.21) implies that the coordinates ζ^i are taken with respect to a natural basis. From Section 2.9 and (11.18),

$$\mathbf{e}_i(\zeta^k, \tau) = \frac{\partial\mathbf{r}}{\partial\zeta^i} \tag{11.22}^\dagger$$

and

$$\gamma_{ij}(\zeta^k, \tau) = \mathbf{e}_i(\tau) \cdot \mathbf{e}_j(\tau)$$

We can connect (11.22) to $\mathbf{M}$ of (11.12). Differentiate (11.22) partially with respect to τ, interchange the order of differentiation, and use $(\partial\mathbf{r}/\partial\tau)_{\tau=t} = \mathbf{v}(\mathbf{x}, t)$, where $\mathbf{x} = x^i\mathbf{e}_i(t)$. Then

$$\left(\frac{\partial\mathbf{e}_i}{\partial\tau}\right)_{\tau=t} = \left(\frac{\partial\mathbf{v}}{\partial\zeta^i}\right)_{\tau=t} = \left(\frac{\partial\mathbf{v}}{\partial x^i}\right) = v^k{}_{,i}\mathbf{e}_k(\mathbf{x}, t) \tag{11.23}$$

Then at $\tau = t$ we can compare (11.12) and (11.23).

$$M^j{}_i = v^j{}_{,i}$$

or, recalling the notation of (5.44),

$$\mathbf{M} = \nabla\mathbf{v} \tag{11.24}$$

We have taken $d_1/d\tau$ of (11.12) to be a time derivative holding material points constant, and will continue below to use that interpretation.

The result (11.24) justifies the notation of (11.13) since, at $\tau = t$, $\mathbf{D}$ and $\mathbf{W}$ now become the rate of stretching and rate of rotation tensors of (5.53) and (5.52), respectively.

11.8. RELEVANCE OF TIME-DEPENDENT BASES TO CONSTITUTIVE EQUATIONS

It is not difficult to see how the material of the preceding section relates to formulation of constitutive equations. To do this, let us return to (11.11) and Oldroyd's criticism of it. On physical grounds we have seen that interpretation of $\partial/\partial t$ as a time derivative with

---

† Unless there is a special reason for doing otherwise, we do not bother to note specifically the ζ^k dependence of $\{\mathbf{e}_i\}$.

respect to space-fixed coordinates is objectionable. Equivalently, that interpretation renders the result incapable of obeying the principle of material objectivity, as may be shown formally or by merely noting that a motion of the observer will change the nature of $\partial/\partial t$ taken with respect to a frame in which the observer is fixed. One is led therefore to interpret $\partial/\partial t$ as a time derivative with respect to a frame associated with the *material*, not with an observer who is fixed in space. (Note, however, that the fundamental measure of length is taken from a yardstick in the observer-fixed frame. Otherwise, we would have no length standard against which to measure deformation.) Several possible frames come to mind:

(a) A frame with a basis $\{\mathbf{e}_i\}$ defined in terms of *material* points of the body. This is an embedded basis which will translate, rotate, and deform along with the material relative to a space-fixed frame.

(b) A frame *reciprocal* to that defined above, the basis for which is given by $\{\mathbf{e}^j\}$, where $\mathbf{e}_i \cdot \mathbf{e}^j = \delta_i^j$.

(c) Since descriptions of the motion written for these two frames obviously satisfy the principle of material objectivity, so must any linear combination of bases $\{\mathbf{e}_i\}$ and $\{\mathbf{e}^i\}$. One combination of particular interest is the basis $\{\mathbf{e}_i(\tau) + g_{ik}\mathbf{e}^k(\tau)\}$, where $g_{ik} = \mathbf{e}_i(t) \cdot \mathbf{e}_k(t)$. From (11.14) and (11.17) we have, at $\tau = t$,

$$\left[\frac{d_1}{d\tau} (\mathbf{e}_i + g_{ik}\mathbf{e}^k) \right]_{\tau=t} = 2\mathbf{W}\mathbf{e}_i \qquad (11.25)$$

and it is seen that this basis shares in the rotation of the material but does not deform with it. We shall find later that this combination offers some special advantages.

Our dilemma with respect to (11.11) remains. We agree that the equation should be formulated with respect to a body-based reference frame. But ultimately, in order to solve boundary-value problems, we will wish to write the result in a space-fixed frame; and we see, for example by comparison of (11.12) and (11.17), that the results will be *different* depending upon whether (a), (b), or some mixture is used as the frame with which $\partial/\partial t$ of (11.11) is associated. One readily sees that the operation of raising and lowering tensor component indices, which was such a trivial matter with time-independent bases, can involve fundamental changes if time-dependent bases are being used.

In the interpretation of (11.11) we have another example of the nonuniqueness which exists when one proceeds from infinitesimal deformations to finite deformations. As long as the constitutive model is derived from a mechanical or molecular model which is rigorous only for infinitesimal deformations, there is no *a priori* way to generalize it. On the other hand, if a specific model, for example a deformable particle embedded in a fluid undergoing shear, is considered and the fluid dynamic problem is solved with proper attention to all tensor components for finite deformations, then the proper form of time derivative arises quite naturally [17], as we shall see in Chapter 13.

11.9. OLDROYD'S FORMULATION OF CONSTITUTIVE EQUATIONS

Because it would appear that body-fixed bases form a natural reference for formulation of constitutive equations, Oldroyd and others, notably Lodge [18], have argued for a system rather different than the one of Noll, Coleman, and Truesdell. The argument runs something like this: if constitutive equations involve the history of deformation of *material* elements, then the equations should be formulated with respect to these elements. That makes the physics of the problem more evident and also removes the necessity of an explicit material objectivity requirement. The requirement is automatically satisfied.

We now proceed to develop some of Oldroyd's ideas by applying concepts from the previous sections. To begin let us consider the time derivative $[\partial(\mathbf{u}(\zeta^i, \tau))/\partial\tau]\zeta^i_{=const}$ of some vector $\mathbf{u}$ as viewed by an observer in the frame with body-fixed basis $\{\mathbf{e}_i(\tau)\}$. However, for computational purposes we wish to express the result at $\tau = t$ in terms of components of $\mathbf{u}$ in the space-fixed frame with basis $\{\mathbf{e}_i(t)\}$ and coordinates $\{x^i\}$. Thus we write

$$\mathbf{u}(\zeta^i, \tau) = \mu^l(\zeta^i, \tau)\mathbf{e}_l(\tau) = u^l(x^i, \tau, t)\mathbf{e}_l(t) \tag{11.26}$$

and require that at $\tau = t$, $\zeta^i = x^i$. Then we ask this question: an observer in the body-fixed frame with basis $\{\mathbf{e}_i(\tau)\}$, which appears constant to an observer in that frame, records a time rate of change $(\partial\mathbf{u}(\zeta^i, \tau)/\partial\tau)_{\tau=t}$ of $\mathbf{u}$. How does one express components of $(\partial\mathbf{u}(\zeta^i, \tau)/\partial\tau)_{\tau=t}$ in terms of quantities defined in the space-fixed frame with basis $\{\mathbf{e}_i(t)\}$? To answer this one can take the body-fixed frame as frame 2 and the space-fixed frame as frame 1, and use the results just developed. Applying (11.12) to (11.26) and setting $\tau = t$,

$$\left\{\left[\frac{\partial\mu^l(\zeta^i, \tau)}{\partial\tau}\right]_{\tau=t} + v^l_{,m}\mu^m\right\}\mathbf{e}_l(t) = \left[\frac{\partial u^l(x^i, \tau, t)}{\partial\tau}\right]_{\substack{\zeta^i = const \\ \tau=t}}\mathbf{e}_l(t)$$

The time derivative on the right-hand side is simply the usual material derivative and can be handled by the chain rule, since the basis is constant with respect to τ. To denote $[\partial\mu^l(\zeta^i, \tau)/\partial\tau]_{\tau=t}$ expressed in terms of components in the space-fixed frame we follow Oldroyd and write this as $\eth u^l(x^i, t)/\eth t$. Then

$$[\partial\mu^l(\zeta^i, \tau)/\partial\tau]_{\tau=t} \equiv \eth u^l(x^i, t)/\eth t = \frac{\partial u^l}{\partial t} + v^m u^l_{,m} - v^l_{,m}u^m \tag{11.27}$$

It should now be clear that this is *not*, in general, the same as a time derivative seen by an observer in a body-fixed frame with basis $\{\mathbf{e}^i(\tau)\}$. Expressing components of that time derivative in a space-fixed frame with basis $\{\mathbf{e}^i(t)\}$, we obtain, using (11.17),

$$\frac{\eth u_l(x^i, t)}{\eth t} = \frac{\partial u_l}{\partial t} + v^m u_{l,m} + v^m_{,l}u_m \tag{11.28}$$

Note that $\eth u_l/\eth t$ is a different time derivative than $\eth u^l/\eth t$.

The analysis is readily extended to second-order tensors. By $\eth a_{ij}/\eth t$ we signify the time derivative of some tensor $\mathbf{A}(\zeta^i, \tau)$ as seen by someone in the body-fixed frame with basis $\{\mathbf{e}^i(\tau)\}$ but expressed at time t in terms of components in a space-fixed frame with basis $\{\mathbf{e}^i(t)\}$ so that $\mathbf{A}(\zeta^k, \tau) = a_{ij}(x^k, \tau, t)\mathbf{e}^i(t) \otimes \mathbf{e}^j(t)$. Then it follows that

$$\frac{\eth a_{ij}}{\eth t} = \frac{\partial a_{ij}}{\partial t} + v^k a_{ij,k} + v^k_{,i}a_{kj} + v^k_{,j}a_{ik} \tag{11.29}$$

or, equivalently, using (11.17)

$$\frac{\eth a_{ij}}{\eth t} = \frac{\partial a_{ij}}{\partial t} + v^k a_{ij,k} + d_i{}^k a_{kj} + d_j{}^k a_{ik} - w_i{}^k a_{kj} - w_j{}^k a_{ik} \tag{11.30}$$

Similarly, for a time derivative in a frame with basis $\{\mathbf{e}_i(\tau)\}$, components expressed in terms of the space-fixed frame with basis $\{\mathbf{e}_i(t)\}$ are

$$\frac{\eth a^{ij}}{\eth t} = \frac{\partial a^{ij}}{\partial t} + v^k a^{ij}_{,k} - v^i_{,k} a^{kj} - v^j_{,k} a^{ik} \tag{11.31}$$

or

$$\frac{\eth a^{ij}}{\eth t} = \frac{\partial a^{ij}}{\partial t} + v^k a^{ij}_{,k} - d^i_{\ k} a^{kj} - d^j_{\ k} a^{ik} - w^i_{\ k} a^{kj} - w^j_{\ k} a^{ik} \tag{11.32}$$

Equations (11.30)–(11.32) are readily generalized to tensors of arbitrary order. For example,

$$\frac{\eth a^{i_1 i_2 \cdots i_m}_{\quad\ j_1 j_2 \cdots j_n}}{\eth t} = \frac{\partial a^{i_1 i_2 \cdots i_m}_{\quad\ j_1 j_2 \cdots j_n}}{\partial t} + v^k a^{i_1 i_2 \cdots i_m}_{\quad\ j_1 j_2 \cdots j_n,k} + v^{i_1}_{\ ,k} a^{k i_2 \cdots i_m}_{\quad\ j_1 j_2 \cdots j_n}$$

$$+ \cdots + v^k_{\ ,j_n} a^{i_1 i_2 \cdots i_m}_{\quad\ j_1 j_2 \cdots k} - v^i_{\ ,k} a^{k i_2 \cdots i_m}_{\quad\ j_1 j_2 \cdots j_n} - \cdots - v^{i_m}_{\ ,k} a^{i_1 i_2 \cdots k}_{\quad\ j_1 j_2 \cdots j_n} \tag{11.33}$$

In Section 10.7 we anticipated the importance of the time rate of change of the metric tensor $\mathbf{\Gamma}(\zeta, \tau)$ with components $\gamma_{ij} = \mathbf{e}_i(\tau) \cdot \mathbf{e}_j(\tau)$. This time rate of change, expressed in the space-fixed frame with basis $\{\mathbf{e}^i(t)\}$ can be written down using (11.29), or by applying (11.12). Remembering that at $\tau = t$, $\gamma_{ij} = g_{ij} = \mathbf{e}_i(t) \cdot \mathbf{e}_j(t)$ and that $g_{ij,k} = 0$, one finds

$$\frac{\eth g_{ij}}{\eth t} = v_{i,j} + v_{j,i} = 2 d_{ij} \tag{11.34}$$

Thus the time rate of change of the body-fixed metric tensor is indeed just twice the local value of the rate of stretch. In fact, one can show that

$$\frac{\eth^n g_{ij}}{\eth t^n} = A^{(n)}_{ij} \tag{11.35}$$

where $A^{(n)}_{ij}$ is the covariant ij component of the nth Rivlin–Ericksen tensor (5.62). From what we have already seen, it is not surprising that $(\eth g^{kl}/\eth t) g_{ki} g_{lj} \neq \eth g_{ij}/\eth t$. In fact,

$$\frac{\eth g^{ij}}{\eth t} = - 2 d^{ij} \tag{11.36}$$

Another way to look at this lack of equivalence is to say that the operation $\eth/\eth t$ does not commute with raising and lowering of indices. Thus, in general

$$\frac{\eth a_{kl}}{\eth t} g^{ik} g^{jl} \neq \frac{\eth a^{ij}}{\eth t} \tag{11.37}$$

11.10. A CHOICE OF TIME DERIVATIVES

Oldroyd's 1950 paper showed very clearly that a unique generalization of time derivatives $\partial/\partial t$ which arise in constitutive equations derived for infinitesimal deformations is not possible. A case in point is the Fröhlich and Sack result (11.11). If we write (11.11) in component form, make the conversion $\partial/\partial t \to \eth/\eth t$, and apply (11.33), it is clear that we are led to different constitutive equations, if we use, say, $\eth t_{ij}/\eth t$ or $\eth t^{ij}/\eth t$. The reason for this is simply that in the former case $\partial/\partial t$ in (11.11) is being taken as referring to a body-fixed frame with basis $\{\mathbf{e}^i(\tau)\}$, while in the latter, the basis is $\{\mathbf{e}_i(\tau)\}$. The time derivatives of these bases, as viewed by an observer in a space-fixed frame, are different, even though either formulation satisfies the principle of material objectivity. Only in the limit, when terms involving products of velocity, stress, or their gradients are of second order, does the distinction become negligible. Then one is in the regime of infinitesimal deformations. Note also how the attractive linearity of (11.11) is lost when the Oldroyd (sometimes called *convected*) derivative $\eth/\eth t$ is used.

The physical significance of components of the metric tensor $\gamma_{ij}(\zeta^k, \tau)$ and their time derivatives is clear from (11.20), (11.21), (11.22), and (11.34). $\gamma_{ij}(\zeta^k, \tau)$ is seen from (11.20) to be a scale factor governing the separation of neighboring material points. Let us now consider the role of the reciprocal to the basis (11.22). In particular, we note first the physical significance of $\gamma^{ij}(\zeta^k, \tau)$, following arguments given by Lodge [18, p. 318].

Consider two adjacent surfaces in a material. These surfaces are defined by the requirement that each surface contains the same material points over all times τ. (The surfaces are *material surfaces*.) Then we can write equations for these surfaces by

$$\sigma(\zeta) = c$$

where c is a parameter that characterizes each surface and, at some time τ, $\zeta = \bar{\zeta}^i \mathbf{e}_i(\tau)$. We wish to form an equation which expresses the distance δh between two neighboring surfaces $\sigma(\zeta) = c$ and $\sigma(\zeta + \delta\zeta) = c + \delta c$. By the chain rule we can write

$$(\nabla\sigma) \cdot \delta\zeta = \delta c$$

where ζ refers to a material point in surface c and $\zeta + \delta\zeta$ to one in $c + \delta c$. Then the distance between surfaces, measured along the normal $\mathbf{n}$ to either surface, is (as $\delta c \to 0$)

$$\delta h = \mathbf{n} \cdot \delta\zeta$$

But

$$\mathbf{n} = \frac{\nabla\sigma}{[(\nabla\sigma) \cdot (\nabla\sigma)]^{1/2}}$$

so that

$$\delta h = \frac{\delta c}{[(\nabla\sigma) \cdot (\nabla\sigma)]^{1/2}} \tag{11.38}$$

The dot product in the denominator can be written in terms of $\gamma^{ij}(\zeta, \tau) = \mathbf{e}_i(\zeta, \tau) \cdot \mathbf{e}^j(\zeta, \tau)$, which is related to $\gamma_{ij}(\zeta, \tau)$ in the usual manner as given in Section 3.6. We have

$$\nabla\sigma \cdot \nabla\sigma = \gamma^{ij} \frac{\partial\sigma}{\partial\zeta^i} \frac{\partial\sigma}{\partial\zeta^j}$$

Then we obtain from (11.38)

$$\left(\frac{\delta c}{\delta h}\right)^2 = \gamma^{ij}(\zeta, \tau) \frac{\partial\sigma}{\partial\zeta^i} \frac{\partial\sigma}{\partial\zeta^j} \tag{11.39}$$

In contrast to (11.20), one infers from (11.39) that $\gamma^{ij}(\zeta, \tau)$ is a scale factor governing separation of adjacent material *surfaces*, and the time rate of change of distance between material surfaces will not, in general, be the same as the corresponding change for material points. Hence the distinction between $\mathfrak{d}g_{ij}/\mathfrak{d}t$ and $\mathfrak{d}g^{ij}/\mathfrak{d}t$.

We mentioned in Section 11.8 that bases formed from linear combinations of $\{\mathbf{e}_i(\tau)\}$ and $\{\mathbf{e}^i(\tau)\}$ are acceptable. In particular from (11.25) it is apparent that the basis $\mathbf{\varepsilon}_i(\tau) = \frac{1}{2}[\mathbf{e}_i(\tau) + g_{ij}\mathbf{e}^j(\tau)]$ shares in the *rotation* of the material but not in its deformation. Consider now time derivatives in this *co-rotational* frame when expressed in a space-fixed frame with basis $\mathbf{\varepsilon}_i(t) = \frac{1}{2}[\mathbf{e}_i(t) + g_{ij}\mathbf{e}^j(t)]$. Given some tensor

$$\mathbf{A} = a^{ij}(\zeta^k, \tau)\mathbf{\varepsilon}_i(\tau) \otimes \mathbf{\varepsilon}_j(\tau) = a^{ij}(x^k, \tau, t)\mathbf{\varepsilon}_i(t) \otimes \mathbf{\varepsilon}_j(t)$$

we denote the co-rotational time derivative expressed in a space-fixed frame by $\mathscr{D}a^{ij}/\mathscr{D}t$. It is easily shown from (11.25) that

$$\frac{\mathscr{D}a^{ij}}{\mathscr{D}t} = \frac{\partial a^{ij}}{\partial t} + v^k a^{ij}_{,k} - w_k{}^i a^{kj} - w^j{}_k a^{ik} \tag{11.40}$$

Analogous to our earlier treatment of frames with a basis $\{\mathbf{e}^i(\tau)\}$, let us now inquire about time derivatives in a frame which at $\tau = t$ is the *reciprocal* to $\mathbf{\varepsilon}_i(\tau)$, i.e.,

$$\mathbf{\varepsilon}^j(\tau) = \frac{1}{2}[\mathbf{e}^j(\tau) + g^{jk}\mathbf{e}_k(\tau)] \tag{11.41}$$

Then one finds, when conversion to a space-fixed frame is effected, that

$$\frac{\mathscr{D}a_{ij}}{\mathscr{D}t} = \frac{\partial a_{ij}}{\partial t} + v^k a_{ij,k} - w_i{}^k a_{kj} - w_j{}^k a_{ik} \tag{11.42}$$

Thus

$$\frac{\mathscr{D}a_{ij}}{\mathscr{D}t} - g_{ik}g_{jl}\frac{\mathscr{D}a^{kl}}{\mathscr{D}t} = 0 \tag{11.43}$$

and we have found in $\mathscr{D}/\mathscr{D}t$ a time derivative which *does* commute with raising and lowering of indices. From (11.40) and (11.42) it is a simple matter to write a general expression for $\mathscr{D}a^{i_1 i_2 \cdots i_m}_{j_1 j_2 \cdots j_n}/\mathscr{D}t$. The commutative property of $\mathscr{D}/\mathscr{D}t$, which is often called the *Jaumann derivative*, is a great convenience. Furthermore, its appropriateness for constitutive equations has been indicated from the natural way in which it appears in some specific models [19].

To sum up, we see that the seemingly innocuous equation (11.11) has burgeoned into numerous possible nonlinear relations between stress and rate of deformation, all of which in the limit of infinitesimal deformation reduce to the parent linear equation. In 1958 Oldroyd suggested that any direct theoretical connection to (11.11) be abandoned, and that additional nonlinear combinations of deformation rate with itself or with the stress be included. The result is [20], when written for a Cartesian coordinate system,

$$\left(1 + \lambda_1 \frac{\mathscr{D}}{\mathscr{D}t}\right) t_{ij} + \mu_0 t_{kk} d_{ij} - \mu_1 (t_{ik}d_{kj} + t_{kj}d_{ik}) + \nu_1 t_{kl}d_{kl}\delta_{ij}$$

$$\tag{11.44}$$

$$= 2\mu\left[d_{ij} + \lambda_2 \frac{\mathscr{D}d_{ij}}{\mathscr{D}t} - 2\mu_2 d_{ik}d_{kj} + \nu_2 d_{kl}d_{kl}\delta_{ij}\right]$$

Oldroyd delineated two special cases of (11.44), which he labeled A and B.

$$\left.\begin{array}{l} \text{Liquid A: } \mu > 0; \quad \lambda_1 = -\mu_1 > \lambda_2 = -\mu_2 \geqslant 0; \quad \mu_0 = \nu_1 = \nu_2 = 0 \\ \text{Liquid B: } \mu > 0; \quad \lambda_1 = \mu_1 > \lambda_2 = \mu_2 \geqslant 0; \quad \mu_0 = \nu_1 = \nu_2 = 0 \end{array}\right\} \quad (11.45)$$

The equations obtained with liquid B predict a positive Weissenberg effect, i.e., when the liquid is sheared between coaxial cylinders the free surface near the inner cylinder rises. This is in agreement with experimental observations of rheologically complex fluids. Liquid A exhibits the reverse effect, i.e., a corresponding depression.

In Section 10.9 we discussed the connection between integral and differential models within the framework of linear viscoelasticity, and some expansions were considered in early sections of this chapter which also related differential to integral constitutive expressions. In his 1950 paper Oldroyd showed how integral expressions, such as (10.32), which take explicit account of memory effects can be properly formulated for finite deformations. The key hypothesis, as with the differential approach, is that relatively simple statements, e.g. (10.32), have some validity if written for a given material point and if, in the case of integration over past time, it is the history of a *material* point, not a spatial point that is recorded. Thus if in a body-fixed reference frame we express the stress tensor $\mathbf{T}$ by $\mathbf{T} = \pi_{ij}(\zeta^k, \tau)\mathbf{e}^i(\tau) \otimes \mathbf{e}^j(\tau)$, (10.32) written with respect to $\{\mathbf{e}^i(\tau)\}$ becomes

$$\pi_{ij}(\zeta^k, t) = \int_{-\infty}^{t} \psi(t-\tau)\frac{\partial\gamma_{ij}(\zeta^k, \tau)}{\partial\tau}\, d\tau \quad (11.46)$$

The expression $\partial\gamma_{ij}(\zeta^k, \tau)/\partial\tau$ in (11.46) has the same significance as in (11.21), and represents the rate of separation of adjacent material points noted by an observer using a distance scale in the $(\mathbf{x}, t)$ reference frame. Calling $\partial\gamma_{ij}(\zeta^k, \tau)/\partial\tau \equiv \gamma_{ij}^{(1)}$, we can write $\mathbf{D}$ with respect to two different bases by

$$\mathbf{D} = \frac{1}{2}\gamma_{ij}^{(1)}\mathbf{e}^i(\zeta^k, \tau) \otimes \mathbf{e}^j(\zeta^k, \tau) = d_{ij}(x'^k, \tau; t)\mathbf{e}^i(x'^k, t)\otimes\mathbf{e}^j(x'^k, t) \quad (11.47)$$

where, as before, we require that at $\tau = t$, $\zeta^k = x^k$. The $\{x'^i\}$ are coordinates associated with the space-fixed basis $\{\mathbf{e}_i(x'^k, t)\}$. These coordinates locate the position at time τ of that material point which is at some particular $\{x^i\} = \{\zeta^i\}$ at $\tau = t$. Thus

$$\mathbf{r} = \zeta^i\mathbf{e}_i(\zeta^k, \tau) = x'^i\mathbf{e}_i(x'^k, t)$$

The relation between these coordinates is shown pictorially in Fig. 11.1[†] Association of $\gamma_{ij}^{(1)}$ with the rate of deformation tensor is in keeping with the interpretation of (11.21) and with (11.34).

Next we apply a straightforward transformation to (11.47), from which we can write for (11.46)

$$t_{ij}(x^m, t) = 2\int_{-\infty}^{t}\psi(t-\tau)\frac{\partial x'^k}{\partial x^i}\frac{\partial x'^l}{\partial x^j}d_{kl}(x'^m, \tau)d\tau \quad (11.48)$$

since at $\tau = t$, $\pi_{ij}(\zeta^k, t) = t_{ij}(x^k, t)$. One can also write the memory integral in terms of $\gamma^{(1)ij}(\zeta^m, \tau) = \gamma_{lk}^{(1)}\gamma^{ki}\gamma^{lj}$.

[†] The coordinates $\{x'^i\}$ are identical in meaning to the $\{\xi^i\}$ used in Section 5.6 and subsequently to describe relative deformation. We have chosen to switch notation to coincide with usage of most of those references relevant to Chapter 5, which use $\{\xi^i\}$, and those pertinent to this section, which use $\{x'^i\}$.

$$\pi^{ij}(\zeta^m, t) = \int_{-\infty}^{t} \psi(t-\tau)\gamma^{(1)ij}(\zeta^m, \tau)d\tau \qquad (11.49)$$

which transforms to

$$t^{ij}(x^m, t) = 2\int_{-\infty}^{t} \psi(t-\tau)\frac{\partial x^i}{\partial x'^k}\frac{\partial x^j}{\partial x'^l}d^{kl}(x'^m, \tau)d\tau \qquad (11.50)$$

In these integrals we have suppressed the dependence of d_{kl} or d^{kl} on t. Instead, it is understood that at $\tau = t$, $x'^i = x^i$. The expressions (11.48) and (11.50) were identified as liquids A' and B', respectively, by Walters [21]. Liquid B' has been used extensively as a constitutive model equation. Walters has noted that the liquids A and B of Oldroyd (11.45) can be recovered as special cases when the following substitution is made in (11.48) and (11.50), respectively:

$$\left.\begin{array}{l} \psi(t-\tau) = \int_{0}^{\infty} \dfrac{\mathcal{N}(t')}{t'} e^{-[(t-\tau)/t']} dt' \\[20pt] \mathcal{N}(t') = \mu\dfrac{\lambda_2}{\lambda_1}\delta(t') + \mu\dfrac{(\lambda_1 - \lambda_2)}{\lambda_1}\delta(t'-\lambda_1) \end{array}\right\} \qquad (11.51)$$

where $\delta(x)$ is the Dirac delta function.

11.11. A COMPARISON OF APPROACHES

There are, of course, many ways to classify the endless variety of constitutive equations which have developed over the years. Most of the equations we have studied, as well as many which we have not, can be associated with one of three continuum approaches to constitutive theory. For convenience we call these approaches:

(i) *Coleman–Noll*. This refers to description of the motion of a simple fluid by means of a functional and characterization of the stress by the history of the deformation.

(ii) *Green–Rivlin*. Here one expresses the stress as an integral expansion of the deformation and its time derivatives over past time.

(iii) *Oldroyd*. The stress and its time derivatives are given in terms of the strain and its time derivatives. In each case "time derivative" refers to one of the variety of convected and/or co-rotational derivatives discussed in the last section.

All of these approaches assume, explicitly or implicitly, that the fluid is isotropic if it has been in a state of no flow for a sufficiently long time, and that the stress at a material point depends upon the motion in an arbitrarily small neighborhood around that point (the principle of local action). One may well expect that these three approaches, though superficially distinct, are actually equivalent. In fact, we have already noted in Section 11.2 that if the Coleman–Noll functional is sufficiently smooth so that arbitrarily high orders of Fréchet differentials can be defined, then the functional can be expanded with the result being equivalent to a Green–Rivlin formulation. An equivalence between Coleman–Noll and Oldroyd formulations has been shown to be plausible [11], but its formal demonstration is not trivial. Lodge and Stark have presented a connection [22] which is based on the premise that Oldroyd's 1950 enunciation of constitutive equations is an expression for the extra stress as a *functional* of the metric tensor, and that the stress and metric tensor are given in convected components.

Although no thorough discussion of either exception is included in this book, it should be emphasized that the two assumptions of local action and rest-state isotropy do isolate our treatment from a number of material models. For example, one can imagine a suspension in which neighboring suspension particles exert a mutual influence on each other, thus transcending the limitation of local action. It is also realistic to imagine fluid models which possess inherent anisotropy, and that this anisotropy must be accounted for by explicit inclusion of a description of orientation into the constitutive model. For representative expositions of these types of constitutive laws one can consult the work of Ericksen [23], Eringen [24], Allen [25] and their coworkers.

11.12. THE BKZ MODEL

An additional approach to constitutive behavior was published in 1963 by Bernstein, Kearsley, and Zapas [26], and evidently a similar procedure was developed by Kaye [27, 28] slightly earlier. Subsequent refinement and application of the BKZ (or KBKZ) model, as it is usually called, has led to considerable success in prediction of rheological behavior of polymer solutions. We present a brief qualitative description.

A central feature of elasticity theory is the energy associated with deformation. In fact one of the methods for arriving at a form for constitutive equations is to consider the energy (the strain energy function) which is stored in a material during a deformation, to relate this function to invariants of some measure of strain, and to connect the energy to the stress by considering a virtual deformation [29]. Application of this procedure to the theory of rubber elasticity is discussed in a readable fashion by Treloar [30, p. 156], and we have briefly referred to a form of this approach in Section 10.13. One way to develop the BKZ model is to consider the history of the Cauchy–Green strain measure $\mathbf{C}_{(t)}$ and its inverse $\mathbf{C}_{(t)}^{-1}$ when they are weighted by terms involving derivatives of a strain energy function with respect to the invariants of $\mathbf{C}_{(t)}^{-1}$. The result can be expressed in component form for a given material point by

$$t_{ij} = \int_0^\infty [m_1(s, I_1(\tau), I_2(\tau))(c_{(t)}^{-1})_{ij} + m_2(s, I_1(\tau), I_2(\tau))c_{(t)ij}]ds \qquad (11.52)$$

where $s = t - \tau$

$$(c_{(t)}^{-1})^{ij} = g^{\alpha\beta} \frac{\partial x^i}{\partial \xi^\alpha} \frac{\partial x^j}{\partial \xi^\beta}; \quad c_{(t)ij} = \frac{\partial \xi^\alpha}{\partial x^i} \frac{\partial \xi^j}{\partial x^\beta} g_{\alpha\beta}$$

and I_1 and I_2 are the first two invariants of $\mathbf{C}_{(t)}^{-1}$. A distinguishing feature of (11.52) is the fact that it is nonlinear and, furthermore, is nonlinear with respect to strain rather than strain rate.

A number of particular forms for m_1 and m_2 have been suggested. For specific details the reader is referred to work of Bogue and coworkers [28, 31].

11.13. A RETURN TO MOLECULAR MODELS

The so-called "molecular models", in which category we include the bead–spring and network models discussed in Chapter 10, have been useful guides for formulation of constitutive equations, and in this section we return to several representative variations of them. Before doing so, it is important to recall the following facts: any molecular model which reduces to a form simple enough to be useful for solutions to flow problems is bound to be an unrealistic picture of true molecular behavior. Also, any truly continuum formulation will lead to phenomenological coefficients, the numerical values of which are unknown. Thus one must resort to empirical evaluation and adjustment of basic models, assuming that the goal is a constitutive equation with some degree of predictive power for certain classes of real materials. If the equation is based on a molecular model, the result should also be helpful in associating flow properties with molecular parameters such as size, shape, and extensibility. In this section we discuss in turn several models which have origins in a molecular theory but which have been modified to a greater or lesser extent to permit better agreement with known experimental results. Since in most cases the final equation has only a tenuous connection with the initial molecular model, the reader is spared the substantial journey required to derive each specific model in detail. As with the last section, we give only a qualitative summary in each case. Those interested in more detail can refer to the cited references.

I. THE NETWORK MODEL

We have already noted that the kinetic theory of rubber elasticity has been a helpful guide for description of polymer solutions. In particular it appears that Yamamoto [32–34] was the first to modify the description of Green and Tobolsky [35] for rubber crosslinks to correspond to concentrated polymer solutions. The physical picture used by Yamamoto was that of a strongly interacting system of polymer solute molecules in which inter-molecular junctions were constantly being formed and broken. Yamamoto's final expression, which results from a "material balance" of network junctions, is a nonlinear theory in that the possibility of shear dependent viscosity is admitted, but the results are in too general a form to be used without additional specialization. Such specialization was given independently by Lodge [36] in his network theory, to which we have referred in Chapter 10, and a brief description of which is provided below. We draw heavily on Lodge's general comparison of network theories [37].

Consider a solution of macromolecules in which each macromolecule, at any instant of time, can be divided into subunits between junction points (or entanglement points) with other molecules. Taking one of these points as an origin one writes an expression for the probability that the next junction point will be a distance r away. This probability P is taken to have the form

$$P \sim \exp\left[\frac{-3r^2}{2nl^2}\right]$$

Here n is the number of freely jointed links, each of length l, in the subunit and can correspond to a measure of monomer units. n is assumed large and the argument of the exponent is assumed small compared to unity. This will be recognized (see Section 10A1) as a Gaussian probability. Though not necessary for all network formulations, the Gaussian assumption is fundamental to the simplest network theories and will be employed here.

In Section 10A1 we related a configuration probability function to Helmholtz free energy by use of the Boltzmann equation for entropy. We do the same thing here. From $S = k \ln P$ we write, to within a constant,

$$S_n = -k \left[\frac{3r^2}{2nl^2} \right]$$

Then if in some macroscopically small element of the material there are N subunits and if the system is isothermal, we can write for the Helmholtz free energy

$$A = kT \sum^{N} \left[\frac{3r^2}{2nl^2} \right] + \text{const}$$

This can be expressed as an integral by writing the result in terms of the concentration of junctions with separation distance r. It is convenient to formulate the expression in terms of the body-fixed coordinates $\{\zeta^i\}$ used in Section 11.7. If we use a distribution function $F(\zeta^k, n, t)d\zeta^\dagger$ to represent the concentration of subunits with n freely jointed links and employ (11.20), one can write

$$A(t) = kT \sum_n \frac{3}{2nl^2} \int_{-\infty}^{\infty} F(\zeta^k, n, t)\gamma_{ij}\zeta^i\zeta^j d\zeta + \text{const} \qquad (11.53)$$

The integrand is assumed to contribute only over distances r which are small with respect to a length scale characterizing appreciable change of any macroscopic variable. Also, by equating the distance between junctions with the distance between material points we have tacitly assumed that the junctions move with the material as it is deformed. This is known as the "affine deformation" assumption [37]. The major task is to find a suitable expression for F, and to do so we need additional assumptions about the physics of junction formation. We first write a junction material balance by assuming that the net change is governed by a characteristic time τ_c. Then if $\psi(\zeta^k, n, \tau; t)d\zeta d\tau$ is the concentration at time t of subunits which were formed between τ and $\tau + d\tau$ and are characterized by $\{\zeta^i\}$ and n, we write

$$\left(\frac{\partial}{\partial t} + \frac{1}{\tau_c} \right) \psi(\zeta^k, n, \tau; t) = 0 \qquad (11.54)$$

so that

$$\psi(\zeta^k, n, \tau; t) = \psi(\zeta^k, n, \tau; \tau) \exp\left(\frac{\tau - t}{\tau_c} \right) \qquad (11.55)$$

The pre-exponential factor is assumed, following precedent from rubber elasticity theory, to be itself a Gaussian distribution function of the form

$$\exp\left[-k_n \gamma_{ij}(\tau)\zeta^i\zeta^j \right]$$

where k_n is a constant. Then

† $d\zeta$ is a shorthand for the element of volume included between ζ^1, ζ^2, ζ^3, and $\zeta^1 + d\zeta^1$, $\zeta^2 + d\zeta^2$, and $\zeta^3 + d\zeta^3$.

$$F = \int_{-\infty}^{t} \psi d\tau \tag{11.56}$$

Substitution of (11.55) into (11.56) and subsequent substitution of F into (11.53) yields

$$A(t) = \frac{1}{2} kT \sum_{n} c_n \int_{-\infty}^{t} \exp\left(\frac{\tau - t}{\tau_c}\right) \gamma_{ij}(t)\gamma^{ij}(\tau)d\tau + \text{const} \tag{11.57}$$

where the c_n are constants. The result is readily extended to the case where τ_c depends on n [37].

The final step is to relate $A(t)$ to the stress by considering an isothermal reversible change of shape during which the material changes its free energy from A to $A+dA$. For example, consider a stress $\mathbf{T}$ acting on a volume element of the material. We equate the change in free energy to the work done per unit volume by the stress as the material undergoes an infinitesimal deformation. It is convenient to form the volume element from a set of body-fixed base vectors $\{\mathbf{e}_i(\tau)\}$. We shall take these vectors to be small enough in magnitude so that the volume formed from them undergoes a spatially homogeneous deformation under a homogeneous stress. Then the deformation is described by the change from one parallelepiped to another, as shown in Fig. 11.2. Consider the work $(dW)_1$ done on the faces with unit normals $\pm\mathbf{n}_1$.

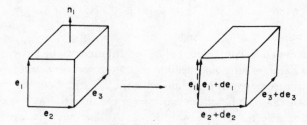

Fig. 11.2. Infinitesimal deformation of a volume element.

$$(dW)_1 = \mathbf{f}_1 \cdot d\mathbf{e}_1 S_1$$

where $\mathbf{f}_1$ is the force acting per unit area on face 1, which has area S_1. From (4.23) we can write, in a body-fixed system,

$$\mathbf{f}_1 = \pi^{j1}(\mathbf{e}_j \otimes \mathbf{e}_l)\,\mathbf{n}_1$$

From Fig. 11.2 it is clear that $\mathbf{n}_1$ is in the direction of $\mathbf{e}_2 \times \mathbf{e}_3$. In fact, from (11.15) we can write

$$\mathbf{n}_1 = \frac{\mathbf{e}^1}{|\mathbf{e}^1|} = \frac{V\mathbf{e}^1}{S_1}$$

where V is the volume of the element. Then the work done per unit volume on faces ± 1 is

$$(dW)_1 = \sum_j \pi^{j1}\mathbf{e}_j \cdot d\mathbf{e}_1$$

The change in free energy is then the sum of similar expressions over the three pairs of faces. We obtain

$$dA = \pi^{ji}\mathbf{e}_j \cdot d\mathbf{e}_i \quad \text{(summation convention in effect)} \tag{11.58}$$

Since the stress is symmetric (11.58) can be rewritten

$$dA = \frac{1}{2}\pi^{ij}(\mathbf{e}_i \cdot d\mathbf{e}_j + \mathbf{e}_j \cdot d\mathbf{e}_i)$$

Finally, using $\gamma_{ij} = \mathbf{e}_i \cdot \mathbf{e}_j$,

$$dA = \frac{1}{2}\pi^{ij}d\gamma_{ij} \tag{11.59}$$

Comparing (11.57) with (11.59),

$$\pi^{ij}(\zeta^k, t) = kT\int_{-\infty}^{t} M(t-\tau)\gamma^{ij}(\tau)d\tau \tag{11.60}$$

where

$$M = \sum_n c_n \exp\left(\frac{\tau-t}{\tau_c}\right)$$

Analogous to the transformation of (11.49) to (11.50) we can express (11.60) in space-fixed coordinates, the result being

$$t^{ij}(x^m, t) = kT\int_0^{\infty} M(s) \frac{\partial x^i}{\partial x'^k} \frac{\partial x^j}{\partial x'^l} g^{kl}(x'^m) \, ds \tag{11.61}$$

where $s = t - \tau$. It is interesting to note that, though both (11.49) and (11.60) imply memory, the former is a summation of *rates* of deformation over past times while the latter is a summation of deformations.

II. THE WISCONSIN IDEAS

An appreciation for the effort and care required to formulate constitutive equations which contain some features of molecular respectability, observer invariance, and experimental reality can be obtained by following the evolution of several generations of equations that have emanated from the students of Professor R. B. Bird at The University of Wisconsin. Each successive equation, as it was published, offered better agreement than its predecessor with existing data. Yet, as more sophisticated techniques for experimental testing have become available, each equation has in turn been shown to have serious shortcomings. As an example, we present here a model which contains an admirable degree of flexibility but which is still far from a universally applicable description of flow behavior of polymer solutions.

The Bird–Carreau model was motivated by a desire to have a nonlinear integral model which in special cases could be reduced to relatively simple molecular models and which, most importantly, could be used to fit a wide variety of rheological data. Bird and Carreau proposed the equation [38]

$$\mathbf{T}(t) = \sum_{p=1}^{\infty} \int_{-\infty}^{t} \frac{\eta_p \exp-[(t-\tau)/\lambda_{2p}]}{(\lambda_{2p})^2[1-4(\lambda_{1p})^2 I_2(\tau)]}\left[\left(1 + \frac{\varepsilon}{2}\right)\bar{\mathbf{\Gamma}} - \frac{\varepsilon}{2}\mathbf{\Gamma}\right]d\tau \qquad (11.62)$$

where $\bar{\Gamma}^{ij} = \dfrac{\partial x^i}{\partial x'^k}\dfrac{\partial x^j}{\partial x'^l}g^{kl}(x'^n) - g^{ij}(x^n)$

$$\Gamma^{ij} = g^{ij}(x^p) - \frac{\partial x'^k}{\partial x^m}\frac{\partial x'^l}{\partial x^n}g_{kl}(x'^p)g^{im}(x^p)g^{jn}(x^p)$$

$I_2(x'^k)$ = second invariant (3.94) of $\mathbf{D}(x'^k)$, the rate of deformation tensor

$\eta_p, \lambda_{1p}, \lambda_{2p}$ = constants

ε = constant

From an analogy to the Rouse theory, Bird and Carreau suggested

$$\eta_p = \frac{\eta_0 \lambda_{1p}}{\sum\limits_{p=1}^{\infty}\lambda_{1p}}; \quad \lambda_{1p} = \lambda_1\left(\frac{1+n_1}{p+n_1}\right)^{\alpha_1}; \quad \lambda_{2p} = \lambda_2\left(\frac{1+n_2}{p+n_2}\right)^{\alpha_2} \qquad (11.63)$$

Then η_0 corresponds to a zero-shear viscosity, and choices of the various constants in (11.63) can be made which cause (11.62) to reduce to one or another of earlier constitutive models. For example:

(a) Rouse model [39] for small amplitude oscillatory shear flow:

$n_1 = n_2 = 0; \quad \lambda_1 = \lambda_2 = \lambda; \quad \alpha_1 = \alpha_2 = 2$

(b) For a different choice of constants Bird and Carreau indicated that (11.62) reduces to the Spriggs model [40] for steady or small amplitude oscillatory shear flow.

The constant ε is a measure of the second normal stress difference N_2 in steady shear flow (see (7.33)).

This model has been quite successful for fitting the viscometric functions in simple shearing flow and for characterizing results of small-amplitude oscillatory experiments. However, as more severe time-dependent experiments such as stress growth and decay have become available, limitations of (11.62) have become increasingly apparent. Carreau, arguing on the basis of a network theory, has suggested an improvement which amounts to a modification of the relaxation time in the exponential by making it a function of $I_2(x'^k)$ and weighting it over past times. Given the proper choice of dependence, an improved fit to transient data can be obtained [41].

III. THE NETWORK RUPTURE THEORY

The Lodge formulation (11.61) of a network model is based upon a Gaussian distribution of subunit lengths and hence is a linear theory in the sense that there is a linear relation between stress and the history of the deformation expressed in body-fixed coordinates. (We note again that (11.61) is not, however, an infinitesimal theory as discussed in some sections of Chapter 10.) One would certainly expect that the assumption of a Gaussian distribution of subunit lengths would become increasingly unrealistic as the shear rate, and the deformation, of the system increased. By accounting for non-Gaussian statistics [30, pp. 99 *et seq.*] one can introduce nonlinearity into the model and hence obtain, among other things, a shear dependent viscosity [37]. Another way to introduce nonlinearity is to postulate that there is a strain limit beyond which the network junction must rupture. This concept has been made precise by Tanner and coworkers [42–44] through introduction of a cutoff time in the lower limit of integrals such as (11.61) or (11.62). The cutoff time can be made a function of, for example, the first invariant (i.e., the trace) of some measure of the strain. An alternative would be to place a step function in the kernel to represent the junction rupture limit. The network rupture model quantifies what Tanner has coined the "straingth" of the system.

IV. THE DUMBBELL MODEL

The reader may have noted a general trend from the preceding discussion. As models have been adjusted to obtain increasing agreement with experimental realities they have also become farther removed from the rigorous theoretical foundations of the model. The pragmatist can in fact justifiably ask if a result such as (11.62) might not best be found through intelligent curve fitting without bothering to force a molecular connection. In a reversal to this trend Bird and coworkers have conducted a vigorous study of the properties of models of dilute polymer solutions in which the macromolecules are modeled by two masses joined by connectors with increasingly exotic properties [45]. These dumbbell models, though unrealistic for quantitative prediction of rheological behavior, have shown a remarkable capacity for qualitative indication of response. Most important, since empirical adjustments of the basic model are not introduced, the effects can be interpreted directly in terms of model parameters. Although one may argue over the molecular significance of these parameters, their dumbbell significance is unmistakable.

The uninitiated may suspect that the simplicity of a dumbbell configuration would provide for relatively simple mathematics. Such is not necessarily the case, however, primarily because of the need for a distribution function to describe dumbbell orientations and the fact that in most cases of interest Brownian motion must be considered. A thorough review of this subject through 1970 has been provided by Bird and coworkers [46].

11.14. CONCLUDING REMARKS

The review of constitutive possibilities given in this chapter has had several purposes. These include an exposition of the following:

(1) The power of continuum formulations which put to use the ideas of material invariance.

(2) The limitations of (1) when one wishes to proceed from constitutive *form* to constitutive equation.

(3) The utility of simple molecular concepts, even if conceived in a one-dimensional and improperly invariant fashion, as a starting point for more general formulations.

(4) The opinion, not heretofore expressed but certainly implicit throughout, that a search for constitutive equations which fit ever more demanding data is ultimately self-defeating. The end result just reintroduces the unwieldiness for which general continuum equations are criticized. Thermodynamicists learned long ago that before one reaches for an equation of state, one should know whether the goal is to design a high-pressure vessel for a gas composed of "awkward" polar molecules,

or to associate the existence of, say, certain critical phenomena with crude measures of molecular properties. Rheologists are not always as clear in formulating the purpose for which a constitutive equation is being chosen.

The present chapter has been illustrative rather than inclusive. Variants of constitutive equations constantly appear, and no attempt has been made here to be encyclopedic, though each of the major categories from which constitutive equations evolve has been discussed. For a balanced review to 1970 the reader is referred to Bogue and White [31].

Though the Bird–Carreau model is of the integral type, most of its antecedents were in differential form [38]. Differential models, evolving from Oldroyd's formulation (Section 11.9) can be very difficult to apply to the momentum equation (4.26) because they are not explicit in the stress. In addition they seem susceptible to more numerical difficulties than their integral counterparts. For example, Tanner and Simmons [47] have documented several cases of numerical instability which were peculiar to differential formulations.

Finally, we remark on the role of strain and strain-rate (or deformation-rate) history. Both seem to be important, and that fact has been utilized in recent formulations [31].

REFERENCES

1. Barnes, H. A., Walters, K., and Townsend, P., *Nature* **224**, 585 (1969).
2. Green, A. E., Rivlin, R. S., *Arch. Rational Mech. Anal.* **1**, 1 (1957/8).
3. Green, A. E., Rivlin, R. S., and Spencer, A. J. M., *Arch. Rational Mech. Anal.* **3**, 82 (1959).
4. Green, A. E., and Rivlin, R. S., *Arch. Rational Mech. Anal.* **4**, 387 (1959/60).
5. Noll, W., *Arch. Rational Mech. Anal.* **2**, 197 (1958/9).
6. Volterra, V., *Theory of Functionals and of Integral and Integro-Differential Equations* (New York: Dover Publications, 1959).
7. Truesdell, C., and Noll, W., The non-linear field theories of mechanics, in *Handbuch der Physik*, Band III/3 (S. Flügge, ed.), (Berlin : Springer-Verlag, 1965).
8. Sokolnikoff, I. S., *Mathematical Theory of Elasticity*, 2nd edn. (New York: McGraw-Hill, 1956).
9. Coleman, B. D., and Noll, W., *Rev. Mod. Phys.* **33**, 239 (1961).
10. Coleman, B. D., and Noll, W., *Arch. Rational Mech. Anal.* **6**, 355 (1960).
11. Walters, K., *Z. Math Physik* **21**, 592 (1970).
12. Oldroyd, J. G., *Proc. Roy. Soc. (London)*, A **200**, 523 (1950).
13. Fröhlich, H., and Sack, R., *Proc. Roy. Soc. (London)*, A, **185**, 415 (1946).
14. Einstein, A., *Ann. Physik* **19** 289 (1906); **34**, 591 (1911).
15. Reed, X. B., PhD thesis, University of Minnesota, 1965.
16. Sokolnikoff, I. S., *Tensor Analysis*, 2nd edn. (New York: Wiley, 1964).
17. Frankel, N. A., and Acrivos, A., *J. Fluid Mech.* **44**, 65 (1970).
18. Lodge, A. S., *Elastic Liquids* (New York: Academic Press, 1964).
19. Leal, L. G., and Hinch, E. J., *J. Fluid Mech.* **55**, 745 (1972).
20. Oldroyd, J. G., *Proc. Roy. Soc. (London)*, A **245**, 278 (1958).
21. Walters, K., *Quart. J. Mech. Appl. Math.* **15**, 63 (1962).
22. Lodge, A. S., and Stark, J. H., *Rheol. Acta* **11**, 119 (1972).
23. Ericksen, J. L., *Koll. Zeit.* **173**, 117 (1960).
24. Eringen, A. C., *J. Math. Mech.* **16**, 1 (1966).
25. Allen, S. J., and DeSilva, C. N., *J. Fluid Mech.* **24**, 801 (1966).
26. Bernstein, B., Kearsley, E. A., and Zapas, L. J., *Trans. Soc. Rheol.* **7**, 391 (1963).
27. Tanner, R. I., *Trans. Soc. Rheol.* **12**, 155 (1968).
28. Chen, I-J., and Bogue, D. C., *Trans. Soc. Rheol.* **16**, 59 (1972).
29. Rivlin, R. S., Large elastic deformations, in *Rheology*, vol. 1, p. 351 (F. R. Eirich, ed.), (New York: Academic Press, 1956).
30. Treloar, L. R. G., *The Physics of Rubber Elasticity*, 2nd edn. (London: Oxford University Press, 1958).
31. Bogue, D. C., and White, J. L., Engineering analysis of non-Newtonian fluids, NATO-AGARDograph No. 144, 1970.
32. Yamamoto, M., *J. Phys. Soc. Japan* **11**, 413 (1956).

33. Yamamoto, M., *J. Phys. Soc. Japan* **12**, 1148 (1957).
34. Yamamoto, M., *J. Phys. Soc. Japan* **13**, 1200 (1958).
35. Green, M. S., and Tobolsky, A. V., *J. Chem. Phys.* **14**, 80 (1946).
36. Lodge, A. S., *Trans. Faraday Soc.* **52**, 120 (1956).
37. Lodge, A. S., *Rheol. Acta* **7**, 379 (1968).
38. Bird, R. B., and Carreau, P. J., *Chem. Eng. Sci.* **23**, 427 (1968).
39. Rouse, P. E., Jr., *J. Chem. Phys.* **21**, 1272 (1953).
40. Spriggs, T. W., *Chem. Eng. Sci.* **20**, 931 (1965).
41. Carreau, P. J., *Trans. Soc. Rheol.* **16**, 99 (1972).
42. Tanner, R. I., and Simmons, J. M., *Chem. Eng. Sci.* **22**, 1803 (1967).
43. Tanner, R. I., *Trans. Soc. Rheol.* **12**, 155 (1968).
44. Tanner, R. I., *AIChE Jl.* **15**, 177 (1969).
45. Warner, H. R., *Ind. Eng. Chem. Fundamentals* **11**, 379 (1972).
46. Bird, R. B., Warner, H. R., and Evans, D. C., *Fortschritte der Hochpolymerenforschung* **8**, 1 (1971).
47. Tanner, R. I., and Simmons, J. M., *Chem. Eng. Sci.* **22**, 1079 (1967).

PROBLEMS

11.1. The nonuniqueness of the Oldroyd derivative $\mathfrak{d}/\mathfrak{d}t$, discussed in Section 11.10, is readily seen when one attempts to generalize the Fröhlich and Sack equation (11.11). In that equation, replace $\partial/\partial t$ by $\mathfrak{d}/\mathfrak{d}t$ and apply the equation to stress and rate-of-deformation tensors written, successively, as covariant, contravariant, and mixed tensor components. Apply these equations to steady, fully developed laminar flow through circular tubes. For each case, answer the following questions:

(a) Does $\sum_i t(ii) = 0$?

(b) Is $N_2 = t(rr) - t(\theta\theta)$ zero?
(c) Is the effective viscosity, $\tau(\kappa)/\kappa$, independent of κ?

11.2. In a rectangular Cartesian coordinate system (x^1, x^2, x^3), consider plane Couette flow, $\mathbf{v} = (\kappa x^2, 0, 0)$. Compare predictions for the two Oldroyd liquids A and B of (11.44) and (11.45). Specifically:

(a) Is $\tau(\kappa)/\kappa$ independent of κ?
(b) To the extent possible, say whether N_1 and N_2, the first and second normal stress differences, are greater than, less than, or equal to zero.

11.3. Consider the classical "start-up" problem for flow in a circular tube, where at $t < 0$ the fluid is at rest, and for $t \geqslant 0$ a constant pressure gradient $-\partial p/\partial z = K$ is imposed. Suppose that the fluid is in laminar flow and is characterized by the covariant form of (11.11). Show that $t(r\theta) = t(\theta z) = t(\theta\theta) = t(zz) = 0$, and that the equations for the velocity and shear stress $t(rz)$ constitute a linear system of equations.

11.4. Ericksen [23] has developed a theory of "transversely isotropic fluids". For such a fluid the stress at any instant of time is determined by the velocity gradient $\nabla\mathbf{v}$ and some preferred direction $\mathbf{n}$, with no distinction being given between $\mathbf{n}$ and $-\mathbf{n}$. Restrictions of material objectivity, addition of a dynamic equation for $\mathbf{n}$, and linearization of equations with respect to the velocity gradient, lead to the constitutive equations, in Cartesian notation,

$$t_{ij} = 2\mu d_{ij} + (\mu_1 + \mu_2 d_{\alpha\beta} n_\alpha n_\beta) n_i n_j + 2\mu_3 (d_{j\alpha} n_\alpha n_i + d_{i\alpha} n_\alpha n_j)$$

$$\frac{Dn_i}{Dt} - w_{i\alpha} n_\alpha = \lambda (d_{i\alpha} n_\alpha - d_{\alpha\beta} n_\alpha n_\beta n_i)$$

$$\text{where } d_{ij} = \frac{1}{2}(v_{i,j} + v_{j,i}); \quad w_{ij} = \frac{1}{2}(v_{i,j} - v_{j,i})$$

and λ, μ, and μ_i are constants.

Write out the equations for t_{ij} and Dn_i/Dt for plane Couette flow, $\mathbf{v} = (\kappa x_2, 0, 0)$. If $\lambda \geqslant 1$ and one requires a time-independent solution as $t \to \infty$, derive expressions for the t_{ij} in terms of the constitutive constants and κ. Note that the results are included in the theory of simple fluids.

CHAPTER 12

NONVISCOMETRIC FLOWS REVISITED: APPLICATIONS

12.1. INTRODUCTION

In our study of flow problems in Chapters 8 and 9 we have in a sense defined the flow in such a way that the tools at hand were sufficient for at least an approximate, if not an exact, solution. Thus we introduced the simple fluid and then proceeded with an exhaustive treatment of viscometric flows, in part because of the elegance with which the simple fluid permits such a treatment. There was a similar motivation for steady extension in Chapter 9. We now become more realistic, from the standpoint of the engineer, and invert the procedure. A flow of some practical interest is posited and one asks whether means are available for describing the flow sufficiently well so that some degree of predictive value is present. The reader is warned at the outset not to expect clear, concise, and reliable procedures for attacking wide varieties of flows of engineering interest. In the opinion of this writer such recipes simply do not exist for non-Newtonian fluids. It will be seen that in many cases our ability to put the fundamental theory of earlier chapters to work is woefully weak. Nevertheless, these fundamentals have helped significantly to permit orderly evaluation of flow problems and to provide guidelines for treatment of complex flows. The subject of this chapter is still evolving. It is certain that in future years we shall become more clever with our adaptations of fundamental theory to engineering practice. For that reason it is not felt obligatory that each fundamental principle enunciated earlier justify its right to the printed page by showing up as a useful tool in current applications.

As with the treatment of constitutive equations in Chapters 10 and 11, an attempt has been made to be representative rather than encyclopedic. Since the subject of boundary-layer theory has been of such central importance in Newtonian fluid dynamics, we discuss its evolution and application to non-Newtonian fluids. The concept of a Deborah number and solid- vs. liquid-like behavior will be discussed and then applied to several flow situations. The phenomenon of drag reduction in turbulent flow and the subject of flow through porous materials are both important and fascinating enough to demand attention. We also include an overview of the current status of hydrodynamic stability and of some other unsteady flows in time and/or space.

12.2. BOUNDARY-LAYER THEORY FOR INELASTIC FLUIDS

(i) GENERAL DEVELOPMENT

To offer some perspective, we recall briefly the plight of fluid mechanics at the beginning of this century. Techniques for solving the equation of motion for an inviscid fluid were rather highly developed. Also, when one could neglect the effect of inertia and thereby reduce description of the motion to a linear equation, several solutions were known. However, a major dilemma was present in the startling discrepancy between predictions of drag, as computed from the inviscid equations, and experimental results with fluids, such as air, of low but nonzero viscosity. Resolution of the paradox was possible because of the realization by Prandtl that no matter how small the viscosity, its influence must be acknowledged in a region near the surface of a body at which the velocity goes to zero. We are now well aware of the role of Reynolds number in determining the boundary-layer thickness, of Prandtl's order-of-magnitude analysis to arrive at the boundary-layer equations, and of the essential role that boundary-layer theory has played in providing meaningful solutions to complex flow problems of engineering interest [1]. More recently our knowledge of the mathematical structure of boundary-layer theory has improved, and we now recognize Prandtl's development as the first approximation, to $O(1/\sqrt{N_{\mathrm{Re}}})$, where N_{Re} is the Reynolds number, in a matched asymptotic expansion [2,3].

Our ability to formulate and even to solve problems of a boundary-layer nature with non-Newtonian fluids is reasonably satisfactory if one can describe the fluid behavior in terms of a constitutive equation for a viscous "generalized Newtonian fluid" (Section 10.3). To incorporate elastic effects we shall need some additional material which is to be discussed in the next section. Hence we defer our treatment of viscoelastic boundary layers.

Although boundary-layer problems are definitely relevant to non-Newtonian fluid mechanics, they are less pervasive than one might expect. This is simply because non-Newtonian materials often exhibit very high viscosities and hence the effective Reynolds number is frequently low.

The first quantitative treatment of non-Newtonian boundary layers is that of Oldroyd, who considered several applications of the boundary-layer equations for a Bingham plastic [4]. Later developments include the description by Acrivos and coworkers of flow past a flat plate [5] and Schowalter's development of similarity solutions [6]. Both of these analyses are for power-law fluids.

Even if one restricts the discussion to steady, laminar, two dimensional boundary-layer flows of power-law fluids, as we do here, an order-of-magnitude analysis along the lines of Prandtl is not exactly parallel to Newtonian fluids because of the fact that the effect of flow on the effective viscosity enters through the second invariant of the rate of deformation tensor, and the flow in the boundary layer is two-dimensional. Fortunately one can, consistent with the boundary-layer assumptions, circumvent this difficulty. Thus we begin with the usual steady two dimensional equations of motion, in the absence of significant gravitational effects, and the continuity equation

$$u \frac{\partial u}{\partial x} + v \frac{\partial u}{\partial y} = -\frac{1}{\rho} \frac{\partial p}{\partial x} + \nabla \cdot (\mathbf{T} \cdot \mathbf{i}) \qquad (12.1)$$

$$u\frac{\partial v}{\partial x} + v\frac{\partial v}{\partial y} = -\frac{1}{\rho}\frac{\partial p}{\partial y} + \nabla\cdot(\mathbf{T}\cdot\mathbf{j})$$

$$\frac{\partial u}{\partial x} + \frac{\partial v}{\partial y} = 0$$

(12.1)

To these we add the constitutive equation for a power-law fluid

$$\mathbf{T} = K(-4I_2)^{(n-1)/2}\,(\nabla\mathbf{v}+\nabla\mathbf{v})^T]$$ (10.10)

where

$$\bar{I}_2 = -\frac{1}{8}\,\mathrm{tr}\{[\nabla\mathbf{v}+(\nabla\mathbf{v})^T]^2\}$$

We assume that far from the surface of the body in question the flow in the x-direction is described by the equation of motion for an inviscid fluid. Application of well-known arguments [1, p. 120] leads to the conclusion that the streamwise pressure gradient in the boundary layer can be found from solution of the inviscid equation of motion along the body surface. Thus if $U(x)$ denotes the x-component of velocity found from the inviscid equation of motion for $x \geqslant 0$, $y=0$, we have

$$U\frac{dU}{dx} = -\frac{1}{\rho}\frac{dp}{dx}$$ (12.2)[†]

with respect to the coordinate system shown in Fig. 12.1. Since the boundary layer is

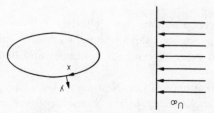

Fig. 12.1. Coordinates for two-dimensional boundary layer.

assumed thin relative to the radius of curvature of the body, the coordinate system of Fig. 12.1 can be treated as if it were rectangular Cartesian. Reduction to dimensionless form and application of an order-of-magnitude argument lead to a natural formulation of the appropriate non-Newtonian Reynolds number $\mathcal{N}_{\mathrm{Re}}$ and the final form of the two-dimensional boundary-layer equation [6].

$$\beta[1-(F')^2] + \alpha FF'' + \frac{d}{d\eta}\,\{[(F'')^2]^{(n-1)/2}\,F''\} = 0$$ (12.3)

Several new variables have been introduced. Lengths x°, etc., are scaled to a characteristic length L and velocities U°, etc., to the characteristic velocity U_∞. Then,

---

[†] Slattery [7] has noted that the assumption of an irrotational flow external to the boundary layer is less direct for non-Newtonian fluids than for their Newtonian counterpart.

$$\eta = \frac{y(\mathcal{N}_{\text{Re}})^N}{Lg(x)}; \quad F'(\eta) = u(x,y)/U(x)$$

$$\mathcal{N}_{\text{Re}} = L^n \rho U_\infty^{2-n}/K; \quad \alpha = \frac{g^n}{(U^\circ)^{n-1}} \frac{d(U^\circ g)}{dx^\circ}$$

$$N = 1/(n+1); \quad \beta = \frac{g^{n+1}}{(U^\circ)^{n-1}} \frac{dU^\circ}{dx^\circ}$$

Equation (12.3) is valid only if the dimensionless velocity $u(x,y)/U(x)$ can be expressed as a function of η alone. The as yet undetermined function $g(x)$ is chosen to convert (12.3), if possible, into an ordinary differential equation. Note the similarity in form of (12.3) to the corresponding equation (9.15) for Newtonian fluids in Schlichting [1, p. 139].

It is not difficult to show that (12.3) can be reduced to an ordinary differential equation in the independent variable η whenever the flow external to the boundary layer is of the form

$$U^\circ = a(x^\circ)^c \tag{12.4}$$

where a is a constant, and one sets

$$g(x) = \left[a^{n-2}(x^\circ)^{c(n-2)+1} \frac{n+1}{1+c(2n-1)} \right]^{1/(n+1)} \tag{12.5}$$

Those familiar with boundary layers for Newtonian fluids will recognize (12.4) as the solution for potential flow past a wedge with included angle $\beta\pi$ (see Fig. 12.2), where $\beta = 2c/(c+1)$. There are other pathological cases for which a similarity transformation to an ordinary differential equation is possible.

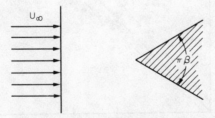

Fig. 12.2. Boundary-layer flow past a wedge.

(ii) THE FLAT PLATE

The first application of non-Newtonian boundary-layer theory for power-law fluids was, naturally enough, the solution for flow past a flat plate [5]. According to the definitions given above for α, β, and $g(x)$, we obtain

$$nF''' + F(F'')^{2-n} = 0 \tag{12.6}$$

with boundary conditions

$$F(0) = F'(0) = 0; \quad F'(\infty) = 1 \tag{12.7}$$

Solutions for velocity profiles and drag coefficients have been tabulated by Acrivos and coworkers over a range of n between 0.1 and 5.0 [5].

(iii) HEAT TRANSFER FROM CYLINDERS

An example of boundary-layer flow past curved bluff objects is the treatment by Shah *et al.* [8] for flow past a cylinder which is submerged in a uniform stream. The work is especially noteworthy in that it represents one of the few reports of local heat transfer coefficients for external flows of non-Newtonian fluids. As would be expected from our knowledge of Newtonian boundary-layer analyses, the theory becomes increasingly deficient as the separation point is reached.

Shah *et al.* performed experiments by inserting a heated cylinder in dilute (0.09% to 0.35%) aqueous solutions of carboxymethylcellulose (CMC) which was recirculating in a water tunnel. Considering the many possibilities for discrepancy between experimental conditions and theoretical assumptions (power-law fluid, standard boundary-layer assumptions, no polymer degradation, no temperature dependence of physical properties), the agreement between theory and experiment is remarkable. Representative results are shown in Fig. 12.3. The authors found that curves of $\mathcal{N}_{\mathrm{Nu}}\mathcal{N}_{\mathrm{Pe}}$ vs. $\mathcal{N}_{\mathrm{Re}}$ showed little de-

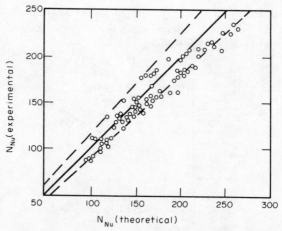

Fig. 12.3. Heat transfer from a cylinder to flowing solutions of carboxymethylcellulose. Comparison of theoretical and experimental results [8]. Dashed lines indicate $\pm 10\%$ deviation.

parture from the Newtonian curve for $n > 0.8$. Here $\mathcal{N}_{\mathrm{Pe}}$ is the usual Peclet number, $c\rho U_\infty L/k$, c being the heat capacity and k the thermal conductivity. The Nusselt number is $\mathcal{N}_{\mathrm{Nu}} = qL/(k\Delta T)$, q being the local heat transfer rate at a given angle from the forward stagnation point and ΔT the temperature difference between cylinder surface and free stream. An example of the theoretical results is shown in Fig. 12.4.

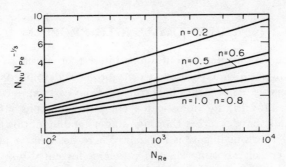

Fig. 12.4. Predicted local heat transfer rates to pseudoplastic fluids from a cylinder at an angle of 60° from the forward stagnation point [8].

(iv) STAGNATION FLOW

Flow impinging on a flat plate ($\beta = 1$ in Fig. 12.2) represents an attractive prototype for study. Classical stagnation flow of an inviscid fluid may be two-dimensional (Fig. 12.5),

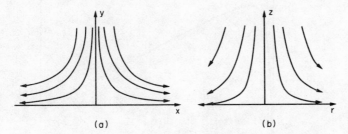

Fig. 12.5. Stagnation flow: (a) two-dimensional, (b) axisymmetric.

$$U = ax; \quad V = -ay \tag{12.8}$$

or axisymmetric,

$$U_r = \frac{ar}{2}; \quad U_z = -az \tag{12.9}$$

thus satisfying identically the continuity equation. The boundary layer interior to these fields has been computed for a bewildering variety of constitutive models, both elastic and inelastic [9–12]. Because of the different techniques and fluid models employed, fruitful comparison of theoretical results has been difficult, and unambiguous comparison with data has been impossible. First, measurements of velocity profiles in the stagnation region are rare [13, 14], and, second, it is not obvious how one fits a value of a in (12.8) and (12.9) to the data. Representative of the theoretical work done with inelastic fluids are the computations of Maiti [11], who used the power-law model for axisymmetric stagnation flow. He also employed the von Karman–Pohlhausen integral method to obtain an approximate solution, and reported that the boundary-layer thickness varies according to $r^{(n-1)/(n+1)}$ as one proceeds outward from the stagnation point.

(v) THE ENTRY-LENGTH PROBLEM

A problem which has attracted continuing interest of workers in Newtonian fluid mechanics is the manner in which a fluid velocity profile is developed as the fluid enters a conduit from a reservoir. Approaches have ranged from early applications of boundary-layer theory [1, p. 176; 15] to sophisticated computer solutions [16, 17], and it now appears that we finally have a rather thorough understanding of this flow—at least when the fully developed Newtonian flow is laminar. There has been a parallel though less-complete development of the same class of problems for entry flow of non-Newtonian fluids, beginning with boundary-layer analyses of power-law fluids in two-dimensional channels and circular conduits [18–20]. Experimental data are available which indicate that if the Reynolds number is sufficiently high so that there is negligible upstream diffusion of vorticity, and if the fluids are not appreciably elastic, then the analysis of Collins and Schowalter is quite acceptable [21]. In Fig. 12.6 the theoretical predictions of Collins

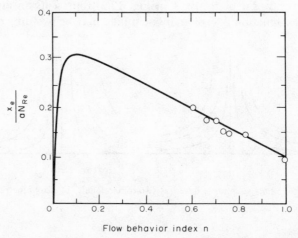

Fig. 12.6. Entrance length x_e as a function of flow behavior index for pseudoplastic power-law fluids. ——= theory of Collins and Schowalter; o=data of Rama Murthy and Boger. N_{Re} is based on the tube diameter, 2a.

and Schowalter are shown for entry lengths of pseudoplastic power-law fluids in tubes along with the experimental results of Rama Murthy and Boger. Entry length x_e is defined as the length required for the centerline velocity to reach 98% of its fully developed value. Reynolds number is based on the pipe diameter $2a$.

(vi) TURBULENT BOUNDARY LAYERS

Those familiar with Newtonian boundary-layer theory will recall that analyses of turbulent boundary layers are necessarily more empirical than those of the laminar counterpart. The situation with non-Newtonian fluids is similar. Skelland [22, pp. 287 *et seq.*] has developed correlations for turbulent boundary layers of power-law fluids flowing over flat plates. Adaptation for flow through tubes shows good agreement with data of Bogue and Metzner [23] using carbopol. This subject is treated at some length in Skelland's book.

12.3. DIMENSIONLESS GROUPS FOR NON-NEWTONIAN FLUIDS

Straightforward dimensional analysis of the sort that has been so successful for complex problems with Newtonian fluids is less obvious in its formulation for non-Newtonian systems because of the ever-present constitutive uncertainties. Truesdell [24] showed how one may incorporate the notion of a natural time into the framework of the theory of simple fluids, and his ideas have been used, with some modification, by others. One of the boldest and also most successful approaches to definition and interpretation of a characteristic time for a fluid has been that of Metzner *et al.* [25]. We summarize their argument below.

Recall the integral expansion of Rivlin and coworkers which was discussed in Section 11.2. In particular, consider the formulation (11.3) in terms of deformation rates. We apply that expression to a fluid at rest which, at time $t=0$, is subjected to the simple shearing flow expressed in rectangular Cartesian coordinates by

$$\mathbf{v} = \{\kappa x_2, 0, 0\} \tag{12.10}$$

Then (11.3) leads to expressions of the following form for the extra stress:

$$\left.\begin{aligned}
t_{12} &= \int_0^\infty \mathcal{N}_1(s)\,\frac{\kappa}{2}\,ds + \cdots \\[2mm]
t_{11} - t_{22} &= \int_0^\infty \mathcal{N}_1(s)s\kappa^2 ds + \cdots \\[2mm]
t_{22} - t_{33} &= -\int_0^\infty \mathcal{N}_1(s)s\kappa^2 ds + \int_0^\infty \int_0^\infty \mathcal{N}_2(s_1, s_2)\left(\frac{\kappa^2}{4}\right) ds_1 ds_2 + \cdots
\end{aligned}\right\} \tag{12.11}$$

The moduli $\mathcal{N}_i$ are of course taken to be well-behaved functions of time so that as $t \to \infty$ the coefficients of κ in (12.11) approach a constant value, and one recovers the familiar expressions for a viscometric flow. We call this "fluid-like" behavior since the properties, for a given flow, are time independent when the rate of deformation has been constant for a sufficiently long time. Contrast this familiar result with the opposite extreme at $t = \Delta t \cong 0$. Now the functions $\mathcal{N}_i$ are expected on physical grounds to have a time dependence which is approximately exponential. Hence for very small times the weighting functions can be approximated by their initial values. Terms of higher order than those shown in (12.11) are taken to be negligible, and (12.11) reduces, approximately, to

$$t_{12} = C_1\kappa\Delta t; \quad t_{11} - t_{22} = C_2(\kappa\Delta t)^2; \quad t_{22} - t_{33} = C_3(\kappa\Delta t)^2 \tag{12.12}$$

Thus the shear stress increases linearly and the normal stresses quadratically with time, a behavior characteristic of elastic solids in simple shear [26, eqn. (44)].

From (12.11) one can infer a characteristic time that should be a measure of the magnitude of normal stress difference $(t_{11} - t_{22})$ relative to shear stress:

$$t_{\text{ch}} = \frac{\displaystyle\int_0^t s\mathcal{N}_1 ds}{\displaystyle\int_0^t \mathcal{N}_1 ds} \tag{12.13}$$

If the flow has been of the form (12.10) for a sufficiently long time, t_{ch} approaches the steady value θ, where

$$\theta = \frac{\displaystyle\int_0^\infty s\mathcal{N}_1 ds}{\displaystyle\int_0^\infty \mathcal{N}_1 ds} \tag{12.14}$$

A dimensionless quantity relating first normal stress difference to shear stress is

$$\mathcal{N}_w = \theta \frac{V}{D} \tag{12.15}$$

where V and D are a characteristic velocity and length, respectively. If one is willing to associate normal stress differences with elastic effects, then $\mathcal{N}_w$ can also be interpreted as a measure of the relative importance of elastic and viscous effects. White [27] has called $\mathcal{N}_w$ the Weissenberg number. Note that if a specific constitutive model, a Maxwell fluid, for example, is used as the starting point [28], then the connection to elasticity is immediately apparent.

The ratio θ/t can be considered as an inverse measure of the time for which the liquid has been tending toward the "fluid-like" behavior of (7.33). As $\theta/t \to 0$, the relevance of the viscometric functions becomes apparent. On the other hand, for $\theta/t \gg 1$ we expect viscometric functions to be of little value, and in fact for sufficiently large θ/t we can expect the fluid to tend toward the "solid-like" behavior of (12.12). These notions have led to the definition of a *Deborah number*, which is the ratio of a characteristic fluid time to a characteristic flow or process time [25]. Clearly there is no unique definition of the two relevant times, and in fact choice of the appropriate parameters has been a source of some controversy. If the Deborah number is to have the significance given above, then the process time must, in some sense, characterize an unsteady flow which is experienced by material points.

The development of the term *Deborah number* is interesting in its own right, and the historically inclined reader is urged to consult references on the subject [28].

If one accepts the ideas set out above, the wide range of behavior of non-Newtonian fluids becomes a little less unmanageable. The Deborah number is a quantitative link joining fluid and flow, a connection which we have emphasized on several previous occasions as being crucially important for analysis of non-Newtonian fluids.

The physical interpretation of $\mathcal{N}_{\text{Deb}}$ was inferred from two rather special extremes; namely, fully developed laminar shearing and sudden imposition of laminar shearing. To extend the interpretation to the whole range of non-Newtonian behavior is dangerous. Because of the simple origins of Weissenberg and Deborah numbers, one should not be surprised at the inadequacy of these two hand-holds for describing all of non-Newtonian fluid mechanics. Nevertheless, the concept has been a fruitful one, as we shall see in subsequent sections. Finally, we note in passing, this example of application of fundamental concepts to explanation of phenomena which have great technological importance. There is no longer an excuse for a show of surprise upon finding that the rheogram of a material fails to provide all of the data necessary for any flow system arbitrarily far removed from viscometric flow.

12.4. BOUNDARY-LAYER THEORY FOR VISCOELASTIC FLUIDS

The considerations of the preceding section make it clear that elastic effects can have a profound influence on fluid response in a region where a material element is undergoing a rapid change in deformation with time; that is, when the Lagrangian time derivative is far removed from zero. Such is often the case near the leading edge of a bluff body past which a fluid is flowing. Thus any application of boundary-layer theory to viscoelastic materials is, of necessity, tentative and conditional upon absence of gross alterations to the flow because of solid-like behavior near a leading edge. If, however, effects of elasticity are "small" in this sense, then one can proceed by reasonably straightforward means to account for the elastic effects which do exist. One finds that even in these cases the boundary-layer behavior can be substantially altered from that discussed in Section 12.2 for a purely viscous fluid.

Although the numerous treatments of viscoelastic boundary layers differ in detail, they all require that the fluid be sufficiently inelastic so that the Deborah number is not large in the regions of flow which are of interest. Then the integral expansion, for example (11.2), can be truncated at an early stage. Beard and Walters [10], for example, began with the Walters model B′ (11.50), but assumed that elastic effects were sufficiently small so that constitutive behavior was adequately represented by an expansion similar to that employed in obtaining (11.9) from (11.1). The essential difference is that by using the Walters model B′, one obtains an expansion in tensors of the Rivlin–Ericksen type, but defined in terms of the inverse of $\mathbf{C}_{(t)}(\tau)$. In contrast to (5.62) *et seq.* one can define tensors [29, 30]

$$\mathbf{B}_n(t) = - \left[\overset{n}{\overline{\mathbf{C}_{(t)}^{-1}(\tau)}} \right]_{\tau=t} \tag{12.16}$$

As with the Rivlin–Ericksen tensors, $\mathbf{A}_n$, the $\mathbf{B}_n$ follow naturally from the Oldroyd formalism. Analogous to (11.35) one can show

$$\frac{\delta^n g^{ij}}{\delta t^n} = - b_n^{ij} \tag{12.17}$$

Beard and Walters' expansion of the model B′ was first truncated to two terms. (Walters subsequently showed the effect of an additional term [31].) In component notation one obtains

$$t^{ij} = \eta_0 b_1^{ij} - K_0 b_2^{ij} \tag{12.18}$$

where η_0 and K_0 are constants. Note the close resemblance to the second-order fluid (10.16) with $\alpha_2 = 0$. Note also that $b_1^{ij} = -2d^{ij}$. For the usual two-dimensional steady flow problem, components of the stress tensor are readily found from (12.18) and can be substituted into the equation of motion. On the rationalization that a *slightly* elastic fluid will produce a boundary layer only slightly altered in its dimensions from a viscous one, Beard and Walters [10] postulated a boundary-layer thickness δ which scales in the classical fashion [1, p. 119]

$$\delta/L = O(\sqrt{N_{\mathrm{Re}}})$$

L being a characteristic length, V a characteristic velocity, ρ the density, and $N_{\mathrm{Re}} = LV\rho/\eta_0$. Then, to first order in the elasticity, the boundary-layer equations can be derived for steady two-dimensional flow. One can easily show, neglecting terms of $O(\delta)$, that the x-component of the equation of motion reduces to

$$u\frac{\partial u}{\partial x} + v\frac{\partial u}{\partial y} = -\frac{\partial p}{\partial x} + \frac{1}{N_{\mathrm{Re}}}\frac{\partial^2 u}{\partial y^2} + k\left[u\frac{\partial^3 u}{\partial x \partial y^2} + v\frac{\partial^3 u}{\partial y^3} + \frac{\partial u}{\partial x}\frac{\partial^2 u}{\partial y^2} - \frac{\partial u}{\partial y}\frac{\partial^2 u}{\partial x \partial y}\right] \quad (12.19)$$

and that $\partial p/\partial y = O(N_{\mathrm{Re}}^{-1})$. Here $\mathbf{v} = (u, v)$. Independent and dependent variables in (12.19) have been made dimensionless as follows:

$$x = \bar{x}/L; \quad y = \bar{y}/L; \quad u = \bar{u}/V; \quad v = \bar{v}/V; \quad p = \bar{p}/\rho V^2$$

where an overbar has been used, momentarily, to indicate dimensional quantities. The effect of elasticity is embodied in the parameter

$$k = \frac{K_0}{\rho L^2} \quad (12.20)$$

The product kN_{Re}, being a measure of importance of elastic relative to viscous forces, can be identified with the Weissenberg number N_{We}. It is apparent from (12.19) that the presence of even a small amount of elasticity can greatly alter the boundary-layer equations, and hence the flow, from the purely viscous case. In his paper on the subject [31] Walters shows how one must be cautious in the ordering procedure when terms of order higher than those shown in (12.19) are considered.

A slightly less restrictive approach to viscoelastic boundary layers was carried out by Denn [32]. His analysis is also confined to finite but small effects of elasticity, and neither his analysis nor that of Beard and Walters is expected to be valid near the leading edge of a bluff body where the Deborah number may be appreciable, and where the customary concept of a fluid boundary layer is open to question.[†]

The constitutive equation used by Denn is an empirical modification of (12.18) which allows for shear dependent viscosity and normal stresses. Denn also used the modified Rivlin–Ericksen tensors $\mathbf{B}_n$ in his formulation, the constitutive approximation being

$$t^{ij} = \eta_0(\bar{I}_2)b_1^{ij} - K_0(\bar{I}_2)b_2^{ij} \quad (12.21)$$

where now, in contrast to (12.18), η_0 and K_0 can be functions of the second invariant $\bar{I}_2$ of $\mathbf{B}_1$. Specifically, Denn chose the power functions

---

† A point which has not as yet received sufficient attention in the literature is the idea that high Deborah number effects near the leading edge of, for example a wedge [33], may have a macroscopically important effect on the boundary layer as a whole.

$$\eta_0(\bar{I}_2) = (-\bar{I}_2)^{(n-1)/2}\,\mu; \quad K_0(\bar{I}_2) = (-\bar{I}_2)^{(s-2)/2}\,\lambda \tag{12.22}$$

where μ and λ are constants. By casting the equation of motion into dimensionless form one arrives at the usual power-law expression for the Reynolds number,

$$\mathcal{N}_{\mathrm{Re}} = \frac{L^n \rho V^{2-n}}{\mu} \tag{12.23}$$

and a Weissenberg number

$$\mathcal{N}_{\mathrm{We}} = \frac{\lambda}{\mu}\,(V/L)^{s-n} \tag{12.24}$$

For problems in which there is no characteristic length, L can be scaled to make the Reynolds number unity. Then the Weissenberg number can be written

$$\varepsilon = \frac{\lambda}{\rho V^2}\left(\frac{\rho V^2}{\mu}\right)^{s/n} \tag{12.25}$$

and it is possible to solve problems in terms of the single parameter ε, which is assumed to be small.

Rather than dealing specifically with the problem of stagnation flow, which was the objective of Beard and Walters, Denn developed a formalism for description of general wedge flows $(U=x^a)$, much as we have done earlier for the purely viscous case. The technique, however, is more difficult to apply since a similarity transformation is not, in general, possible. A hierarchy of ordinary differential equations is obtained by defining a stream function

$$u = \frac{\partial \psi}{\partial y}; \quad v = -\frac{\partial \psi}{\partial x}$$

and relating it to $f(x, \eta)$ through

$$f(x, \eta) = x^{-[(2an-a+1)/(n+1)]}\,\psi(x, y) \tag{12.26}$$

with

$$f(x, \eta) = \sum_{m=0} \varepsilon^m x^{m[s(3a-1)/(n+1)-2a]} f_m(\eta) \tag{12.27}$$

where

$$\eta = x^{[(2a-an-1)/(n+1)]}\,y \tag{12.28}$$

Equations (12.26) and (12.27) are substituted into the dimensionless form of the boundary layer equation for u. This equation, the analogue to (12.19), is

$$u\frac{\partial u}{\partial x} + v\frac{\partial u}{\partial y} - U\frac{dU}{dx} - n\left(\frac{\partial u}{\partial y}\right)^{n-1}\frac{\partial^2 u}{\partial y^2} + \varepsilon\left[(3-2s)\left(\frac{\partial u}{\partial y}\right)^{s-1}\frac{\partial^2 u}{\partial x \partial y}\right.$$

$$+ \left(\frac{\partial u}{\partial y}\right)^{s-2}\left(u\frac{\partial^3 u}{\partial x \partial y^2} + v\frac{\partial^3 u}{\partial y^3} + \frac{\partial u}{\partial x}\frac{\partial^2 u}{\partial y^2}\right) \tag{12.29}$$

$$\left. + (s-2)\left(\frac{\partial u}{\partial y}\right)^{s-3}\frac{\partial^2 u}{\partial y^2}\left(u\frac{\partial^2 u}{\partial x \partial y} + v\frac{\partial^2 u}{\partial y^2} + 2\frac{\partial u}{\partial x}\frac{\partial u}{\partial y}\right)\right] = 0$$

Denn [32] has given the ordinary differential equations which result for f_0 and f_1, the boundary conditions being the usual which follow from no slip at the solid surface and a necessity for matching the inviscid solution as $\eta \to \infty$. The hope is expressed that these solutions should give insight into boundary-layer behavior for weakly elastic fluids, in the sense that $\varepsilon \ll 1$.

The expansion (12.27) has been shown to be inappropriate for $s \geqslant n+1$ [34]. Nevertheless, solutions based on (12.27) for $s < n+1$ illustrate the high sensitivity of the results to the elastic parameter s. An example is flow past a flat plate aligned at zero incidence to the flow. The equation for f_0 is of course just the viscous power-law result, (12.6) and (12.7), suitably altered for the slightly different definition of dependent variable

$$n(n+1)f_0''' + f_0(f_0'')^{2-n} = 0 \qquad (12.30)$$

with $f_0(0) = f_0'(0) = 0$; $f_0'(\infty) = 1$. Denn [32] and Serth [34] have reported the following effects of elasticity on wall shear stress for $n=1$:

s	$f_1''(0)$
1.25	0.129
1.50	−0.082
1.75	−0.260

Thus it is seen that elasticity can, according to the predictions, either increase or reduce wall drag. Equally interesting is the variation of f_1' with η. An example, taken from Denn's paper, is given in Fig. 12.7.

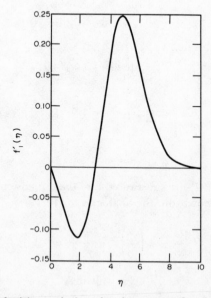

Fig. 12.7. Effect of elasticity on the boundary layer over a flat plate. $n=1$, $s=1.5$ [32].

One cannot help but be suspicious of the validity of such drastic oscillations. Unfortunately, exhaustive and careful measurements do not exist for boundary-layer behavior of viscoelastic fluids which are well-characterized rheologically. Hermes and Fredrickson [35] have studied the flow of several aqueous solutions of carboxymethylcellulose (CMC) in an approximation to uniform flow past a semi-infinite flat plate. Their results show that fluid elasticity does have a substantial effect on the velocity profile in the boundary layer. Their work also casts doubt upon the validity of analyses which do not allow for important memory effects near the leading edge of the plate. Included in the doubtful category would be second-order fluid models. Unfortunately, the data of Hermes and Fredrickson cannot be compared directly with the analysis of Denn because elastic parameters of the CMC solutions were not measured.

Because of the more difficult analysis required, the pipe-entry problem, so exhaustively treated for Newtonian and viscous fluids, has not been subjected to a comparable analysis for viscoelastic fluids. Entry behavior has been studied experimentally by Boger and Rama Murthy [36]. Employing a series of Methocel solutions which exhibited varying degrees of normal stress in a rheogoniometer, they compared the pressure drop in the entry region to the predictions of Collins and Schowalter [20] for viscous fluids. Measured pressure drop decreased monotonically with increasing elasticity (as measured by normal stress), approaching zero for the most elastic solutions tested. However, later work by Boger and Rama Murthy [37] indicates that entry behavior may be far too complicated to admit description by a simple correlation with first normal stress difference.

Most of the interest in turbulent boundary layers of viscoelastic fluids has centered on the phenomenon of drag reduction. We treat this topic separately in Section 12.11.

12.5. FLOWS IN WHICH INERTIAL EFFECTS ARE SMALL

We consider very briefly some of the progress made at the opposite extreme of the Reynolds number spectrum from that of Section 12.4. One is immediately reminded of the rich literature of flows at low Reynolds numbers in Newtonian fluid mechanics [38]. The successes there are a consequence of linearity of the creeping motion equations, a luxury which unfortunately does not carry over to creeping flows of non-Newtonian systems. We shall see later, in our discussion of unsteady flows, that nonlinear behavior can give rise to some interesting new phenomena which may have both practical and fundamental significance. In the present section we consider only flows which are steady in the Eulerian sense.

Though it is not often stated explicitly, the classical steady laminar flows in circular tubes or between parallel planes constitute flows in which the inertial term $\rho \ddot{\mathbf{x}}$ in (4.26) is of no consequence. Furthermore, symmetry offered by the flow permits *a priori* specification of the stress distribution. We have seen in Chapter 7 how an expression for the velocity distribution is then readily obtained for fluids with very general constitutive properties. Slight alteration of the flow geometry, for example, flow through a pipe with rectangular cross-section, greatly complicates the nature of the problem. Several workers have shown, however, that variational techniques can be applied to the solution of these problems if the constitutive behavior is approximated by a sufficiently simple expression [39–42]. Schechter [43], for example, has described the use of a variational technique to determine the velocity profile for laminar flow of a power-law fluid through ducts with noncircular cross-section.

As soon as nonparallel walls are considered, the flow solution is greatly complicated. Flow under conditions of negligible elasticity was studied by Sutterby [44]. His analysis is a superposition of solutions for two asymptotic cases where viscous and inertial effects, respectively, dominate. A more complete solution for power-law fluids is due to Oka and Takami [45]. Far more interesting phenomena occur, however, when one deals with fluids which exhibit elastic effects. Giesekus has reported some remarkable experiments showing a variety of pathlines in the conical approach to a nozzle [46]. Photographs of the limiting case of a 180° cone are shown in Fig. 12.8. Similar results have been found by Schümmer [47] in cones with smaller angles.

The role of inertia as a cause for secondary flow in conical flows of Newtonian fluids is well documented, the chief theoretical results being due to Ackerberg [48]. However, the secondary flow, and ultimate instability, observed with viscoelastic fluids appears to be quite distinct from the former. For example, the paths of streamlines predicted by Ackerberg are compared in Fig. 12.9 with those which can result, according to a study of Langlois and Rivlin [49, 50], for a second-order fluid. Computations of conical flow of a second-order fluid have also been reported by Kaloni [51].

The degree of complexity, already substantial for some of the internal flows just cited, becomes especially formidable when one attempts to analyze external flows past bluff objects. Slattery and coworkers have published a series of papers dealing with the creeping flow of a power-law fluid past a rigid sphere (see, for example, [52]). We mention in passing that power-law calculations have been extended into the regime $5 < N_{Re} < 40$ (Reynolds number based on sphere radius) by Nakano and Tien for flow of a power-law fluid past a Newtonian drop [53]. They employed Galerkin's method to obtain approximate solutions to the appropriate differential equations. Good agreement was obtained with experimentally observed values of the drag coefficient for a carboxymethylcellulose (CMC) solution.

Extension to viscoelastic fluids and inclusion of fluid inertia, by means of matched asymptotic expansions, was performed independently by Caswell and Schwarz [54] and by Giesekus [55]. The procedure, as can be seen from study of either of these papers, is extremely laborious and susceptible to error [56]. A double perturbation expansion is performed: one to account for terms beyond the Newtonian in, say, a third-order representation (11.9) of fluid behavior, and another to represent a small but finite effect of fluid inertia. Following a series of papers by several authors on this subject, Caswell [56] has indicated that, to the degree of approximation used in the analyses, the solutions are equivalent irrespective of the particular constitutive model which may have been used.

There is a practical reason for wishing to know the basic theory for slow flow of spheres in non-Newtonian liquids. It should afford a convenient means for determining the "lower-limiting" or "zero-shear" viscosity, an important parameter for fluid characterization which was discussed in connection with some of the elementary models of Chapter 10. Cygan and Caswell [57] obtained data for spheres falling through dilute polyisobutylene solutions at terminal velocities of the spheres as low as 10^{-2} cm/s. Curiously, the data do not show the parameter dependence that is predicted from their theory (simplified to exclude inertial effects). According to theory one expects the departure from zero-shear viscosity to be quadratic in the net hydrodynamic drag force on the particle. In fact, the data showed an extremely close match to a departure *linear* in the drag force.

The insights into flow behavior which may be obtained from creeping flow solutions appear to be limited. Results of broad applicability have not been forthcoming and it is unlikely that significant results beyond those already achieved will be found. An exception may be some of the new information that is available when unsteady flows are considered. We consider this topic in Section 12.8.

Fig. 12.8. Flow approaching an orifice in a planar surface. Fluid is a 2% aqueous solution of polyacrylamide. Flow rates are (a) 50, (b) 100, (c) 200, (d) 400, (e) 500, (f) 750 cm³/min. Similar results were observed in conical approaches to the orifice. From Giesekus [46].

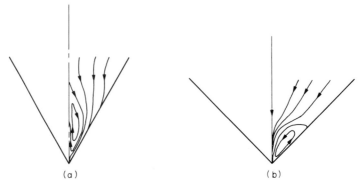

Fig. 12.9. Sketch of profiles showing (a) the effect of inertia on flow of a Newtonian fluid in a cone [48]; (b) the flow predicted for a second-order fluid in a cone [46, 49].

12.6. EXTENSIONAL FLOWS

In Section 9.4 we discussed at some length the theory of steady uniform extensions and made reference to the difficulties associated with experimental realization of the flow. We now return to the subject of extensional flow from a rather more practical point of view.

We have already indicated that knowledge of the viscometric functions for a non-Newtonian fluid provides no *a priori* insight into the behavior of that fluid in extensional flow. In contrast to Newtonian fluids there is no fundamental connection between the two flows, such as the Trouton viscosity 3μ of (9.37) for Newtonian fluids. Since many of the most important processing steps associated with molten polymers, e.g., fiber spinning and film blowing, involve a high degree of material *stretching* as opposed to shearing, the practical motivation for interest in extensional flows is clear.

Another strong motivation is provided by the intriguing predictions of a host of simplified constitutive equations. Consider, for example, the convected Maxwell model. This is one of the simplest constitutive models that allows for elastic behavior and also displays correct invariance properties. Its origin is readily derived from a generalization of the simple Maxwell model (10.22)

$$t^{ij} + \theta \frac{\eth t^{ij}}{\eth t} = 2\mu d^{ij} \tag{12.31}$$

Consider application of (12.31) to a uniaxial extension, defined in rectangular Cartesian coordinates by

$$\left.\begin{aligned} d^{11} &= d \\ d^{22} &= d^{33} = -\frac{1}{2}d \\ \text{all other } d^{ij} &= 0 \end{aligned}\right\} \tag{12.32}$$

Then the extensional viscosity $\bar{\mu}$, defined by

$$\bar{\mu} = \frac{t^{11} - t^{33}}{d} \tag{12.33}$$

is readily seen to be

$$\bar{\mu} = \frac{3\mu}{(1-2\theta d)(1+\theta d)} \tag{12.34}$$

Note that as d is increased from zero the extensional viscosity will increase from its Newtonian value of 3μ, approaching infinity as $d \to (2\theta)^{-1}$. If we wish to define a Deborah number for this flow,

$$\mathcal{N}_{\mathrm{Deb}} = \theta d$$

we see that $\mathcal{N}_{\mathrm{Deb}} = \frac{1}{2}$ is a critical value at which, according to (12.34), an infinite axial stress is required to achieve the stretching (12.32). The literature is filled with speculations over the meaning of this result, if indeed it should be taken seriously in the first place [58, 59].

The first measurements in which care was taken to approach a state of steady elongation appear to be those of Ballman [60]. He conducted a tensile test with polystyrene at temperatures sufficiently high so that the material could be considered liquid-like. However, the viscosity was still high enough $(O(10^7 \text{ poise}))$ to prevent serious distortion from the action of gravity. The work of Ballman reveals two important difficulties with extensional flow experiments. The first is the problem of achieving steady uniform extension, unencumbered by spatial inhomogeneities and memory effects from startup. In Ballman's case this meant limiting measurements to extension rates d well below $10^{-1}\ \mathrm{s}^{-1}$. Second, one immediately sees how presence of a gravitational field can make it virtually impossible to stretch "liquid" columns with viscosities in the neighborhood of one poise.

An ingenious variant to the experiment described above is the so-called "tubeless siphon", shown diagrammatically in Fig. 12.10, which was apparently first reported by

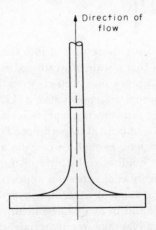

Fig. 12.10. The "tubeless siphon" for studying extensional flows.

Fabula [61]. Later measurements with this device have been published by Astarita [62]. Astarita measured the requisite forces by cutting the siphon at an appropriate point and measuring the change in weight registered on a balance upon which the fluid reservoir rested. From this, and a knowledge of the extension rate, extensional viscosity could be calculated. The chief drawback with this technique is that just mentioned above: Is a given fluid element responding as though it has been in steady extensional flow for all past time? Unless the rates of extension are extremely low the method is probably not a good approximation to steady uniform extension.

(We note that such applications as fiber spinning and film blowing are certainly not good approximations to steady extension either. It is not implied here that measurements made under conditions other than steady extension are of no utility. However, the distinct difference between them and the conditions for which the analysis of Chapter 9 was performed should be clearly recognized.)

Recent attempts to measure extensional flow behaviour have been numerous. We cite in particular the work of Vinogradov *et al.* [63] and of Meissner [64].

We return now to further discussion of some of the theoretical aspects of extensional flow. It is interesting that the prediction of a catastrophic increase in extensional viscosity, though shared by many rheological models with respectable credentials (various forms of Maxwell, Oldroyd, and network models; the Bird–Carreau model) is not universally predicted. A notable exception is the network rupture model of Tanner [59, 65]. In fact, this model has been used with some success to describe the data of Ballman [66]. The exact consequences of this fit are, however, not completely clear [67, 68].

A useful clarification of the significance of (12.34) has been given by Denn and Marrucci [69], who noted the importance of the transient portion of an extension experiment. Suppose, for example, we apply the convected Maxwell model (12.31) to a fluid which is at rest for $t < 0$ and at $t \geqslant 0$ follows (12.32). Then the expression for axial stress $(t^{11} - t^{33})$ of a cylinder stretched in the 1-direction becomes

$$t^{11} - t^{33} = \frac{3\mu d}{(1 - 2\theta d)(1 + \theta d)} - \frac{2\mu d}{1 - 2\theta d} \exp\left[\frac{(2\theta d - 1)t}{\theta}\right]$$

$$- \frac{\mu d}{1 + \theta d} \exp\left[-\frac{(1 + \theta d)t}{\theta}\right]$$

$$(12.35)$$

This equation provides significantly more insight into the problem than the abbreviated version (12.34). Note first of all that a state of steady extension is predicted to be *impossible* if $N_{\text{Deb}} > \frac{1}{2}$. Furthermore, if $N_{\text{Deb}} = \frac{1}{2}$ one predicts from (12.35) that an infinite time is required before the state of infinite stress, shown also in (12.34), is achieved. Representative curves computed from (12.35), are shown graphically in Fig. 12.11. One immediately sees that the time since startup is of critical importance. To date, tests in uniform uniaxial extension conducted over times long enough to provide a definitive test of (12.35) have not been performed. However, qualitative features of the analysis seem reasonable, and consistency with some results for biaxial extension appears to be established.

One can readily show that a uniaxial *compression* experiment is not equivalent to the extension described above [70], and in fact can be identified with a *biaxial extension*. This flow configuration has certain advantages for materials testing, not the least of which is the fact that biaxial stretching is important in polymer processing, examples being film and bottle forming. Although interest in biaxial extension is by no means new [71], the work of Denson and colleagues [72] has been a stimulus for recent activity. Also, there seems to exist a possibility for achieving steady-state conditions at higher stretch rates, with materials which border on liquid-like behavior, than has been possible in the uniaxial experiments just described [73].

It is noteworthy that an analysis similar to that which leads to (12.35) can be carried

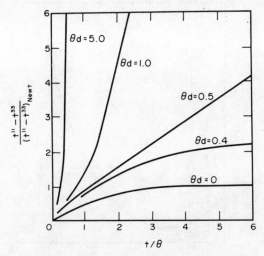

Fig. 12.11. Axial stress of a cylinder in uniaxial extension, relative to Newtonian stress, plotted as a function of time relative to the relaxation time.

out for biaxial extension of a convected Maxwell fluid. In this case the net axial stress $(t^{11} - t^{33})$ is, for $d^{11} = d^{22} = d$; $d^{33} = -2d$,

$$t^{11} - t^{33} = \frac{6\mu d}{(1 - 2\theta d)(1 + 4\theta d)} - \frac{2\mu d}{(1 - 2\theta d)} \exp\left[\frac{(2\theta d - 1)t}{\theta}\right]$$
$$- \frac{4\mu d}{(1 + 4\theta d)} \exp\left[-\frac{(1 + 4\theta d)t}{\theta}\right] \tag{12.36}$$

Once more the extensional viscosity, $\bar{\mu}_b = (t^{11} - t^{33})/d$, approaches infinity at sufficiently large times as $\mathcal{N}_{\text{Deb}} \to \frac{1}{2}$. It is interesting to note, however, that at asymptotically large times, for $\mathcal{N}_{\text{Deb}} < \frac{1}{2}$, $\bar{\mu}_b$ goes through a minimum at $\mathcal{N}_{\text{Deb}} = \frac{1}{8}$ as d is increased from zero. This is shown in Fig. 12.12. The qualitative occurrence of a minimum is consistent with experiments on low molecular weight polyisobutylenes [73]. Note from Fig. 12.12 and eqn. (12.36) that the Trouton ratio $(\bar{\mu}_b/\mu)_{\theta = 0}$ is six for biaxial extension.

We have seen that steady extension experiments discussed thus far are limited to stretch rates below approximately 1 s⁻¹, and in most cases are far below that. Furthermore, according to the predictions of several rheological models, it is futile to hope for steady-state conditions beyond a critical value of stretch rate which, for some materials tested, corresponds to about 1 s⁻¹ [73]. The region of practical interest, however, may be nearer to 10^3 s⁻¹, with corresponding values of $\theta d > 1$. Thus it appears that detailed study of the transient aspects of extensional flow is in order. Work in this direction for biaxial extension has been reported by White [74], who obtained a more general form of (12.36).

Consequences of the peculiar effects of stress buildup and/or high Deborah number are not limited to flows which are purely extensional in nature. We cite just a few examples in passing. A phenomenon known as the 'Uebler effect'' has been the basis of an approximate analysis by Metzner [75] and involves extensional flow. The effect was observed by workers at Delaware when using small bubbles as tracer particles to follow pathlines of a viscoelastic fluid as it accelerated in flowing from a reservoir to a constricted opening [76].

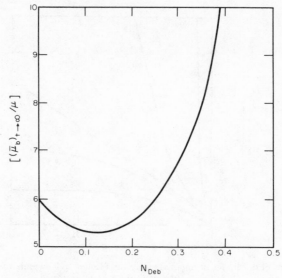

Fig. 12.12. Steady-state value of extensional viscosity according to predictions of convected Maxwell model (12.36).

Large bubbles (diameter $\cong 0.5$ cm) occasionally reached a point, just prior to the constriction, at which they abruptly stopped even though fluid continued to pass around them. After a few moments the large bubbles would again accelerate and pass into the constriction. Metzner has attributed the phenomenon to development of tensile stresses around the bubble, noting that near a critical N_{Deb} these stresses can become very large indeed.

Metzner and Astarita [33] have also described the effects of sudden deceleration near a stagnation point, and have noted that the resulting large stresses and "solid-like" behavior can have serious consequences for both momentum and heat transfer. We have already noted this in connection with viscoelastic boundary layers (Section 12.4). Metzner and Metzner [77] have used approximate analyses of converging flow in cones to quantify extensional flows at the high stretch rates which characterize industrial operations. More work in this area is critically needed. It does seem, as shown in (12.36) and other work cited above, that problems of practical interest must contain time as a significant variable. Put in the framework of Chapter 9, it is unlikely that we can look at elastic effects in industrially important extensional flows without recognizing that the flows do not have a constant stretch history.

The technologically important process most closely associated with extensional flow is probably fiber spinning. This operation involves far more than fluid extension, and is severely complicated by factors which have thus far been neglected. In dry spinning, a fluid element is withdrawn continuously from a spinneret at high speed, simultaneously stretched, cooled in gaseous surroundings, solidified, and then wound on a takeup reel. The time lapse for a given fluid element from moment of exit from the spinneret to arrival at the takeup reel may be less than one second, and stretch rates may exceed 10^2 s^{-1}. The first one to attempt a reasonably complete analysis of the spinning process was Ziabicki (see, for example, [78, 79]). A comprehensive statement of the governing equations for the problem has been given by Ohzawa *et al.* [80]. Solutions to the thread-drawing problem which account explicitly for thermal effects have been presented by Pearson and Shah

[81]. They have shown that thermal and stability considerations can, in the range of industrially important conditions, be dominant.

12.7. FLOW THROUGH POROUS MEDIA

This subject has a long and somewhat confusing history even when one restricts the fluid to be incompressible and Newtonian. The additional complications caused by non-Newtonian constitutive behavior are far from being fully understood, but it is clear that elastic effects can have important consequences. Therefore, attempts to describe the flow by exact analogy to Newtonian fluids are, at best, only partially successful.

Flow of Newtonian fluids through porous materials at Reynolds numbers well below unity are described by Darcy's law [82], a linear relation between pressure drop and flow rate, the form of which can be inferred from the Navier–Stokes equation. In many applications it is convenient to follow the convention well known from tube flow and to express the result in terms of a friction factor. In standard engineering treatments a connection between friction factor and Reynolds number is often made by modeling a porous medium as a collection of long capillary tubes, in each of which fully developed laminar flow occurs. A similar treatment for power-law fluids [83] leads to

$$f = \frac{1}{\mathcal{N}_{Re}} \quad (\mathcal{N}_{Re} \ll 1) \tag{12.37}$$

The friction factor is defined here by

$$f = \frac{\Delta P D_p \varepsilon^3 \rho}{L G^2 (1 - \varepsilon)} \tag{12.38}$$

and the Reynolds number by

$$\mathcal{N}_{Re} = \frac{D_p G^{2-n} \rho^{n-1}}{150 H (1 - \varepsilon)} \tag{12.39}$$

where ΔP is the pressure drop over a thickness L of packed bed, G is the superficial mass velocity, D_p is the particle diameter, ε is the void fraction in the bed, ρ is the fluid density, and H is a "viscosity parameter" defined by

$$H = \frac{K}{12} \left[9 + \frac{3}{n} \right]^n (150 \, k \varepsilon)^{(1-n)/2}$$

k is the bed permeability and, by a semiempirical argument, is related to bed geometry by [84, pp. 196 *et seq.*],

$$k = \frac{D_p^2 \varepsilon^3}{150 (1 - \varepsilon)^2}$$

Here K and n are the usual power-law constants from (10.7). These rather awkward definitions permit a proper collapse to the Blake–Kozeny equation familiar from Newtonian technology [84, pp. 196 *et seq.*]. If the power law was a satisfactory measure of non-Newtonian response, one would hope that beyond the range of Reynolds number for which (12.37) is adequate Newtonian correlations, such as the Ergun equation [84, pp. 196 *et seq.*] with suitable modification of defining parameters to coincide with those just given, would be useful. In fact, the problem appears to be significantly more complex— even in those cases where complicating factors of polymer degradation, adsorption on surfaces of porous material, and the like, can be ignored.

Although anomalous data still wait for a rational explanation, an important factor has been cited by Marshall and Metzner [85] which indicates once more the crucial role played by the Deborah number, or some equivalent measure of solid- vs. liquid-like behavior. As a crude model Marshall and Metzner considered flow of a fluid element through a porous material to be analogous to flow through a series of frustra of cones with adjoining vertices. Thus a given fluid element experiences a sequence of alternating accelerative and decelerative flows. Then if V is a mean value for the fluid velocity in this flow it can be argued that a Deborah number can be defined by

$$\mathcal{N}_{\text{Deb}} = A \frac{V}{D_p} \theta$$

where A reflects the details of the geometry, but should be a constant of order unity, and D_p is a length scale representative of the particle size. The characteristic fluid time can be associated with the convected Maxwell model (12.31). On the basis of earlier discussions of the significance of the Deborah number, one might expect radical departures from the behavior predicted from (12.37) as $\mathcal{N}_{\text{Deb}} \to 1$. Results obtained by Marshall and Metzner for some low Reynolds number experiments ($\mathcal{N}_{\text{Re}} < 0.05$) are shown in Fig. 12.13. The constant θ was estimated from viscometric flow data.

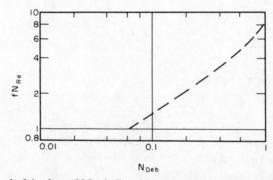

Fig. 12.13. Trend of the data of Marshall and Metzner [85] for flow through porous media.

Several comments concerning these results are in order. First, the phenomenon of an anomalously high effective viscosity is not unique to the data of Marshall and Metzner. For example, Dauben and Menzie [86] noted an increase in effective viscosity with polyethylene oxide solutions at sufficiently high flow rates. Second, the analytical arguments put forward by Marshall and Metzner, though plausible, are primarily qualitative. Hence the low value of $\mathcal{N}_{\text{Deb}}(\mathcal{N}_{\text{Deb}} \cong 0.06)$ at which they notice departure from power-law behavior should not be assigned undue significance. More recently, Wissler [87] has performed calculations to describe converging flow by allowing for elastic effects through a perturbation of the Newtonian calculation. Results were expressed in a form which permits comparison with Marshall and Metzner's data. Agreement is good.

But the picture, as we said earlier, is still not complete. For example, Griskey and co-workers [88] have carried out experiments of flow of molten polymer, in this case a polyethylene "Alathon 10", through porous media. They found good agreement with predictions for a power-law fluid at $\mathcal{N}_{\text{Deb}}$ up to 0.19. Furthermore, phenomena not accounted for by any of the descriptions presented above also exist. An especially interesting example is the presence of a "velocity gap" reported by James and McLaren [88a]. They found, in experiments with flow of polyethylene oxide solutions through a bead matrix, that regions of inherent instability can exist. As the flow rate was increased, a point was reached where the pressure loss actually *decreased* with increase in flow. As a result, the throughput jumped to some higher value at which the flow again stabilized, the new pressure drop being below that which existed prior to the jump.

12.8. UNSTEADY FLOWS

If a fluid displays elastic behavior, one would expect that its response to time-dependent forces would be distinctly different from predictions based upon a Newtonian assumption. We have already seen from our discussion of oscillatory testing that moduli defined in terms of amplitude and phase relations between stress and strain were among the early criteria to be established for evaluation of viscoelastic response (Section 10.5). Even a constitutive law which leads to linear equations, and hence is amenable to elementary analysis, leads to prediction of a number of interesting phenomena which have been qualitatively verified by experiment. One example is the analysis by Etter and Schowalter [89] of tube flow. They used a constitutive law in which the covariant form of the Oldroyd convected derivative (11.29) was applied to (11.11). The fluid was assumed to have a velocity field of the form

$$\mathbf{v} = \{0, 0, v_z(r, t)\} \tag{12.40}$$

Relevant components of the constitutive law then reduce to a linear form. Physical components of the stress in a cylindrical coordinate system are found to be:

$$
\begin{aligned}
&t(r\theta) = t(\theta z) = t(\theta\theta) = t(zz) = 0 \\
&t(rz) + \lambda_1 \partial t(rz)/\partial t = \mu[\partial v_z/\partial r + \lambda_2 \partial^2 v_z/\partial r \partial t] \\
&t(rr) + \lambda_1[\partial t(rr)/\partial t + 2t(rz)\partial v_z/\partial r] = 2\mu\lambda_2(\partial v_z/\partial r)^2
\end{aligned}
\tag{12.41}^{\dagger}
$$

The nonlinear inertial terms in the equation of motion are identically zero. Consider the predictions which such an admittedly crude constitutive assumption (note that the fluid does not have a shear dependent viscosity in viscometric flow) provide for the so-called "startup" problem, defined by the conditions

$$
\begin{aligned}
v_z = 0; \quad &dp/dz = 0 \quad \text{for } t \leqslant 0 \\
&dp/dz = -P \quad \text{for } t > 0
\end{aligned}
\tag{12.42}
$$

Centerline velocity profiles and volumetric flow rate are shown for a given set of parameters in Fig. 12.14. Note that both overshoot and undershoot of the final fully developed values are possible. Oscillatory approaches to fully developed values of stress and velocity have in fact been reported for real materials [90, 91]. Because of the symmetry in the equations and boundary conditions the inverse problem of "shutdown" is readily obtained from the solution described above. These results are also shown in Fig. 12.14.

A second tube flow of some illustrative value is that in which the pressure gradient oscillates sinusoidally with frequency ω.

---

† Note that with this form of constitutive equation $\Sigma t(ii)$ is not, in general, zero.

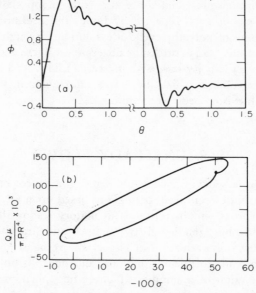

Fig. 12.14. Unsteady flow through a tube of radius R. Response to a step change in pressure gradient for a fluid (see (12.41)) with $\mu = 0.86$ poise, $\lambda_1 = 0.17$ s, $\lambda_2 = 7.8 \times 10^{-2}$ s. (a) Centerline velocity. $\theta = t\mu/\rho R^2$ and $\phi = (4\mu/PR^2)(v_z)_{r=0}$. (b) Volumetric flow rate Q and wall shear stress. $\sigma = t(rz)/PR$.

$$dp/dz = -\mathbf{R}(Pe^{i\omega t}) \tag{12.43}$$

the symbol $\mathbf{R}$ signifying that only the real part of its argument has physical significance. After initial transients have decayed, the volumetric flow rate undergoes a periodic over-shoot and undershoot relative to the flow rate of a corresponding Newtonian fluid at the same value of instantaneous wall stress [89]. A plot of volumetric flow rate $Q(t)$ against wall stress $t(rz)(R, t)$ exhibits a hysteresis loop which is characteristic of model parameters. Similar phenomena have been reported by others [92–94].

Though the fluid responses outlined above show some new and interesting features, several of which have been corroborated by experiment, the essential nonlinearity of rheologically complex fluids is missing from the analysis. This nonlinearity is the source for new phenomena which appear to have both practical and fundamental value, and which we wish to consider next. A thorough study of unsteady flows of nonlinear fluids has been conducted by Walters and his associates. We review some of their findings below.

Appreciation of the phenomenon of flow enhancement is readily gained by considering first a nonlinear *viscous* fluid in laminar flow through a circular tube and under the in-fluence of an axial pressure gradient

$$dp/dz = -\mathbf{R}[P(1 + \varepsilon e^{i\omega t})] \tag{12.44}$$

the constitutive equation being that of a generalized Newtonian fluid

$$t_{rz} \equiv \tau = \mu(\kappa)\kappa \tag{12.45}$$

where $\kappa = dv/dr$ and we use v here to signify the axial velocity. Assuming the amplitude ε of the oscillation to be small, Walters and coworkers have given an approximate solution by expanding velocity, shear rate, and stress in a power series in ε, and neglecting terms of $0(\varepsilon^3)$ [95]. One finds that the expansion takes the form

$$v = v_0(r) + \varepsilon v_1{}^{(2)}e^{i\omega t} + \varepsilon^2[v_2{}^{(1)}(r) + v_2{}^{(2)}(r)e^{i\omega t} + v_2{}^{(3)}e^{2i\omega t}] \tag{12.46}$$

with similar expressions for the stress and shear rate. An essential point is that the non-linearity of the problem introduces harmonics of the fundamental frequency ω of the oscillatory forcing function $-Pe e^{i\omega t}$. Furthermore, the oscillation gives rise to a new contribution to the *steady* component of the flow. It is this steady contribution which leads to a time-average flow different from that for steady flow with pressure gradient $-P$. Substitution of expansions such as (12.46) into the equation of motion yields a series of equations which can be solved numerically for the velocity functions $v_j{}^{(i)}$. Walters compared the experimentally observed increase in flow rate for a 1.5% aqueous polyacrylamide solution with predictions based upon viscometer measurements of the solution. Results for amplitude $\varepsilon = 0.2$ are shown in Fig. 12.15. Although quantitative agreement does not

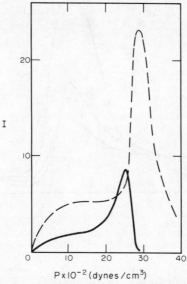

Fig. 12.15. Predicted (———) and experimental (– – – –) curves for enhancement of flow of a 1.5% aqueous solution of polyacrylamide. The ordinate I is the percentage enhancement of mean flow rate over that which should occur for $\varepsilon = 0$. In the case shown $\varepsilon = 0.2$, $\omega = 0.14$ Hz.

exist, there is a surprisingly good correspondence between general shape of experimental and theoretical curves. One notes in particular the substantial resonance effect which, if parameters are properly chosen, can give rise to an average flow rate substantially in excess of that due to a steady pressure gradient $-P$.

A more complete description in which the purely viscous fluid is a special case has been given by Barnes *et al.* [96]. Once more an expansion of the form (12.46) to $O(\varepsilon^2)$ has been used. Though the formal development is in terms of an integral constitutive model, it is shown to be consistent with the Oldroyd contravariant form of (11.11). Results of sample calculations are shown for the latter model, which is

$$t^{ij} + \lambda_1 \frac{\delta t^{ij}}{\delta t} + \mu_0 t_k{}^k d^{ij} = 2\eta_0 \left[1 + \lambda_2 \frac{\delta}{\delta t} \right] d^{ij} \tag{12.47}^\dagger$$

† As in (12.41), the stress for this model is not traceless.

Note that this model, in contrast to (12.41), provides for both elasticity and a shear dependent viscosity in steady flow. Barnes *et al.* computed the percentage enhancement in flow due to the fluctuation, i.e.,

$$I = \frac{100(Q - Q_s)}{Q_s}$$

Q referring to the volumetric flow rate and Q_s denoting the steady flow associated with time-independent pressure gradient $-P$. Representative results are shown in Fig. 12.16.

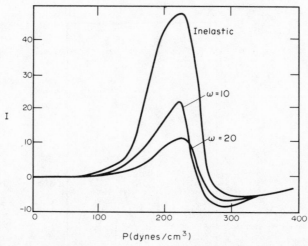

Fig. 12.16. Predicted flow enhancement for a viscoelastic fluid in pulsating pipe flow. Parameters are $\eta_0 = 10$ poise, $\lambda_1 = 0.06$ s, $\lambda_2 = 0.01$ s, $\mu_0 = 1.0$ s, $R = 0.25$ cm, $\epsilon = 0.25$. Frequency ω is in radians per second.

Of particular interest is the fact that maximum enhancement is predicted for the inelastic case, which corresponds to the limit, for an elastic fluid, of zero frequency. Note also that both positive and negative values of I are possible, and that the relatively simple inelastic model appears adequate to permit prediction of the conditions for I_{max}.

A qualitative explanation for the substantial change from the steady-state value of throughput is not hard to find. The timewise variation in pressure gradient causes a corresponding change in shear rate, which in turn results in a change in the effective viscosity. However, since the relation between pressure gradient and effective viscosity (and hence throughput) is nonlinear, there is no assurance that the mean throughput will correspond to the mean pressure gradient $-P$.

One should also take note of the energy necessary to effect the flow enhancement shown in, for example, Fig. 12.16. It is not obvious that the energy requirement of a pump designed to produce the pulsation (12.44) will be favorable relative to that required for the same average delivery under steady conditions. Barnes *et al.* [96] claim that the per-centage increase in energy can be positive *or* negative depending upon the value of flow parameters. As with their previous work, experimental results have confirmed the essential features of the theoretical predictions. One concludes that there may be substantial practical advantages to operation of pumping systems in an unsteady mode. It is also clear, however, that "tuning" of each situation is required to ensure optimum benefit—and in fact to avoid detrimental results which are inferior to steady-state operation.

At the time of this writing the value of nonlinear analysis of unsteady flows, as well as the attendant analytical difficulties, are just beginning to be appreciated. Dodge and Krieger [97], for example, have noted the possible effect of inertia in several viscometer geometries. This point is treated explicitly by Goldstein and Schowalter [98] in an analysis of oscillatory Couette flow, with or without a superposed parallel transverse steady flow. Most work is based upon a negligible effect of density. The degree of validity of such an assumption depends, of course, on the fluid in question and on the operating conditions [99]. If inertial effects can be neglected in, say, flow between parallel planes, then the stress is homogeneous, and the velocity profile is given without recourse to the equation of motion; only the constitutive relation is relevant. However, in the more general situation it is seen that density can have an important effect on the results.

In closing we note the importance of unsteady flows as a severe test of constitutive behavior. The equation developed by Bird and Carreau (eqn. 11.62)) has been successful in fitting several types of rheological data. Carreau has shown, however, that (11.62) is inadequate for quantitative fits to certain types of time dependent data [91]. He has also indicated some adjustments which appear to offer the greatest hope for success with his earlier model.

12.9. MORE ON SECONDARY FLOWS

In Section 12.5 we saw that fluid elasticity can have a profound effect on the nature o certain secondary flows. For example, the secondary flow patterns in a conical converging duct were discussed, and the limiting case of flow approaching an orifice in a flat plate was illustrated in Fig. 12.8. In the present section a few of the many possible examples of secondary flow are selected for further description.

(i) FLOW THROUGH NONCIRCULAR CONDUITS

One of the first and most fascinating predictions of secondary flow is due to Green and Rivlin [100] and to Ericksen [101]. They showed, using a Reiner–Rivlin constitutive equation (Section 10.2), that rectilinear flow through straight tubes with noncircular cross-section is not, in general, possible when one applies the usual pressure gradient dp/dz in the axial direction. This is true because, for the constitutive equation used, satisfaction of the axial component of the equation of motion does not lead to trivial satisfaction of the other two components. Green and Rivlin showed that, for the particular case of a pipe with elliptical cross-section, the equations of motion could be satisfied by adding cells of secondary flow symmetric about the axes of the ellipse as shown in Fig. 12.17. The subject is also developed in Fredrickson's book [102, p. 82]. Work on the same problem has been reported by Giesekus [103] and by Rivlin and Sawyers [104], who have cited experimental confirmation of the secondary flow.

In addition to its inherent interest the "elliptical tube" problem provides a cautionary lesson to those who may be tempted to believe that they have obtained a solution to the equation of motion when in fact they have only satisfied one component of the equation. This pitfall awaits the unwary in several areas of non-Newtonian fluid mechanics.

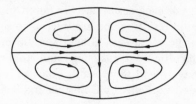

Fig. 12.17. Secondary flow cells for a Reiner–Rivlin fluid flowing through a tube with elliptical cross-section.

(ii) CONE-AND-PLATE FLOW

It will be recalled from Chapter 8 that cone-and-plate geometry is one of the mainstays of modern viscometry. In Section 8.5 we noted that the requirements for a viscometric flow could only be satisfied if one ignored the centrifugal term in the equation of motion. Thus, as centrifugal effects become appreciable one expects a corresponding secondary flow to develop. Recall also, however, that the analysis in Chapter 8 was based on a small angle between cone and plate so that one could approximate the flow field as one of constant shear rate. Substantial theoretical and experimental evidence shows that for appreciable cone angles secondary flows, which are of an elastic rather than an inertial nature, are to be expected. The effects have been beautifully documented by Giesekus [105, 106], who also derived theoretical predictions, and by Miller and Hoppmann [107]. In addition, Giesekus has shown the secondary flow which one can expect from a sphere rotating in a large mass of fluid.

Naturally, one is concerned about the effect that secondary flow, which only goes to zero asymptotically with decreasing cone angle, may have upon viscometric measurements—particularly measurements of normal stress. The subject has not been settled conclusively. For a good discussion of the possible influence of pressure-hole error, inertial effects, and boundary conditions on cone-and-plate viscometry, the reader is referred to a paper by Olabisi and Williams [108].

(iii) A NEARLY VISCOMETRIC FLOW

In Section 9.5 we developed in some detail Pipkin and Owen's treatment of nearly viscometric flows [109]. This formalism has provided an opportunity for analysis of a number of systems in which a strong primary flow is modified by a much weaker secondary flow, the result being a nearly viscometric flow in the sense of Pipkin and Owen. As an example of such a flow we choose the disk-and-cylinder system shown in Fig. 12.18. The cylinder, filled with fluid, has radius R and height H. It is fitted at the top with a disk which rotates at constant angular velocity Ω. This system has been considered theoretically and experimentally by Kramer and Johnson [110] and by Hill [111]. To obtain a tractable set of equations and also to be in a position to assign numerical values to flow parameters, they considerably simplified their description of constitutive behavior from the 13 material functionals which Pipkin and Owen cite as a maximum for this system.

A starting point for their analysis is the constitutive relation (7.11) for a simple fluid.

$$\mathbf{T}(t) = \mathbf{S}(t) + p(t)\mathbf{I} = \underset{s=0}{\overset{\infty}{\mathscr{H}}} \left[\mathbf{C}_{(t)}(s) \right] \tag{7.11}$$

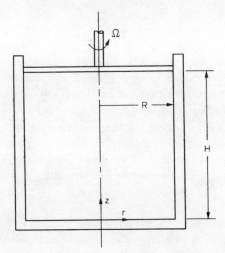

Fig. 12.18. The disk-and-cylinder configuration.

Following the formalism of Section 9.5 the functional $\mathscr{H}$ is expanded about the value $\overline{\mathbf{C}}_{(t)}(s)$ associated with a primary velocity field $\overline{\mathbf{v}}$, which is viscometric. The perturbation in the right relative Cauchy–Green strain tensor from its viscometric value is measured by

$$\mathbf{E}_{(t)}(s) = \mathbf{C}_{(t)}(s) - \overline{\mathbf{C}}_{(t)}(s) \tag{12.48}$$

as discussed in Section 9.5, and the velocity perturbation is

$$\mathbf{v}_1(\mathbf{x}) = \mathbf{v}(\mathbf{x}) - \overline{\mathbf{v}}(\mathbf{x}) \tag{12.49}^\dagger$$

where $|\mathbf{v}_1| \ll |\overline{\mathbf{v}}|$. The simplest possible nontrivial relation of $\mathbf{E}_{(t)}(s)$ to $\mathbf{v}_1(\mathbf{x})$ is a linear dependence of $\mathbf{E}_{(t)}(s)$ on the deformation rate (i.e., the first Rivlin–Ericksen tensor) of the perturbed flow. Thus

$$\mathbf{E}_{(t)}(s) \cong -s\mathbf{A}_{(1)1} \tag{12.50}$$

where $\mathbf{A}_{(1)1} = \nabla \mathbf{v}_1 + (\nabla \mathbf{v}_1)^T$. Furthermore, we know from Section 9.5 that $\mathscr{H}$ of (7.11) can be expanded around its viscometric value by a functional $\delta\mathscr{H}$ which is linear in $\mathbf{E}_{(t)}(s)$. The linear functional can be represented as an integral linear in $\mathbf{E}_{(t)}(s)$, so that

$$\delta\mathscr{H}_{ij} = \int_0^\infty K_{ij}^{kl}(\overline{\mathbf{C}}_{(t)}(s))E_{(t)kl}(s)\,ds \tag{12.51}$$

Substituting (12.50) we see that

$$\delta\mathscr{H}_{ij} = \overline{K}_{ij}^{kl}A_{(1)1kl} \tag{12.52}$$

where

$$\overline{K}_{ij}^{kl} = -\int_0^\infty K_{ij}^{kl}(\overline{\mathbf{C}}_{(t)}(s))s\,ds$$

† Cf. (12.55).

so that

$$\mathbf{T} = \overline{\mathbf{T}} + \overline{\mathbf{K}}\mathbf{A}_{(1)1} \tag{12.53}$$

where $\overline{\mathbf{T}}$ is the extra stress associated with the basic viscometric flow. As pointed out by Kramer and Johnson, the rather drastic restrictions already imposed on constitutive behavior imply that the basic flow itself is "slow" in the sense that a small departure from it can be described by (12.50). This is admittedly a rather restrictive assumption.

Kramer and Johnson next connect the material function $\overline{\mathbf{K}}$ to a familiar constitutive equation by invoking a somewhat generalized form of (11.62). Using (11.62), for example, one can express the integral in terms of the basic plus disturbance flows. Then, to first order in the disturbance, (11.62) can be used to obtain

$$\overline{K}^{kl}_{ij} = \mu(\kappa)\delta^k_i\delta^l_j \tag{12.54}$$

Here $\mu(\kappa)$ is the shear dependent viscosity, found from (11.62), for a viscometric flow. The result (12.54) is significant because it collapses the Pipkin–Owen formalism to only *one* material function, that being the fluid shear viscosity. It is surprising, and probably also unrealistic, that the secondary flow should be governed solely by the shear dependent viscosity. That is not to say, however, that elasticity does not enter into the problem. Kramer and Johnson approximated the flow by

$$\left.\begin{array}{l} v(r) = \mathcal{N}_R^2 v_1(r) \\ v(\theta) = \mathcal{N}_R \bar{v}(\theta) + \mathcal{N}_R^2 v_1(\theta) \\ v(z) = \mathcal{N}_R^2 v_1(z) \end{array}\right\} \tag{12.55}$$

where the basic flow, a function of r and z, is given by $\mathcal{N}_R \bar{v}(\theta)$, the symbols in parentheses here denoting physical components. Note that the velocity has been expanded in terms of a Reynolds number $\mathcal{N}_R = R^2\Omega\rho/\eta_0$. (It is surprising that a Reynolds number was chosen as the expansion parameter, since presumably elastic rather than inertial effects are of primary interest here.) In $\mathcal{N}_R$, η_0 is the zero-shear viscosity of the fluid. Then the expansion (12.55) is to $\mathrm{O}(\mathcal{N}_R^2)$, where $\mathcal{N}_R \ll 1$. Oldroyd [112] has shown that the basic flow is viscometric. One can readily employ $\mathcal{N}_R \bar{v}(\theta)$ to express the kinematics in the viscometric form (7.24) to (7.26), using a transformation similar to that given in (7.61) for curvilineal flows. If one defines

$$\kappa = (\kappa_1{}^2 + \kappa_3{}^2)^{\frac{1}{2}} \tag{12.56}$$

where

$$\kappa_1 = \partial\bar{v}(\theta)/\partial z \quad \text{and} \quad \kappa_3 = r\frac{\partial}{\partial r}\left(\frac{\bar{v}(\theta)}{r}\right)$$

then components of the stress for the basic flow take the form

$$\left.\begin{array}{l} \bar{t}(r\theta) = \mathcal{N}_R \kappa_3 \mu(\kappa) \\ \bar{t}(\theta z) = \mathcal{N}_R \kappa_1 \mu(\kappa) \\ \bar{t}(rz) = \mathcal{N}_R^2 \kappa_3 \kappa_1 \mathcal{N}_2(\kappa) \\ \bar{t}(rr) - \bar{t}(zz) = \mathcal{N}_R^2 [\kappa_3^2 - \kappa_1^2]\mathcal{N}_2(\kappa) \\ \bar{t}(\theta\theta) - \bar{t}(zz) = \mathcal{N}_R^2 [\kappa^2(\mathcal{N}_1(\kappa) + \mathcal{N}_2(\kappa)) - \kappa_1^2 \mathcal{N}_2(\kappa)] \\ \bar{t}(rr) - \bar{t}(\theta\theta) = \mathcal{N}_R^2 [\kappa_3^2 \mathcal{N}_2(\kappa) - \kappa^2(\mathcal{N}_1(\kappa) + \mathcal{N}_2(\kappa))] \end{array}\right\} \tag{12.57}$$

Here $\mathcal{N}_1$ and $\mathcal{N}_2$ are of course the viscometric functions defined by (7.33).

Predictions based on these foundations were compared with experiments performed with Newtonian fluids and with a solution of polyacrylamide. Agreement of velocity data with theory could only be claimed to within about 40%. However, when one considers the severity of the theoretical assumptions and the difficulty of obtaining accurate data, the results can be considered a hopeful sign. It should be noted that the agreement with theoretical predictions was substantially better with the procedure described above than with calculations performed from a second-order model. The latter, however, also predicts correct trends. These are quite dramatic, showing that elasticity can reverse the sense of rotation of secondary flow, a phenomenon noted earlier with other flows (Fig. 12.19).

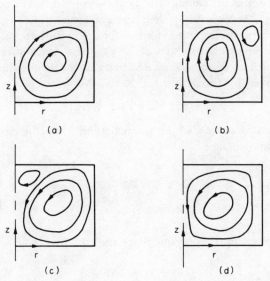

Fig. 12.19. Secondary flow in the disk-cylinder system according to Kramer and Johnson [110]. The effect of elasticity as predicted from a second-order model. $N_2=0$, $N_1(\kappa)/\kappa^2=$(a) 0, (b) 0.025 ρR^2, (c) 0.05 ρR^2, (d) ρR^2.

12.10. HYDRODYNAMIC STABILITY

Perusal of any recent volume of the *Journal of Fluid Mechanics* provides ample documentation for the statement that the subject of hydrodynamic stability is undergoing a period of high activity. The field has also become so specialized and difficult that an understanding of current work requires appreciable prior study of the subject. Such is the case with theories of hydrodynamic stability of Newtonian fluids. Our knowledge of stability of non-Newtonian fluids is much less well developed, and this may in fact account for the lack of coverage given to the subject in texts and review articles dealing with hydrodynamic stability. Although applications of stability theory for non-Newtonian systems are evolving, and although the significance of many of the results is still equivocal, the subject has achieved a stature and importance which warrants inclusion in this volume. We concentrate our attention on linear stability theory, mentioning only in passing some of the developments with nonlinear aspects, though it is the latter that are sure to assume increasing importance in coming years.

Study of stability phenomena with non-Newtonian fluids has an importance over and above the obvious desire to learn about transitions from one mode of flow to another. We shall see below that the stability behavior in some flow situations is highly sensitive to certain rheological parameters. This fact provides an additional tool for evaluation, through observation of stability phenomena, of rheological characteristics.

The two standard references to linear hydrodynamic stability theory are the treatise by Chandrasekhar [113] and the more abbreviated monograph by Lin [114]. The two works are complementary, in that the latter is concerned to a large degree with stability of pressure driven flow through conduits, a subject not covered in the former.

The essential idea used to formulate problems is simple enough. One considers a basic flow, presumed to be steady, which satisfies the appropriate differential equations and boundary conditions. The basic flow is then perturbed from its steady state, and differential equations and boundary conditions are formed for the perturbations, a crucial assumption being that any terms involving products of the perturbations and/or their derivatives are small enough to be neglected when compared to terms containing the perturbed quantities to degree one. Consider, for the moment, velocity. We write

$$\mathbf{v}(\mathbf{r}, t) = \bar{\mathbf{v}}(\mathbf{r}) + \tilde{\mathbf{v}}'(\mathbf{r}, t) \tag{12.58}$$

where $\bar{\mathbf{v}}(\mathbf{r})$ is the basic flow and $\tilde{\mathbf{v}}'$ the perturbation. Then the equation of motion and the continuity equation yield

$$\left. \begin{array}{c} \rho \left(\dfrac{\partial \tilde{\mathbf{v}}'}{\partial t} + \bar{\mathbf{v}} \cdot \nabla \tilde{\mathbf{v}}' + \tilde{\mathbf{v}}' \cdot \nabla \bar{\mathbf{v}} \right) = - \nabla p' + \nabla \cdot \tilde{\tau}' \\[2mm] \nabla \cdot \tilde{\mathbf{v}}' = 0 \end{array} \right\} \tag{12.59}$$

The density has been assumed constant. Since the equation is first order in time we can write

$$\tilde{\mathbf{v}}'(\mathbf{r}, t) = \mathbf{v}'(\mathbf{r})e^{i\sigma t} \tag{12.60}$$

In general σ will be complex, $\sigma = \sigma_r + i\sigma_i$. One seeks solutions to (12.59), with appropriate boundary constraints, to find conditions which divide $\sigma_i > 0$ from $\sigma_i < 0$, under which conditions the disturbance $\tilde{\mathbf{v}}'$ will, respectively, decay or grow in time. The result will of course depend on the details of the disturbance $\tilde{\mathbf{v}}'(\mathbf{r}, t)$, which is not uniquely defined by the information given thus far. Here again the linearity of the problem permits simplification because any disturbance $\tilde{\mathbf{v}}'(\mathbf{r}, t)$ can be synthesized from a superposition of Fourier components. Thus

$$\mathbf{v}'(\mathbf{r}) = \int \mathbf{v}_{\mathbf{k}}(\mathbf{k}, \mathbf{r})e^{i\mathbf{k} \cdot \mathbf{r}}d\mathbf{k} \tag{12.61}$$

the integration being over all values of vector-valued wave number $\mathbf{k}$. We need consider only that Fourier component which is most susceptible to instability, because one is usually interested in conditions which bound a region within which *all* Fourier components will decay in time.

Let us consider several examples:

(i) FLUID HEATED FROM BELOW

(a) *Newtonian Fluid.* Details of the method of linearized stability analysis are admirably demonstrated by the classical problem of a Newtonian fluid held between two horizontal plates in a gravitational field. The lower plate is maintained at a temperature T_1 greater than the upper plate, which is at T_2 (Fig. 12.20). One seeks those conditions for which

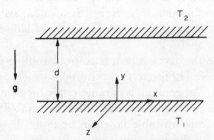

Fig. 12.20 Configuration for a fluid heated from below.

the buoyancy forces tending to initiate motion are just balanced by the viscous opposition to motion and the diffusive capacity for distribution of thermal energy. The relevant equations are

$$\nabla \cdot \mathbf{v} = 0; \quad \rho \frac{D\mathbf{v}}{Dt} = \rho \mathbf{g} + \nabla \cdot \mathbf{S}; \quad \frac{DT}{Dt} = \kappa_T \nabla^2 T \tag{12.62}$$

where $\kappa_T = k/(\rho c_p)$ is the thermal diffusivity, k is the thermal conductivity, c_p is the heat capacity, and $\mathbf{g} = -\mathbf{j}g$ is the acceleration due to gravity.

Let us first consider the basic solution to (12.62) for the case where the planes $y = 0, d$, are held motionless. Boundary conditions and equations are obeyed for heat transfer by simple conduction, i.e.,

$$\bar{\mathbf{v}} = \mathbf{0}; \quad \bar{T} = T_1 - \beta y \tag{12.63}$$

where $\beta = (T_1 - T_2)/d$ is the (constant) magnitude of the temperature gradient.

Proceeding further we specify an equation of state for the density

$$\rho = \rho_1 [1 + \alpha \beta y] \tag{12.64}$$

where $\alpha = 1/V(\partial V/\partial T)$, the coefficient of volumetric expansion. Then the pressure field is obtained from (12.62b), with $\mathbf{S} = -p\mathbf{I}$.

$$p = p_1 - \rho_1 g y \left(1 + \frac{1}{2} \alpha \beta y \right) \tag{12.65}$$

It is this rest solution that is now perturbed in the manner of (12.58). When the perturbed quantities are substituted into the governing differential equations (12.62) and the equations are linearized, the result is a set of equations for the disturbance quantities:

$$\left. \begin{array}{c} \nabla \cdot \tilde{\mathbf{v}}' = 0 \\[2mm] \rho_1 \dfrac{\partial \tilde{\mathbf{v}}'}{\partial t} = \tilde{\rho}' \mathbf{g} + \nabla \cdot \tilde{\mathbf{S}}' \\[2mm] \dfrac{\partial \tilde{T}'}{\partial t} + \tilde{\mathbf{v}}' \cdot \nabla \bar{T} = \kappa_T \nabla^2 \tilde{T}' \end{array} \right\} \tag{12.66}$$

To arrive at (12.66) we have assumed fluid physical properties to be independent of temperature, with of course the exception of density. However, we have made the customary assumption that the temperature dependence of density is important only in the gravitational term of the equation of motion. This is the "Boussinesq assumption" and is discussed by Chandrasekhar [113, p. 16] and, with some care, by Mihaljan [115]. For our purposes we accept its validity.

The next critical point in the development is proper handling of the amplification factor σ in (12.60). We wish to solve (12.66) to find conditions which divide $\sigma_i > 0$ from $\sigma_i < 0$, as explained following (12.60). In general, a parameter for the solution will be σ_r. However, it turns out that there are a number of classical linearized stability problems for which the conditions dividing stable from unstable flow can be found by setting $\sigma_r = 0$. This situation, sometimes described by the term "principle of exchange of stabilities", is valid for the problem at hand, as has been shown by Chandrasekhar [113, p. 24], and permits substantial simplification. Then we can find conditions for marginal stability by setting $\sigma_r = \sigma_i = \sigma = 0$, and, according to (12.60), $\tilde{\mathbf{v}}' = \mathbf{v}'$.

Next we specialize the description to a disturbance of a particular Fourier component, in accord with the discussion following (12.61). We express the disturbance in separable form with respect to the three coordinate directions. Furthermore, since the fluid is infinite in extent in the x–z plane, there are no boundary conditions to contend with for x or z. One is thus led to try a disturbance of the form

$$\left.\begin{aligned}
u' &= U(\xi) \exp\left[i(k_x x + k_z z)\right] \\
v' &= V(\xi) \exp\left[i(k_x x + k_z z)\right] \\
w' &= W(\xi) \exp\left[i(k_x x + k_z z)\right] \\
T' &= \theta(\xi) \exp\left[i(k_x x + k_z z)\right] \\
p' &= P(\xi) \exp\left[i(k_x x + k_z z)\right]
\end{aligned}\right\} \tag{12.67}$$

where $\xi = y/d$. The functions U, V, W, θ, and P are not all independent but are related to each other, and to k_x and k_z, through (12.66). Substitution into (12.66), written for a Newtonian fluid, and elimination of the pressure leads to

$$\left.\begin{aligned}
(D^2 - a^2)^2\, V(\xi) &= \frac{ga}{\nu}\, d^2 a^2 \theta \\[2em]
(D^2 - a^2)\theta &= -\frac{\beta}{\kappa_T}\, d^2 V(\xi)
\end{aligned}\right\} \tag{12.68}$$

where $\nu = \mu/\rho$, the kinematic viscosity; $D = d/d\xi$; $k^2 = k_x^2 + k_z^2$; and $a = kd$, the dimensionless wave number of the disturbance.

It is a simple matter to eliminate θ and to obtain the sixth-order equation

$$(D^2 - a^2)^3 V(\xi) = -N_{\mathrm{Ra}} a^2 V(\xi) \tag{12.69}$$

where $N_{\mathrm{Ra}} = (g\alpha\beta d^4)/(\kappa_T \nu)$ is the celebrated Rayleigh number, which governs the stability of this problem.

One now looks for nontrivial solutions to (12.68) which meet the boundary conditions

$$\left.\begin{aligned}
\theta &= V(\xi) = 0 \quad \text{at } \xi = 0, 1 \\
DV(\xi) &= 0 \qquad\quad \text{at } \xi = 0, 1
\end{aligned}\right\} \tag{12.70}$$

Solutions to (12.69) which meet the boundary conditions (12.70) are possible for only certain combinations of a^2 and $\mathcal{N}_{\rm Ra}$. The lowest Rayleigh number for which marginal stability is possible corresponds to a disturbance which is symmetric about the midplane between the two boundaries. The relation between $\mathcal{N}_{\rm Ra}$ and wave number for this particular type of disturbance is shown in Fig. 12.21. Assuming that a general disturbance would

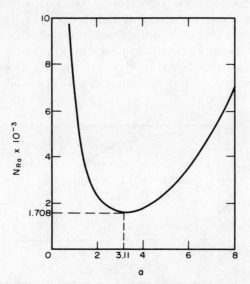

Fig. 12.21. An example of the relation between Rayleigh number, $\mathcal{N}_{\rm Ra}=g\alpha\beta d^4/(\kappa_T\nu)$, and dimensionless wave number $a=kd$ which divides stable and unstable conditions.

contain some contribution from $a=3.11$, one concludes that the critical value of Rayleigh number is 1708. Good agreement with experiment has been found. The interested reader is referred to Chandrasekhar for details [113, ch. 2].

We have noted the great simplification which is possible when the principle of exchange of stabilities is valid. One can then seek states of marginal stability by searching for conditions under which a solution to the governing equations is possible for $\sigma=0$. A second possible simplification concerns the dimensionality of the disturbance. In the example cited above the effect of k_x and k_z could be combined, because of the symmetry of the problem, into a single wave number k. It turns out that in problems with less symmetry one can sometimes still deal with two-dimensional disturbances since it can, in certain cases, be proved that these will be less stable than three-dimensional disturbances. This idea is embodied in what is usually called "Squires' theorem". Though it has wide applicability in Newtonian flows, use of the theorem in problems of non-Newtonian hydrodynamic stability is severely restricted [116].

It is interesting, finally, to contemplate the effect of a superposed shearing motion on the problem just considered. It has been found [117, 118] that as long as one is interested in the minimum Rayleigh number below which the system must be stable to *any* infinitesimal disturbance, there is no effect of shearing, and the no-flow solution is still valid.

(b) *Non-Newtonian Fluid.* We dealt with the Newtonian problem at some length because it provides a straightforward illustration of many key features of hydrodynamic stability. We can now discuss some results of the comparable non-Newtonian problem with relative ease. In particular because of the coupling between rheology and flow, one would wish to consider the problem just mentioned, i.e., plane Couette flow with heating from below, for non-Newtonian fluids. Not surprisingly, the flow can have a profound effect on the mode of heat transfer between the two plates. McIntire and Schowalter [119] carried out a linearized stability analysis for both a Bird–Carreau model (11.62) and a second-order fluid (10.16), following a procedure analogous to that given above for Newtonian fluids. The results show that, in particular, the sign of the second normal stress difference N_2 can appreciably stabilize or destabilize the flow as the difference is negative or positive, respectively. Also, for the parameters used in the illustration the choice between Bird–Carreau and second-order fluid models is of no consequence. Hence the results shown in Fig. 12.22 are given in terms of a second-order model. The results presented are for the

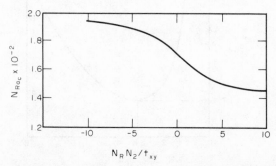

Fig. 12.22. Effect of second normal stress difference N_2 on critical Rayleigh number N_{Ra_c}. $N_R = \rho_1 d^2 \kappa / \alpha_0$ (κ=shear rate, α_0=shear viscosity from (10.16)), $N_{Ra} = \rho_1 g \alpha \beta d^4 / (\kappa_T \alpha_0)$. Results taken from [119].

range of parameters in which the first normal stress difference does not have a noticeable effect on the results. A more thorough treatment of the problem is also available [120].

One should not conclude from these results that the choice of constitutive model is of no consequence in linear stability theory. The second-order model is notorious for its unreliability in time-dependent problems. Furthermore, if one chooses a model which does not admit both first and second normal stress differences, important features of the problem can be overlooked.

Consideration of oscillatory instabilities ($\sigma_i = 0$; $\sigma_r \neq 0$) renders the problem appreciably more difficult. Also, recall that no theorem equivalent to Squires' theorem is generally valid. Hence a complete analysis of the problem must simultaneously deal with $\sigma_r \neq 0$ and with three-dimensional disturbances. Oscillatory disturbances have been considered not only for the problem at hand [120], but for the case where the fluid is initially quiescent [121], and for plane or cylindrical Couette flow of several constitutive models with no temperature gradient present [122, 123]. However, the case of three-dimensional disturbances has not been solved.

(ii) CYLINDRICAL COUETTE FLOW

This configuration was discussed in Section 8.6 as one of the laminar shear flows with which rheological measurements could conveniently be made. We shall see here that the flow can be considerably more complex than that given by (8.36).

With Newtonian fluids an instability can be caused by inertial effects which are present because of the rotational nature of the flow. Taylor was the first to provide a definitive theoretical and experimental description of the viscous and centrifugal forces which control stability of the system [113, ch. 7; 124]. The nature of the flow is governed by the Taylor number, defined for flow under conditions such that $(R_2 - R_1)/R_1 \ll 1$ by

$$T = \frac{2\rho^2 M \Omega_1^2 R_1 \delta^3}{\mu^2} \tag{12.71}$$

where

$$\delta = R_2 - R_1 \quad \text{and} \quad M = (\Omega_1 - \Omega_2)/\Omega_1$$

Other new notation is explained in Fig. 12.23.

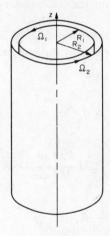

Fig. 12.23 Flow between rotating cylinders.

Instability phenomena are more varied and complex when the fluid is not Newtonian. One can gain a good appreciation of the nature and variety of secondary flows which occur by reading papers of Giesekus [125] and Hayes and Hutton [126]. The most general theoretical analysis of the problem is that of Goddard and Miller [127], referred to in the discussion of nearly viscometric flows, Section 9.6. However, association of their results with experiment is difficult because of the appearance of parameters for which no experimental values exist.

A bewildering variety of analyses of the cylindrical Couette flow problem has been published [127a]. Probably the simplest which also shows correspondence with experimental results is that of Denn and coworkers. In their initial treatment the constitutive model is essentially that of a second-order fluid [128]. Analogous to the description given for the problem of a fluid heated from below, small perturbations about a stationary state were considered. In this case the "stationary state" is a laminar shear flow between the cylinders of Fig. 12.23. Then

$$\bar{v}_\theta = \frac{r(R_2^2 \Omega_2 - R_1^2 \Omega_1)}{(R_2^2 - R_1^2)} + \frac{1}{r} \frac{R_2^2 R_1^2 (\Omega_1 - \Omega_2)}{(R_2^2 - R_1^2)} \tag{12.72}$$

Ginn and Denn assumed two-dimensional axisymmetric disturbances. Following (12.60) and subsequent equations one can write

$$
\left.
\begin{aligned}
v_r' &= -\frac{\alpha_0 k}{\rho \delta} u'(x) \cos (kz/\delta) \\
v_\theta' &= -k\Omega_1 R_1 M v'(x) \cos (kz/\delta) \\
v_z' &= \frac{\alpha_0}{\rho \delta} \frac{du'(x)}{dx} \sin (kz/\delta)
\end{aligned}
\right\}
\tag{12.73}
$$

where $x = (r - R_1)/R_1$; k is the dimensionless wave number of the disturbance (we have suppressed the subscript k, used in (12.61), on the left-hand side of (12.73)); v_r', etc., refer to physical components of the velocity disturbance with wave number k; $u'(x)$, $v'(x)$ are dimensionless quantities representing the x-dependence of perturbations; and α_0 is the second-order constant (see (10.16)).

In this analysis it was assumed that overstability is not important, and σ_r was set to zero. The governing differential equations for marginal stability ($\sigma_i = 0$) were found by Ginn and Denn to be

$$
\left.
\begin{aligned}
\left(\frac{d^2}{dx^2} - k^2\right) v' &= -u' + (P_1 + 2P_2)\left(\frac{d^2}{dx^2} - k^2\right) u' \\
\left[\frac{d^2}{dx^2} - k^2\right]^2 u' &= Tk^2\left[(1-Mx)v' - P_2 M \frac{R_1}{\delta}\left(\frac{d^2}{dx^2} - k^2\right) v' + 2(P_1 + P_2)M \frac{dv'}{dx}\right]
\end{aligned}
\right\}
\tag{12.74}
$$

along with boundary conditions

$$
u' = \frac{du'}{dx} = v' = 0 \text{ at } x = 0 \text{ and } x = 1
$$

P_1 and P_2 are related to first and second normal stress differences.

$$
P_1 = \frac{N_1}{2\rho \delta^2 \kappa^2} = -\frac{\alpha_1}{\rho \delta^2}; \quad P_2 = \frac{N_2}{2\rho \delta^2 \kappa^2} = \frac{(\alpha_2 + 2\alpha_1)}{2\rho \delta^2}
\tag{12.75}
$$

To find conditions of marginal stability one now searches for those combinations of parameters which permit a solution to the system of equations and boundary conditions. For given values of P_1 and P_2, that wave number k is found for which T at $\sigma = 0$ is a minimum. This defines a critical Taylor number. Results calculated by Ginn and Denn are shown in Fig. 12.24. One immediately notes the striking effect of normal stress differences on critical Taylor number relative to the results for a Newtonian fluid. As N_1 increases from zero the flow is destabilized. As N_2 becomes positive or negative for a given value of N_1, the flow is destabilized or stabilized, respectively. The result is qualitatively reminiscent of the coupled heat transfer and flow problem discussed earlier.

Though the analysis was admittedly based upon crude assumptions, impressive consistency was found with experimental determination of critical Taylor number for several dilute polymer solutions [129]. The value k_c of wave number k at which $T = T_c$ is, however, in rather poor agreement with experiment, the predicted cell spacing being too large.

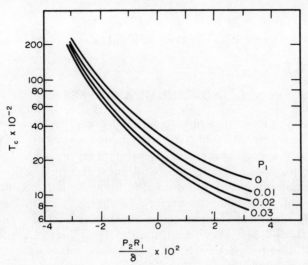

Fig. 12.24. Critical Taylor number T_c for rotational Couette flow of a viscoelastic fluid. From Ginn and Denn [128].

Thus k_c from theory is smaller than that observed. This discrepancy has not been removed by use of a more sophisticated constitutive equation [130]. Of course, there is no assurance that the value of k_c predicted from linear theory must coincide with the cell spacing of the secondary flow which is observed in experiments, since the observed disturbance is of finite amplitude. Furthermore, reduction of the data to isolate values of $\mathcal{N}_1$ and $\mathcal{N}_2$ is not a straightforward process. Jones and coworkers [127a] have conducted experiments and analysis to show that small experimental errors can cause large errors in the evaluation of the second-order constitutive coefficients α_i of (10.16). One must also remember that the second-order model with constant coefficients is known to be an inadequate description of most non-Newtonian materials over a range of shear rates.

(iii) POISEUILLE FLOW

Plane and cylindrical Poiseuille flow have been popular prototypes for analysis of Newtonian hydrodynamic stability, and a corresponding interest has developed for non-Newtonian fluids. In contrast to the examples already discussed, linear theory has not been successful for predicting critical conditions at which one observes transition from laminar to turbulent flow with Newtonian fluids. Nonlinear aspects of the problem are evidently important because of inherent differences from the flows discussed earlier. Nevertheless, linear analyses offer insights into the nature of disturbance propagation and provide the beginnings from which one can proceed to a nonlinear analysis.

The plane Poiseuille problem has been treated in numerous papers. One example is that of Porteous and Denn [131]. They used both second-order and convected Maxwell models. In agreement with other workers it was found that an increase in the absolute value of fluid elasticity, expressed in terms of an elasticity number which, for a second-order fluid, is

$$E = -\frac{\alpha_1}{\rho L^2}$$

destabilized the flow. Here L is a characteristic length and α_1 is defined in (10.16).

(iv) NONLINEAR ANALYSES

Following the development of techniques for nonlinear analysis of Newtonian stability problems, nonlinear effects with non-Newtonian fluids are currently of interest. Samples of two widely different approaches are adaptations of the energy method to non-Newtonian flows [132] and expansions about the results of linear theory. The former provides information about sufficient conditions for stability while the latter is an expansion about necessary conditions for instability. An example of the expansion method is the work on plane Poiseuille flow, published almost simultaneously by McIntire and Lin [133] and by Porteous and Denn [134]. They found that the destabilization predicted from linear theory is partially offset by a countertrend toward stabilization which develops as finite disturbances are considered.

12.11. DRAG REDUCTION

Of all the unusual phenomena observed with flow of non-Newtonian systems, none has caused more scientific and economic speculation than drag reduction. Described briefly, this term usually refers to the substantial reduction in drag which occurs under conditions of turbulent flow when minute amounts of a soluble polymer, for example polyethylene oxide, are dissolved in a solvent, say water, flowing at a fixed flow rate. Additive concentrations of less than 50 parts per million (by weight) have reduced frictional drag by more than 30%. The economic as well as the scientific appeal of such a result is obvious. Drag reduction is sometimes referred to as the "Toms phenomenon", in honor of the person who gave an early report of this remarkable effect [135]. Interest in the subject has been high since the work of Hoyt and Fabula was published in 1966 [136]. The fundamental factors which control drag reduction have been extremely elusive, and we are still not close to a satisfactory description or explanation. Authoritative reviews appear at regular intervals [137–140], and in view of the tentative nature of the subject at this date these reviews and their sequels will be the best source for relatively concise descriptions of research on drag reduction. The material presented in this section is limited to flow through straight conduits with circular cross-section. It is hoped, however, that an overview will acquaint the reader with the phenomenon, will point out some aspects of the subject which are likely to have permanent value, and will indicate a few speculative ideas that currently have appeal.

(i) THE PHENOMENON

Some appreciation of the bewildering variety of data available can be seen in Fig. 12.25. If one is content with a crude description of the phenomenon, the features listed below (for all of which there are exceptions) are typical:

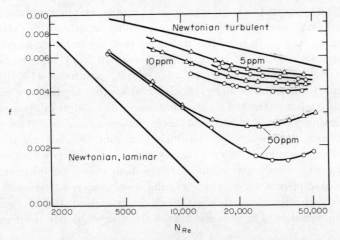

Fig. 12.25. Representative drag reduction data [141]. Data were obtained in a 0.248 in diameter pipe with polyethylene oxide (weight-average molecular weight=500,000). Data indicated by ($\triangle$) were obtained approximately 2 ft downstream from data indicated by ($\bigcirc$). Friction factor $f = 2\tau_w/(\rho v^2_{ave})$. $N_{Re} = Dv_{ave}\rho/\mu$. See [84, p. 186] for a description of friction factor correlations for Newtonian fluids.

(1) Less than 200 ppm of a high molecular weight polymer with little branching is added. Frequently, so little is used that in conventional measurements the solution displays no measurable normal stress differences and the viscosity is not appreciably changed from that of the (Newtonian) solvent.

(2) On a conventional friction factor–Reynolds number plot [84, p. 186], the laminar and transition regimes are unaffected by the additive. However, at some point beyond the conventional transition Reynolds number of 2100 for tube flow, the effect of polymer additive is noticed, and the drag coefficient drops below the standard correlation for Newtonian fluids.

(3) Increasing additive concentration increases drag reduction, but only to a point. Beyond this the drag may actually increase, presumably because of increased solution viscosity.

(4) Although the severity of the effect is a matter of some dispute, pipe diameter is a relevant parameter. Several workers have indicated that the Toms phenomenon is less effective in large pipes than in small [138]. It is also likely that the degree of polymer agglomeration is important [142].

(ii) ATTEMPTS TO EXPLAIN DRAG REDUCTION

In view of the fact that drag reduction does not become prominent in laminar flow, one is immediately attracted to the notion that the time-dependent flow of turbulent eddies is coupled with a characteristic relaxation time of the polymer molecules in such a way that energy is stored rather than dissipated. It has been difficult to quantify this idea, however, for lack of a relaxation time which characterizes the usual polydisperse solutions [138]. Kundu and Lumley [143] have noted that one must, first of all, be careful to make the right comparison of time scales. Beyond that, prior fluid history can have profound effects on relaxation time and hence on drag-reducing capability [141, 144].

It is natural also to suspect that an interaction between polymer molecule and hydrodynamic structure may occur because of corresponding length scales determined from polymer size and eddy size. Lumley has indicated that the polymer length scale is too small, however, the ratio of it to eddy size being of order 10^{-2} [137]. He has argued that a characteristic time is more appropriate as a correlating parameter [140]. Nevertheless, a successful scheme for correlating drag reducing action relies heavily on characterization of the polymer through a characteristic length, namely, the radius of gyration [145].

Another interesting aspect of flow in turbulent eddies is their *extensional* nature [140]. Vortex filaments are continuously being stretched as they entwine and mix with other filaments. It has been postulated that the high resistance of polymer solutions to extension (Section 12.6) accounts for the reduction of energy dissipation (see, for example, Gordon [146]). The associated phenomenon of "vortex inhibition", whereby resistance to stretching appears to inhibit vortex formation during tank drainage of dilute polymer solutions, lends credence to the importance of extensional flow in the drag reduction mechanism [147].

(iii) ATTEMPTS TO CORRELATE DRAG REDUCTION

No single method has been developed which can satisfactorily correlate all drag reduction data. This is due, at least in part, to the fact that the true rheological and molecular parameters of the test fluid are often not known in sufficient detail. Probably the most successful and comprehensive correlation of drag reducing activity has been that developed by Virk and his coworkers [145, 148]. Since their scheme also provides some insight into the features of turbulent flow of drag-reducing solutions through tubes, we outline the major components of this approach.

Based primarily on their own work with pipe flow of polyethylene oxide solutions, but consistent also with data obtained from other aqueous polymer solutions, Virk *et al.* proposed the following guidelines for organizing results of drag reduction experiments:

(1) In laminar flow the friction factor–Reynolds number relation is that of the Newtonian solvent, since the additive is assumed to be so dilute that it does not affect viscometric flow behavior.
(2) Transition to turbulence takes place, for most systems, according to the predictions for the Newtonian solvent.
(3) Either at transition or at a critical value of wall shear stress τ_w^*, whichever is reached at a higher Reynolds number, drag reduction will commence.
(4) With increasing Reynolds number drag reduction will increase, up to a so-called maximum drag reduction asymptote.
(5) Velocity profiles in turbulent flows with drag-reducing additives present are consistent with a three-layer model.

We now expand on these features, following the ideas put forward by Virk in 1971 [148]. Velocity profiles in turbulent tube flows are traditionally correlated in terms of a viscous sublayer immediately adjacent to the wall, a buffer layer which serves as a region of transition, and a fully turbulent core. The result is a universal velocity profile with viscous sublayer given by

$$u^+ = y^+ \tag{12.76}$$

and the fully turbulent core by

$$u^+ = 2.5 \ln y^+ + 5.5 \tag{12.77}$$

If one is willing to disregard the buffer layer, the two regions are seen to divide at $u^+=y^+=$ 11.6. The notation is conventional [84, p. 164], with

$$u^+ = u/u_\tau$$
$$y^+ = yu_\tau/\nu$$
$$u_\tau = \sqrt{\tau_w/\rho}, \text{ the "friction velocity"}$$
$$u = \text{time-averaged velocity at some radial position } r$$
$$y = R-r$$
$$R = \text{pipe radius}$$

Behavior for drag-reducing solutions can be quite different, as shown in Fig. 12.26. Outside of the viscous sublayer it is seen that the data are bounded by solvent behavior and the so-called "ultimate profile", indicative of the maximum drag reduction asymptote. From the results presented in Fig. 12.26, Virk hypothesized that, in general, a turbulent

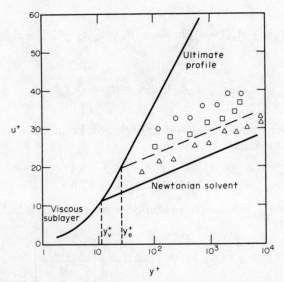

Fig. 12.26. Velocity profiles of drag reducing solutions after Virk [148]. Characters represent typical data Dashed line is the Newtonian plug for a typical solution.

profile will exhibit the characteristics shown with the dashed line on the figure. Near the wall one notes the customary sublayer. This is joined by a second *elastic sublayer*, which is characterized by the ultimate profile, and at some critical point, $y^+=y_e^+$, the elastic sublayer is connected with an outer region called the "Newtonian plug" because it has the same slope as the turbulent core in the purely Newtonian example. Since the ultimate profile, according to Virk's hypothesis, is a universal phenomenon, the location of the outer limit of the elastic sublayer y_e^+ is seen to be crucial in determining the degree of drag reduction. The inner boundary of the elastic sublayer $y^+=y_v^+$ is fixed by the intersection of (12.76) and (12.77). As drag reduction decreases $y_e^+ \to y_v^+$ and the Newtonian velocity profile is observed. At the other extreme, $y_e^+ \to R^+$, where $R^+=Ru_\tau/\nu$, the velocity profile is the ultimate profile, and the maximum drag reduction asymptote is reached.

These ideas were quantified by dividing turbulent flow behavior into three regimes, the boundaries being determined by the location of y_e^+. Thus

$$\ln(y_e^+/y_v^+) = 0 \quad \text{for } (R^+/R^{+*}) < 1 \tag{12.78}$$

$$\ln(y_e^+/y_v^+) = \psi \ln(R^+/R^{+*}) \tag{12.79}$$

$$y_e^+ = R^+ \quad \text{for } (R^+/R^{+*}) > (R^{+*}/y_v^+)^{((1/\psi)-1)} \tag{12.80}$$

where R^{+*} is the value of R^+ at onset of drag reduction. Thus (12.78) represents conditions below onset, and the profile is represented, approximately, by (12.76) to $y^+ = y_v^+$ and (12.77) at $y^+ \geqslant y_v^+$. Beyond onset but for values of the parameter $\psi < 1$ the elastic sublayer behavior shown in Fig. 12.26 and (12.79) obtains. For $\psi > 1$ condition (12.79) is replaced by (12.80), the maximum drag reduction asymptote. Average tube velocity in the three cases (12.78) to (12.80), respectively, is

$$u_{\text{ave}}^+ = A_n \ln R^+ + B_n - \frac{3}{2} A_n \tag{12.81}$$

$$u_{\text{ave}}^+ = [A_m \psi + A_n(1-\psi)] \ln R^+ + B_n - \frac{3}{2} A_n - (A_m - A_n)\psi \ln R^{+*} \tag{12.82}$$

$$u_{\text{ave}}^+ = A_m \ln R^+ + B_m - \frac{3}{2} A_m \tag{12.83}$$

where the contribution from $y^+ < y_v^+$ has been neglected. From available data Virk suggests the following values for empirical constants:

$$A_n = 2.5; \quad B_n = 5.5$$
$$A_m = 11.7; \quad B_m = -17.0$$

A crude correlation for ψ, obtained from results for several random coiling linear polymers in aqueous systems, is

$$\psi \cong (2.3 \times 10^{-14}) \left(\frac{A_n}{A_m - A_n}\right) \left[\frac{N_{\text{Av}} C N^3}{M}\right]^{\frac{1}{2}} \tag{12.84}$$

where N_{Av} is Avogadro's number, C is the weight fraction of polymer, M is the polymer molecular weight, and N is the number of backbone chain links in the polymer. The onset value R^{+*} is obtained from the empiricism

$$\frac{u_\tau^* R_g}{\nu} \cong 0.008$$

for random coiling molecules in aqueous solvents. R_g is the macromolecular radius of gyration. Knowing the relation between u^+ and y^+ for Newtonian fluids, and finding y_e^+ from (12.79), (12.81)–(12.83) can be used to infer velocity profiles.

The results presented above can be readily expressed also in terms of a friction factor–Reynolds number relation. For example, the condition for onset is given by

$$(N_{\text{Re}} f^{\frac{1}{2}})^* \cong 2\sqrt{2} \, (0.008)(R/R_g) \tag{12.85}$$

and the maximum asymptote is given by

$$f^{-\frac{1}{2}} = 19.0 \log_{10}(N_{\text{Re}} f^{\frac{1}{2}}) - 32.4 \tag{12.86}$$

where $f = 2\tau_w/(\rho U^2)$, $N_{\text{Re}} = 2RU/\nu$, and U is the average linear velocity.

Below onset the Newtonian friction factor relation in turbulent flow is operative. Virk uses the correlation

$$f^{-\frac{1}{2}} = 4.0 \log_{10}(N_{\text{Re}} f^{\frac{1}{2}}) - 0.4 \tag{12.87}$$

Since u_{ave}^+ and R^+ are proportional to $f^{-\frac{1}{2}}$ and $(N_{\text{Re}} f^{\frac{1}{2}})$, respectively, it is readily demonstrated [148] that the regime characterized by an elastic sublayer is given by a curve which intersects (12.87), when plotted as $f^{-\frac{1}{2}}$ against $\log(N_{\text{Re}} f^{\frac{1}{2}})$, at the onset point (12.85), but which has a slope of $4.0(1 + \Delta)$, where

$$\Delta \cong (2.3 \times 10^{-14}) \left[\frac{N_{\text{Av}} C N^3}{M} \right]^{\frac{1}{2}} \tag{12.88}$$

Details of the profile hypothesized by Virk and coworkers are oversimplified. In particular, careful measurements do not substantiate that in the buffer region the velocity profile universally follows the ultimate profile [148a].

(iv) OTHER FACTORS

We mentioned the problem of adequate characterization of test solutions. A continuing problem is the fact that drag-reducing activity is substantially affected by the technique used for additive dissolution and by the length of time that the drag reducer has been operative in a flow system. This aging phenomenon has been a major deterrent to widespread application of materials such as polyethylene oxide [138]. The dependence of activity on age and mixing history has not been fully explained. It does seem that there is more to the phenomenon than molecular degradation of the additive. In fact, substantial evidence exists to indicate that aggregation of polymer in some form of supermolecular structure has a strong effect on drag-reducing activity. The existence—and destruction—of aggregates may also explain the elusive "diameter effect" referred to earlier. For example, it has been found [142, 149] that aging or vigorous mixing during dissolution has a stronger effect on large than on small diameter tubes. It has been postulated that the rupture of supermolecular aggregates reduces the additive to a size which is ineffectual for drag reduction in the large tube but is still adequate with respect to the appropriate length scale for the smaller tube.

12.12. HEAT TRANSFER

In Section 4.8 a conservation equation for energy was presented, but it was stated there that little would be said in this book about transport of thermal energy in non-Newtonian systems. This is a subject for which both theory and experiments are in short supply. The case of purely viscous fluids is well understood, and the subject has developed by analogy to theory and empiricism for the Newtonian case. Conventional engineering heat transfer problems for these fluids are set out in detail in Skelland's book [22, ch. 10], to which the interested reader is referred. Christiansen and coworkers have included the effect of temperature variation on the constitutive equation, an example being heat transfer to non-Newtonian fluids in laminar flow through tubes [150].

The more interesting situations concern the role of viscoelasticity on heat and mass transfer. Given the unexpected action of dilute polymer additives on momentum transport, one would expect anomalies also for heat transfer. Experimenters have found that drag-reducing additives can reduce heat transfer, but the effect is not in complete analogy to the momentum transport phenomenon. Examples of heat transfer studies with visco-elastic fluids include those cited by Sidahmed and Griskey in a review of work relevant to their mass transfer experiments [151], a discussion of boundary-layer effects by Metzner and Astarita [33], and a study of hot-film anemometer response by Serth and Kiser [152].

In a radically different approach, Ultman and Denn have speculated that anomalous heat transfer in dilute polymer solutions may be due to an abrupt change in the nature of the flow at a critical condition [153, 154]. This condition is determined by an ordering parameter which contains the velocity of propagation of shear waves in the fluid. Good agreement is obtained between the theory and some heat-transfer data for cylindrical probes in dilute polymer solutions, but a more thorough comparison with data is needed.

12.13. REMARKS ON POLYMER PROCESSING

Though the title of this chapter contains the word "applications", little has been said about direct application of non-Newtonian fluid mechanics to engineering problems of polymer processing. We refer here to such important topics as extruder design, injection molding, film blowing, and the like. For an engineer these are of course the ultimate reasons for wishing to understand non-Newtonian fluid mechanics. It should be clear that no unequivocal methodology exists for approaching these complicated combinations of momentum, heat, and mass transfer. Various aspects of polymer processing have been covered in book form [155, 156], and the literature of the field continues to grow. Each problem has its special features; the wise engineer is the one who knows which aspects of a particular problem are amenable to generalization, which are unimportant, and which are peculiar to the problem at hand. To do that one should, at the very least, have a knowledge of certain types of well-defined nonviscometric flows of non-Newtonian fluids. It is with that task that this chapter has been primarily concerned.

Application of principles to processing problems then becomes rather specific. We give here just one example which illustrates how, by judicious approximations, one can perform analyses that have some value. The example chosen is polymer extrusion. The extrusion process bears little resemblance to the relatively straightforward viscometric flows at constant temperature which have been considered in earlier chapters. In fact, it is far more complicated than any of the other flow programs discussed in this chapter. Typically, a polymer simultaneously undergoes temperature change, phase change, and is subjected to a complicated and nonhomogeneous history of stress variation. Even if the constitutive behavior of the material were completely known, the process is so complex that exact solution of the boundary-value problem would, at best, be formidable. However, availability of large computers has permitted one to ascertain the relative importance of large numbers of variables over a range of operating conditions.

A schematic diagram of a simple single-screw extruder is shown in Fig. 12.27. Solid material enters one end of the screw apparatus from the hopper shown in the figure. In a

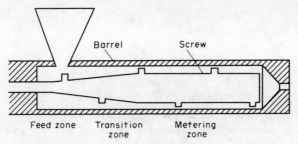

Fig. 12.27. Schematic sketch of a single-screw extruder.

transition zone the polymer begins to melt as a result of conductive heat transfer from the barrel, convection in the polymer liquid, and heat generation due to friction. Finally, a fully melted and highly viscous polymer is transported through the metering section. Heat generation can be appreciable in this section, so there is often a close coupling between energy and momentum transfer. Pearson and his colleagues have outlined analyses of the complete extrusion process [157], but we confine our remarks here to their description of the metering section.

An inelastic power-law description of constitutive behavior was used [158]. Thus (10.10) becomes

$$K[-4I_2]^{(n-1)/2} = \mu_{\text{eff}} = C[I_2]^{(n-1)/2} e^{-b(T-T_0)} \tag{12.89}$$

where $C=$ constant; I_2 is the second invariant of the rate of strain tensor (see Section 3.18; b is the temperature coefficient, a constant; and T_0 is the reference temperature.

Other physical properties, including density, are taken to be constant. Relatively, these are probably acceptable assumptions. Governing equations then are (4.15), (4.26), (4.30), (7.10), and (7.12), with the constitutive history reduced to (10.10) in the form (12.89). Judicious combinations of these equations and expression in dimensionless form lead one to a series of relevant dimensionless groups. Boundary conditions are of course also important for setting geometric as well as dynamic similarity. Of particular importance is the distance between flights w, and the channel height h (Fig. 12.28a). It is assumed that

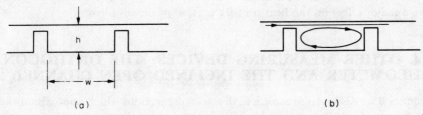

Fig. 12.28. (a) Primary geometric factors in screw extrusion. (b) Secondary flow due to clearance between flight and barrel.

$h/w \ll 1$, and that the gap between flight and barrel is small and of secondary if not negligible importance. Then if we also assume effects of inertia and gravity to be negligible, the problem is governed by five dimensionless groups:

Griffith number $N_G = b\mu_{eff}V^2/k$
Brinkman number $N_{Br} = \mu_{eff}V^2/(k\Delta T)$
Peclet number $N_{Pe} = Vh/\kappa$
Graetz number $N_{Gz} = Vh^2/(\kappa L)$
Aspect ratio $A = w/h$

where V is the characteristic velocity; e.g., circumferential velocity of rotating screw; k is the thermal conductivity; ΔT is the characteristic temperature difference between walls and melt; L is the characteristic length scale along fluid streamlines; $\kappa = k/(\rho c)$, the thermal diffusivity; ρ is the density; and, c is the heat capacity.

One looks for regions where one or more of the dimensionless groups above indicate a term of negligible importance. Since in any event the solutions are numerical, we mention here some qualitative notions concerning the dimensionless groups, but the reader must consult the cited literature to find results of specific calculations.

The Griffith number is a gauge of the importance of temperature dependence of viscosity. If solutions to the flow problem based on a temperature independent viscosity are to be valid, Pearson cites that a constraint $N_G < \frac{1}{2}$ should be imposed. In practice, values of N_G between zero and 200 can be expected [158].

The Brinkman number is a measure of thermal effects due to viscous heat generation relative to those caused by conduction from the surroundings. For realistic situations the number can vary by several orders of magnitude.

Peclet numbers are usually large, indicating that bulk transport of thermal energy in the flow direction by convection exceeds that by conduction.

The Graetz number can be thought of as an adjusted Peclet number which converts the dimension for conduction to the cross-stream direction. Thus N_{Gz} is a measure of the importance of bulk convection relative to conduction across streamlines. In realistic situations the value of N_{Gz} can vary widely. Pearson cites a range from 10^{-1} to 10^4.

The discussion above, and indeed almost all attempts to quantify the extrusion process, have assumed that the space between screw and barrel can be "unwrapped", and the flow approximated by straight streamlines in a channel. Martin [159] has allowed for the effect of the small gap between screw flight and barrel, and the fact that this gap permits communication between fluid on either side of a flight. The result is a complicated secondary flow in which there is not only leakage between the flight clearance separating channels, but overturning within a given channel, as shown in Fig. 12.28b. The flow is still predominantly parallel to the flight walls. Nevertheless, the secondary flow indicated can have a major effect on the heat transfer portion of the problem.

12.14. OTHER MEASURING DEVICES: THE ORTHOGONAL RHEOMETER AND THE INCLINED OPEN CHANNEL

In Chapter 8 we discussed at some length the configurations which approximate laminar shearing flows and which are used in practice to measure the viscometric functions. Use of some nonviscometric flows was described. We now take up some additional nonviscometric flows which have been used to characterize constitutive behavior. To fix ideas we restrict the discussion to two rather different examples of nonviscometric flow.

The orthogonal rheometer is a rheometrical tool which has enjoyed considerable popularity following its development by Maxwell and his students [160, 161]. The apparatus is shown schematically in Fig. 12.29. Test fluid is rotated and sheared between

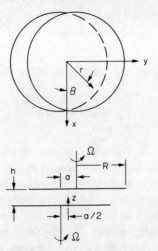

Fig. 12.29. The orthogonal rheometer.

two disks, each rotating with angular speed Ω and separated by a gap h. The disks are eccentric to each other, the top disk having its center of rotation displaced from the lower by a distance a along the y-axis.

In addition to the description cited, numerous analyses of the flow in the orthogonal rheometer have appeared [162–164]. Huilgol [165] has shown that the flow is one of constant stretch history (see Section 9.2). Primary interest in the flow stems from its close, but not complete, similarity to the oscillatory flows discussed in Chapter 10 in connection with linear viscoelasticity. Abbott and Walters [166] have obtained a solution of the Navier–Stokes equation, including inertia, for a Newtonian fluid in an orthogonal rheometer. It is noteworthy that the flow in any plane $z=$ constant is a solid-body rotation about an axis between $y=-a/2$ and $y=a/2$. In the limit of negligible inertia the locus of these centers of rotation is the straight line $z=(h/a)(y+a/2)$, but in general, the locus is more complicated. (Fig. 12.30). The analysis leading to Fig. 12.30 takes no account of edge effects. Boundary conditions are given by

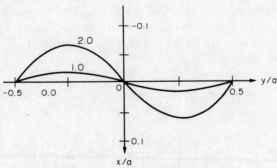

Fig. 12.30. Locus of centers of rotation for orthogonal rheometer flow of a Newtonian fluid. From Abbott and Walters [166]. The parameter is dh, where $d=(\rho\Omega/2\mu)^{\frac{1}{2}}$.

$$z=0:\quad \left.\begin{array}{l} v_x = -\ \Omega y - \tfrac{1}{2}a\Omega \\ v_y = \Omega x \\ v_z = 0 \end{array}\right\} \qquad (12.90)$$

$$z = h: \quad \begin{aligned} v_x &= -\Omega y + \tfrac{1}{2}a\Omega \\ v_y &= \Omega x \\ v_z &= 0 \end{aligned} \right\} \tag{12.91}$$

Then it is natural to assume that the velocity field has the form

$$\begin{aligned} v_x &= -\Omega y + g(z) \\ v_y &= \Omega x + k(z) \\ v_z &= 0 \end{aligned} \right\} \tag{12.92}$$

To proceed with an analysis for non-Newtonian fluids, a constitutive model must be posited. One can see from comparison of results obtained with different forms of constitutive equations, that the results, to first order, are not sensitive to the details of the constitutive model. As an example, consider the integral expression given by (11.62). From the assumed form of the velocity field (12.92) components of the strain tensors $\mathbf{\Gamma}$ and $\bar{\mathbf{\Gamma}}$ can be found and substituted into (11.62). The unknown functions $g(z)$ and $k(z)$ are related to the stress by an expansion in the parameter, assumed small, (a/h). Thus:

$$g(z) = \sum_{m=1}^{\infty} (a/h)^m g_m(z) \left. \vphantom{\sum_{m=1}^\infty} \right)$$

$$k(z) = \sum_{m=1}^{\infty} (a/h)^m k_m(z) \left. \vphantom{\sum} \right\} \tag{12.93}$$

$$t_{ij}(z) = \sum_{m=1}^{\infty} (a/h)^m (t_{ij})_m \left. \vphantom{\sum} \right)$$

Equating coefficients of equal powers of (a/h), one obtains, for the first-order terms,

$$\begin{aligned} (t_{xz})_1 &= \mu' g_1'(z) - \mu'' k_1'(z) \\ (t_{yz})_1 &= \mu' k_1'(z) + \mu'' k_1'(z) \end{aligned} \right\} \tag{12.94}$$

The coefficients μ' and μ'' have exactly the same meaning as the real and imaginary coefficients of the complex viscosity μ^* of (10.20), defined for small amplitude *oscillatory* motion where, in that case, one associates Ω with the oscillatory frequency. (Note that the oscillatory flow, in contrast to orthogonal rheometer flow, is not one of constant stretch history.) Then

$$\mu' = \int_0^\infty m(s, 0) \frac{\sin \Omega s}{\Omega}\, ds = \sum_{p=1}^{\infty} \frac{\eta_p}{1 + \Omega^2 \lambda_{2p}^2} \left. \vphantom{\int} \right\}$$

$$\mu'' = \int_0^\infty m(s, 0) \left[\frac{1 - \cos \Omega s}{\Omega} \right] ds = \sum_{p=1}^{\infty} \frac{\eta_p \lambda_{2p} \Omega}{1 + \Omega^2 \lambda_{2p}^2} \tag{12.95}$$

where $m(s, I_2)$ is defined by writing (11.62) in the form

$$\mathbf{T}(t) = \int_0^\infty m(s, I_2(s)) \left[\left(1 + \frac{\varepsilon}{2} \right) \bar{\mathbf{\Gamma}} - \frac{\varepsilon}{2} \mathbf{\Gamma} \right] ds$$

As mentioned earlier, the form of the integral expressions in (12.95) is not specific to the details of the integral constitutive model [166].

If one also expands the pressure with respect to the rest pressure at the origin by

$$p = \frac{1}{2} \rho(x^2+y^2) - \rho gz + \sum_{m=1}^{\infty} (a/h)^m p_m \qquad (12.96)$$

then pressure and velocity fields can be readily found to first order. It is convenient to define

$$G_1 = g_1 - ik_1; \quad K_1 = k_1 + ig_1 \qquad (12.97)$$

Then from substitution of (11.62), (12.92), (12.94), and (12.96) into the equation of motion one obtains, after some manipulation,

$$\left.\begin{array}{l} G_1 = \left[\dfrac{\Omega h}{2\sinh(bh)}\right] [\sinh(bz) + \sinh(b(z-h))] \\ \\ K_1 = iG_1 \end{array}\right\} \qquad (12.98)$$

where

$$b^2 = -i\Omega\rho/\mu^* \quad \text{and} \quad \mu^* = \mu' - i\mu''$$

In the limit $bh \to 0$, real and imaginary parts of the complex viscosity can readily be obtained from measurement of x- and y-components of force, F_x and F_y, respectively, on, say, the upper disk. From integration of normal stress over the disk one obtains

$$F_x = -(\pi R^2 a\Omega/h)\mu'; \quad F_y = -(\pi R^2 a\Omega/h)\mu'' \qquad (12.99)$$

Solutions correct to $O(a/h)^2$ have also been obtained [164]. As one would expect, characterization correct to this order requires new rheological coefficients.

In Section 8.8 we discussed the vexing problem of pressure-hole error, and the effect which this error had on early measurements of the second normal stress difference N_2. Tanner has published results which show very clearly that most earlier reports of N_2 were not correct with respect to sign [167]. He drew upon an analysis by Wineman and Pipkin [168], who had analyzed the flow of, essentially, a third-order fluid flowing by gravity down a tilted trough. It was found that the shape of the free surface was influenced by the sign and magnitude of N_2, the surface being convex upward for N_2 negative and concave upward for positive N_2. Tanner's experiments were performed in a channel with semicircular cross-section inclined at an angle α to the horizontal as shown in Fig. 12.31. In all cases studied, the center of the free surface rose above the level at the walls, indicating $N_2 < 0$.

Since the full analysis of this flow is very complex, we simply indicate here a first-order analysis. The assumption that, in this instance, small causes will yield small effects is based on the specific approximation scheme used by Wineman and Pipkin. Consider, first of all, rectilinear flow in the semicircular channel shown in Fig. 12.31. If $N_2 = 0$, the height of the free surface $h(x) = 0$, and all components of the equation of motion are satisfied by a viscometric rectilinear flow. Tanner's basic assumption is that the surface perturbation $h(x)$ is small, i.e., $h(x)/R \ll 1$, and that this perturbation has no effect, to first order, on the rectilinear velocity field. Then the fluid, in effect, "acts as its own pressure gauge", and $h(x)$ can be associated with N_2. To first order, flow in the semicircular cross-section is assumed to be viscometric, and the normal stress $s(\phi\phi)$ at the

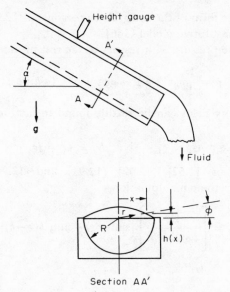

Fig. 12.31. Measurement of $\mathcal{N}_2$ by means of an open channel [167].

plane $\phi=0$ is assumed to be balanced by the hydrostatic head $\rho g h(x) \cos \alpha$ of fluid above the plane $\phi=0$. This is equivalent to expressing the velocity field for the flow by an expansion in powers of $\varepsilon = h_w/R$ $(h_w \equiv h(0))$ about a base viscometric flow $\mathbf{k}w^{(0)}$. Thus

$$\mathbf{v} = \mathbf{k}(w^{(0)} + \varepsilon w^{(1)}) + \varepsilon^2 \mathbf{v}^{(2)} + \mathbf{O}(\varepsilon^3) \qquad (12.100)$$

Substitution of (12.100), along with a corresponding expansion of the stress, into the equation of motion expressed in cylindrical coordinates ((A7.4)–(A7.6)), and isolation of coefficients of ε^0, provide the equation necessary for the viscometric base flow:

$$\frac{\partial p^{(0)}}{\partial r} = \frac{1}{r}\frac{d}{dr}(rt^{(0)}(rr)) - \frac{t^{(0)}(\phi\phi)}{r} - \rho g \cos \alpha \sin \phi \qquad (12.101)$$

$$\frac{\partial p^{(0)}}{\partial \phi} = -\rho g r \cos \alpha \cos \phi \qquad (12.102)$$

$$\frac{\partial p^{(0)}}{\partial z} = \rho g \sin \alpha + \frac{1}{r}\frac{d}{dr}(rt^{(0)}(rz)) \qquad (12.103)$$

where we have used the fact that $\mathbf{T}^{(0)} = \mathbf{T}^{(0)}(r)$. The last of these equations readily integrates to

$$t^{(0)}(rz) = \frac{1}{2}\rho g r \sin \alpha \qquad (12.104)$$

Equations (12.101) and (12.102) can be integrated to give

$$p^{(0)} = t^{(0)}(rr) + \int_0^r [t^{(0)}(rr) - t^{(0)}(\phi\phi)]\frac{d\xi}{\xi} - \rho g r \cos \alpha \sin \phi + C$$

But since $s^{(0)}(\phi\phi) = t^{(0)}(\phi\phi) - p^{(0)}$, one can write

$$- s^{(0)}(\phi\phi) = N_2 + \int_0^r N_2 \frac{d\xi}{\xi} - \rho g r \cos \alpha \sin \phi + C$$

At $r=R$ we can, neglecting interfacial tension, equate $s^{(0)}(\phi\phi)$ to atmospheric pressure, which is taken as zero. Then

$$- C = N_2(R) + \int_0^R N_2(\xi) \frac{d\xi}{\xi}$$

and

$$- s^{(0)}(\phi\phi) = N_2(r) - N_2(R) - \int_r^R N_2(\xi) \frac{d\xi}{\xi} \tag{12.105}$$

At the origin

$$- s^{(0)}(\phi\phi) = - t^{(0)}(\phi\phi) + p^{(0)} = p^{(0)} = \rho g h_w \cos \alpha$$

because, to within the viscometric approximation, the shear rate is zero at the origin. Then (12.105) becomes

$$\rho g h_w \cos \alpha = - N_2(R) - \int_0^R N_2(\xi) \frac{d\xi}{\xi} \tag{12.106}$$

But N_2 can be considered a function of shear stress through the relation (12.104). Hence one finds from differentiation of (12.106) with respect to wall shear stress τ_w,

$$\frac{d}{d\tau_w} (\tau_w N_2) = - \frac{d}{d\tau_w} (\tau_w \rho g h_w \cos \alpha) + \rho g h_w \cos \alpha$$

which can be integrated to

$$N_2(\tau_w) = - \rho g h_w \cos \alpha + \frac{1}{\tau_w} \int_0^{\tau_w} \rho g h_w \cos \alpha \, d\tilde{\tau}_w \tag{12.107}$$

It is convenient to interpret the integral in (12.107) as an integration of data obtained from a series of experiments run at different angles of inclination β between $\beta=0$ and $\beta=\alpha$. Then

$$\tilde{\tau}_w = \frac{1}{2} \rho g R \sin \beta$$

and (12.107) becomes

$$N_2(\tau_w) = - \frac{\rho g}{\sin \alpha} \int_0^{h_w(\alpha)\cos \alpha} \sin \beta \, d[h_w(\beta) \cos \beta] \tag{12.108}$$

From this equation one can obtain estimates of N_2 with relative simplicity over a limited range of shear rates. Tanner has done this [167], and a sample of his results is shown in Fig. 12.32. In all cases N_2 was found to be negative (i.e., of different sign than N_1) and about an order of magnitude smaller in numerical value than N_1.

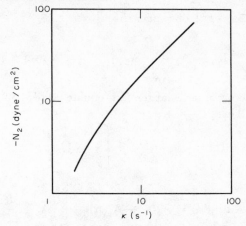

Fig. 12.32. Second normal stress differences from tilted trough experiments for 6.8% polyisobutylene in cetane at 24°C [167].

Subsequent work by Kuo and Tanner [169] showed how the analysis could be refined to include interfacial tension. It was found that interfacial tension could be important, but that the degree of importance depends upon both the fluids used and the conditions of the experiment.

REFERENCES

1. Schlichting, H., *Boundary-Layer Theory*, 6th edn. (New York: McGraw-Hill, 1968).
2. Cole, J. D., *Perturbation Methods in Applied Mathematics* (Waltham, Massachusetts: Blaisdell Publishing Co., 1968).
3. Van Dyke, M., *Perturbation Methods in Flud Mechanics* (New York: Academic Press, 1964).
4. Oldroyd, J. G., *Proc. Camb. Phil. Soc.* **43**, 383 (1947).
5. Acrivos, A., Shah, M. J., and Petersen, E. E., *AIChE Jl.* **6**, 312 (1960).
6. Schowalter, W. R., *AIChE Jl.* **6**, 24 (1960); **10**, 597 (1964).
7. Slattery, J. C., *Chem. Eng. Sci.* **17**, 689 (1962).
8. Shah, M. J., Petersen, E. E., and Acrivos, A., *AIChE Jl.* **8**, 542 (1962).
9. Rajeswari, G. K., and Rathna, S. L., *Z. ang. Math. Phys.* **13**, 43 (1962).
10. Beard, D. W., and Walters, K., *Proc. Camb. Phil. Soc.* **60**, 667 (1964).
11. Maiti, M. K., *Z. ang. Math. Phys.* **16**, 594 (1965).
12. Davies, M. H., *Z. ang. Math. Phys.* **17**, 189 (1966).
13. Leider, P. J., and Lilleleht, L. U., *Trans. Soc. Rheol.* **17**, 501 (1973).
14. Smith, L. S., PhD thesis, Princeton University, 1974.
15. Goldstein, S., *Modern Developments in Fluid Dynamics*, vol. 1, pp. 299–308 (London: Oxford University Press, 1938).
16. Vrentas, J. S., Duda, J. L., and Bargeron, K. G., *AIChE Jl.* **12**, 837 (1966).
17. Christiansen, E. B., Kelsey, S. J., and Carter, T. R., *AIChE Jl.* **18**, 372 (1972).
18. Bogue, D. C., *Ind. Eng. Chem.* **51**, 874 (1959).
19. Collins, M., and Schowalter, W. R., *AIChE Jl.* **9**, 98 (1963); **10**, 597 (1964).
20. Collins, M., and Schowalter, W. R., *AIChE Jl.* **9**, 804 (1963).
21. Rama Murthy, A. B., and Boger, D. V., *Trans. Soc. Rheol.* **15**, 709 (1971).
22. Skelland, A. H. P., *Non-Newtonian Flow and Heat Transfer*, (New York: Wiley, 1967).
23. Bogue, D. C., and Metzner, A. B., *Ind. Eng. Chem. Fund.* **2**, 143 (1963).
24. Truesdell, C., *Phys. Fluids* **7**, 1134 (1964).
25. Metzner, A. B., White, J. L., and Denn, M. M., *AIChE Jl.* **12**, 863 (1966).
26. Rivlin, R. S., Large elastic deformations, in *Rheology*, vol. 1, p. 351 (F. R. Eirich, ed.), (New York: Academic Press, 1956).
27. White, J. L., *J. Appl. Pol. Sci.* **8**, 2339 (1964).
28. Bogue, D. C., and White, J. L., Engineering analysis of non-Newtonian fluids, AGARDograph No. 144, Advisory Group for Aerospace Research and Development, Paris, 1970.

29. White, J. L., *J. Appl. Pol. Sci.* **8**, 1129 (1964).
30. Feinberg, M. R., PhD thesis, Princeton University, 1967.
31. Walters, K., *Z. ang. Math. Phys.* **21**, 276 (1970).
32. Denn, M. M., *Chem. Eng. Sci.* **22**, 395 (1967).
33. Metzner, A. B., and Astarita, G., *AIChE Jl.* **13**, 550 (1967).
34. Serth, R. W., *AIChE Jl.* **19**, 1275 (1973).
35. Hermes, R. A., and Fredrickson, A. G., *AIChE Jl.* **13**, 253 (1967).
36. Boger, D. V., and Rama Murthy, A. V., *AIChE Jl.* **16**, 1088 (1970).
37. Boger, D. V., and Rama Murthy, A. V., *Rheol. Acta* **11**, 61 (1972).
38. Happel, J., and Brenner, H., *Low Reynolds Number Hydrodynamics* (Englewood Cliffs, New Jersey: Prentice-Hall, 1965).
39. Pawlowski, J., *Kolloid Z.* **138**, 6 (1954).
40. Bird, R. B., *Phys. Fluids* **3**, 539 (1960).
41. Johnson, M. W., Jr., *Phys. Fluids* **3**, 871 (1960).
42. Johnson, M. W., Jr., *Trans. Soc. Rheol.* **5**, 9 (1961).
43. Schechter, R. S., *AIChE Jl.* **7**, 445 (1961).
44. Sutterby, J. L., *Trans. Soc. Rheol.* **9**, 227 (1965).
45. Oka, S., and Takami, A., *Japanese J. Appl. Phys.* **6**, 423 (1967).
46. Giesekus, H., *Rheol. Acta* **7**, 127 (1968).
47. Schümmer, P., *Rheol. Acta* **6**, 192 (1967).
48. Ackerberg, R. C., *J. Fluid Mech.* **21**, 47 (1965).
49. Langlois, W. E., and Rivlin, R. S., Steady flow of slightly viscoelastic fluids, Tech. Report No. 3 to Dept. of the Army (Ordnance Corps) on Contract DA-19-020-ORD-4725, Brown University, Division of Appl. Math., Providence, RI, 1959.
50. Giesekus, H., *Rheol. Acta* **8**, 411 (1969).
51. Kaloni, P. N., *J. Phys. Soc. Japan* **20**, 610 (1965).
52. Hopke, S. W., and Slattery, J. C., *AIChE Jl.* **16**, 224, 317 (1970).
53. Nakano, Y., and Tien, C., *AIChE Jl.* **14**, 145 (1968).
54. Caswell, B., and Schwarz, W. H., *J. Fluid Mech.* **13**, 417 (1962).
55. Giesekus, H., *Rheol. Acta* **3**, 59 (1963).
56. Caswell, B., *Chem. Eng. Sci.* **25**, 1167 (1970).
57. Cygan, D. A., and Caswell, B., *Trans. Soc. Rheol.* **15**, 663 (1971).
58. Astarita, G., *Ind. Eng. Chem. Fund.* **6**, 257 (1967).
59. Tanner, R. I., *Trans. Soc. Rheol.* **12**, 155 (1968).
60. Ballman, R. L., *Rheol. Acta* **4**, 137 (1965).
61. Fabula, A. G., Nonsteady elongational flow experiments with polymer solutions, paper presented at meeting of Society of Rheology, Washington, DC, October 1967.
62. Astarita, G., *Chem. Eng. J.* **1**, 57 (1970).
63. Vinogradov, G. V., Radushkevich, B. V., and Fikhman, V. D., *J. Pol. Sci.* Pt. A-2, **8**, 1 (1970).
64. Meissner, J., *Rheol. Acta* **8**, 78 (1969); **10**, 230 (1971).
65. Tanner, R. I., *AIChE Jl.* **15**, 177 (1969).
66. Tanner, R. I., and Ballman, R. L., *Ind. Eng. Chem. Fund.* **8**, 588 (1969).
67. Marrucci, G., *Ind. Eng. Chem. Fund.* **4**, 514 (1970).
68. Tanner, R. I., *Ind. Eng. Chem. Fund.* **9**, 688 (1970).
69. Denn, M. M., and Marrucci, G., *AIChE Jl.* **17**, 101 (1971).
70. Wankat, P. C., *Ind. Eng. Chem. Fund.* **8**, 598 (1969).
71. Treloar, L. R. G., *Trans. Faraday Soc.* **40**, 59 (1944).
72. Joye, D. D., Poehlein, G. W., and Denson, C. D., *Trans. Soc. Rheol.* **16**, 421 (1972); **17**, 287 (1973).
73. Maerker, J. M., and Schowalter, W. R., *Rheol. Acta* **13**, 627 (1974).
74. White, J. L., Theoretical considerations of biaxial stretching of viscoelastic fluid sheets with application to plastic sheet forming, Polymer Science and Engineering Report, No. 11, Dept. of Chemical and Metallurgical Engineering, The University of Tennessee, Knoxville, Tennessee.
75. Metzner, A. B., *AIChE Jl.* **13**, 316 (1967).
76. Metzner, A. B., Uebler, E. A., and Chan Man Fong, C. F., *AIChE Jl.* **15**, 750 (1969).
77. Metzner, A. B., and Metzner, A. P., *Rheol. Acta* **9**, 174 (1970).
78. Ziabicki, A., in *Man-Made Fibers*, vol. 1, pp. 13, 169 (H. F. Mark, S. M. Atlas, and E. Cernia, eds.), (New York: Interscience, 1967).
79. Ziabicki, A., and Kedzierska, K., *J. Appl. Pol. Sci.* **2**, 14, 23 (1959); **6**, 111, 361 (1962).
80. Ohzawa, Y., Nagano, Y., and Matsuo, T., *Proc. 5th Int. Cong. on Rheol.*, vol. 4, p. 393 (S. Onogi, ed.), (Tokyo: University of Tokyo Press, 1970).
81. Pearson, J. R. A., and Shah, Y. T., *Trans. Soc. Rheol.* **16**, 519 (1972).
82. Richardson, J. G., in *Handbook of Fluid Dynamics*, Section 16 (V. L. Streeter, ed.), (New York: McGraw-Hill, 1961).
83. Christopher, R. H., and Middleman, S., *Ind. Eng. Chem. Fund.* **4**, 422 (1965).

84. Bird, R. B., Stewart, W. E., and Lightfoot, E. N., *Transport Phenomena* (New York: John Wiley, 1960).
85. Marshall, R. J., and Metzner, A. B., *Ind. Eng. Chem. Fund.* **6**, 393 (1967).
86. Dauben, D. L., and Menzie, D. E., *J. Petroleum Technology* **19**, 1065 (1967).
87. Wissler, E. H., *Ind. Eng. Chem. Fund.* **10**, 411 (1971).
88. Siskovic, N., Gregory, D. R., and Griskey, R. G., *AIChE Jl.* **17**, 281 (1971).
88a. James, D. F., and McLaren, D. R., *J. Fluid Mech.* **70**, 733 (1975).
89. Etter, I., and Schowalter, W. R., *Trans. Soc. Rheol.* **9**, 351 (1965).
90. Kapoor, N. N., Kalb, J. W., Brumm, E. A., and Fredrickson, A. G., *Ind. Eng. Chem. Fund.* **4**, 186 (1965).
91. Carreau, P. J., *Trans. Soc. Rheol.* **16**, 99 (1972).
92. Pipkin, A. C., *Phys. Fluids* **7**, 1143 (1964).
93. Vela, S., Kalb, J. W., and Fredrickson, A. G., *AIChE Jl.* **11**, 288 (1965).
94. Jones, J. R., and Walters, T. S., *Rheol. Acta* **6**, 240 (1967).
95. Barnes, H. A., Townsend, P., and Walters, K., *Nature* **224**, 585 (1969).
96. Barnes, H. A., Townsend, P., and Walters, K., *Rheol. Acta* **10**, 517 (1971).
97. Dodge, J. S., and Krieger, I. M., *Rheol. Acta* **8**, 480 (1969).
98. Goldstein, C., and Schowalter, W. R., *Rheol. Acta* **12**, 253 (1973).
99. Macdonald, I. F., Marsh, B. D., and Ashare, E., *Chem. Eng. Sci.* **24**, 1615 (1969).
100. Green, A. E., and Rivlin, R. S., *Quart. Appl. Math.* **14**, 299 (1956).
101. Ericksen, J. L., *Quart. Appl. Math.* **14**, 318 (1956).
102. Fredrickson, A. G., *Principles and Applications of Rheology*, (Englewood Cliffs, New Jersey: Prentice-Hall, 1964).
103. Giesekus, H., *Rheol. Acta* **4**, 85 (1965).
104. Rivlin, R. S., and Sawyers, K. N., *Annual Review of Fluid Mechanics*, vol. 3, p. 117 (M. Van Dyke *et al.*, eds.), (Palo Alto: Annual Reviews, Inc., 1971).
105. Giesekus, H., *Proc. 4th Int. Congr. on Rheology*, Part 1, p. 249 (E. H. Lee and A. L. Copley, eds.), (New York: Interscience, 1965).
106. Giesekus, H., *Rheol. Acta* **6**, 339 (1967).
107. Miller, C. E., and Hoppmann, W. H., II, *Proc. 4th Int. Congr. on Rheology*, Part 2, p. 619 (E. H. Lee, ed.), (New York: Interscience, 1965).
108. Olabisi, O., and Williams, M. C., *Trans. Soc. Rheol.* **16**, 727 (1972).
109. Pipkin, A. C., and Owen, J. R., *Phys. Fluids* **10**, 836 (1967).
110. Kramer, J. M., and Johnson, M. W., Jr., *Trans. Soc. Rheol.* **16**, 197 (1972).
111. Hill, C. T., *Trans. Soc. Rheol.* **16**, 213 (1972).
112. Oldroyd, J. G., *Proc. Roy. Soc.* A, **283**, 115 (1965).
113. Chandrasekhar, S., *Hydrodynamic and Hydromagnetic Stability* (Oxford: Oxford University Press, 1961).
114. Lin, C. C., *The Theory of Hydrodynamic Stability* (Cambridge: Cambridge University Press, 1955).
115. Mihaljan, J. M., *Astrophysical J.* **136**, 1126 (1962).
116. Lockett, F. J., *Int. J. Eng. Sci.* **7**, 337 (1969).
117. Gallegher, A. P., and Mercer, A. McD., *Proc. Roy. Soc.* A, **286**, 117 (1965).
118. Deardorff, J. W., *Phys. Fluids* **8**, 1027 (1965).
119. McIntire, L. V., and Schowalter, W. R., *Trans. Soc. Rheol.* **14**, 585 (1970).
120. McIntire, L. V., and Schowalter, W. R., *AIChE Jl.* **18**, 102 (1972).
121. Sokolov, M., and Tanner, R. I., *Phys. Fluids* **15**, 534 (1972).
122. Beard, D. W., Davies, M. H., and Walters, K., *J. Fluid Mech.* **24**, 321 (1966).
123. Giesekus, H., and Bhatnagar, R. K., *Rheol. Acta* **10**, 266 (1971).
124. Taylor, G. I., *Phil. Trans. Roy. Soc.* A, **223**, 289 (1923).
125. Giesekus, H., *Progress in Heat and Mass Transfer*, vol. 5, pp. 187–194 (W. R. Schowalter *et al.*, eds.), (Oxford: Pergamon Press, 1972).
126. Hayes, J. W., and Hutton, J. F., *Progress in Heat and Mass Transfer*, vol. 5, pp. 195–210 (W. R. Schowalter *et al.*, eds.), (Oxford: Pergamon Press, 1972).
127. Goddard, J. D., and Miller, C., A study of the Taylor–Couette stability of viscoelastic fluids, ORA Project 06673, University of Michigan, Ann Arbor, Michigan, 1967.
127a. Jones, W. M., Davies, D. M., and Thomas, M. C., *J. Fluid Mech.* **60**, 19 (1973).
128. Ginn, R. F., and Denn, M. M., *AIChE Jl.* **15**, 450 (1969).
129. Denn, M. M., and Roisman, J. J., *AIChE Jl.* **15**, 454 (1969).
130. Sun, Z. S., and Denn, M. M., *AIChE Jl.* **18**, 1010 (1972).
131. Porteous, K. C., and Denn, M. M., *Trans. Soc. Rheol.* **16**, 295 (1972).
132. Feinberg, M. R., and Schowalter, W. R., *Ind. Eng. Chem. Fund.* **8**, 332 (1969).
133. McIntire, L. V., and Lin, C. H., *J. Fluid Mech.* **52**, 273 (1972).
134. Porteous, K. C., and Denn, M. M., *Trans. Soc. Rheol.* **16**, 309 (1972).
135. Toms, B. A., *Proc. 1st Int. Congr. Rheology*, Part 2, p. 135 (Amsterdam: North-Holland Publishing Co., 1948).
136. Hoyt, J. W., and Fabula, A. G., *Proc. Fifth Symp. on Naval Hydrodynamics*, 1964, p. 947 (Washington, DC, Office of Naval Research, 1966).

137. Lumley, J. L., *Annual Review of Fluid Mechanics*, vol. 1, p. 367 (W. R. Sears, and M. Van Dyke, eds.), (Palo Alto: Annual Reviews, Inc., 1969).
138. Gadd, G. E., *Encyclopedia of Polymer Science and Technology*, vol. 15, p. 224 (H. F. Mark, chm., editorial board), (New York: Interscience–Wiley, 1971).
139. Hoyt, J. W., *Trans. ASME Jl. Basic Eng.* **94D**, 258 (1972).
140. Lumley, J. L., *J. Pol. Sci.: Macromolecular Reviews* **7**, 263 (1973).
141. Paterson, R. W., and Abernathy, F. H., *J. Fluid Mech.* **43**, 689 (1970).
142. Ellis, H. D., *Nature* **226**, 352 (1970).
143. Kundu, P. K., and Lumley, J. L., *Phys. Fluids* **16**, 953 (1973).
144. Brennan, C., and Gadd, G. E., *Nature* **215**, 1368 (1967).
145. Virk, P. S., Merrill, E. W., Mickley, H. S., Smith, K. A., and Mollo-Christensen, E. L., *J. Fluid Mech.* **30**, 305 (1967).
146. Gordon, R. J., *J. Appl. Pol. Sci.* **14**, 2097 (1970).
147. Balakrishnan, C., and Gordon, R. J., *Nature, Phys. Sci.* **231**, 177 (1971).
148. Virk, P. S., *J. Fluid Mech.* **45**, 417 (1971).
148a. Reischman, M. M., and Tiederman, W. G., *J. Fluid Mech.* **70**, 369 (1975).
149. Elliott, J. H., and Stowe, F. S., Jr., *J. Appl. Pol. Sci.* **15**, 2743 (1971).
150. Christiansen, E. B., Jensen, G. E., and Tao, F. S., *AIChE Jl.* **12**, 1196 (1966).
151. Sidahmed, G. A., and Griskey, R. G., *AIChE Jl.* **18**, 138 (1972).
152. Serth, R. W., and Kiser, K. M., *AIChE Jl.* **16**, 163 (1970).
153. Ultman, J. S., and Denn, M. M., *Trans. Soc. Rheol.* **14**, 307 (1970).
154. Ultman, J. S., and Denn, M. M., *Chem. Eng. J.* **2**, 81 (1971).
155. Pearson, J. R. A., *Mechanical Principles of Polymer Melt Processing* (Oxford: Pergamon Press, 1966).
156. Tadmor, Z., and Klein, I., *Engineering Principles of Plasticating Extrusion* (New York: Van Nostrand Reinhold Co., 1970).
157. Martin, B., Pearson, J. R. A., and Yates, B., On screw extrusion: Part I, Steady flow calculations, Report No. 5, University of Cambridge, Dept. of Chemical Engineering Polymer Processing Research Center, 1969.
158. Pearson, J. R. A., *Progress in Heat and Mass Transfer*, vol. 5, pp. 73–87 (W. R. Schowalter *et al.*, eds.), (Oxford: Pergamon Press, 1972).
159. Martin, B., Numerical studies of steady-state extrusion processes, PhD dissertation, University of Cambridge, 1968.
160. Maxwell, B. and Chartoff, R. P., *Trans. Soc. Rheol.* **9**, 41 (1965).
161. Blyler, L. L., Jr., and Kurtz, S. J., *J. Appl. Pol. Sci.* **11**, 127 (1967).
162. Bird, R. B., and Harris, E. K., Jr., *AIChE Jl.* **14**, 758 (1968); **16**, 149 (1970).
163. Gordon, R. J., and Schowalter, W. R., *AIChE Jl.* **16**, 318 (1970).
164. Goldstein, C., and Schowalter, W. R., *Trans. Soc. Rheol.* **19**, 1 (1975).
165. Huilgol, R. R., *Trans. Soc. Rheol.* **13**, 513 (1969).
166. Abbott, T. N. G., and Walters, K., *J. Fluid Mech.* **40**, 205 (1970).
167. Tanner, R. I., *Trans. Soc. Rheol.* **14**, 483 (1970).
168. Wineman, A. S., and Pipkin, A. C., *Acta Mechanica* **2**, 104 (1966).
169. Kuo, Y., and Tanner, R. I., *Rheol. Acta* **13**, 443 (1974).

PROBLEMS

12.1. A 1% aqueous solution of carboxymethylcellulose (CMC) flows past a flat plate at zero incidence. The plate is 0.6 cm long (i.e., along the direction of flow), and the total force measured per unit width of one side of the plate is 1.26×10^3 dyne/cm. Approach velocity U_∞ of the CMC ($\rho \simeq 1$ g/cm^3) is 108 cm/s.

A rheogram of the CMC solution is approximated by the power-law constants $n = 0.5$ and $K = 24.8 \dfrac{\text{g}}{\text{cm(s)}^{2-n}}$

(data adapted from Hermes and Fredrickson [35]).

For a power-law fluid, Acrivos *et al.* [5] have given the following expression for the drag force per unit area τ_w at a distance x from the leading edge of the plate:

$$\frac{\tau_w}{\rho U_\infty^2} = c(n) \left[N_{\text{Re}_x} \right]^{\frac{-1}{1+n}}$$

where $c(0.5) = 0.58$ and $N_{\text{Re}_x} = \dfrac{\rho U_\infty^{2-n} x^n}{K}$

Compare the measured drag of the real fluid with the prediction from an inelastic model. Note that 1% CMC shows appreciable elasticity.

12.2. It is desired to approximate the stretching of a soft polymer filament by a convected Maxwell model and eqn. (12.35). For the polymer used $\mu = 10^4$ poise and $\theta = 0.5$ s. A constant stretch rate $d = 10^2$ s^{-1} is applied to the filament, initially a cylinder at rest with a diameter of 0.4 cm. How long will it take to reduce the filament diameter by a factor of two? According to the prediction of the model, what force, in grams, must be applied to the filament at that time?

12.3. A fluid characterized by the Bird–Carreau model (11.62) is made to oscillate between two parallel planes. Fluid velocity is given by

$$\mathbf{v} = (0, 0, w(y, t))$$

The lower plane (at $y=0$) is at rest, and the upper plane (at $y=h$) moves in the z-direction with velocity $V_0 \cos \omega t$. One wishes to predict the velocity profile and corresponding stresses.

Assume that velocities and stresses can be expanded in terms of a parameter, assumed small, $\alpha = V_0/(\omega h)$ [78]. Thus,

$$w = \sum_{l=1}^{\infty} \alpha^l w_l$$

$$t_{yz} = \sum_{l=1}^{\infty} \alpha^l t_l$$

$$t_{ii} = \sum_{l=1}^{\infty} \alpha^{2l} t_{ii}(2l)$$

(a)   Show that, to first order, the description is that of a linear viscoelastic fluid; i.e.,

$$t_1 = \mathbf{R}[t_1^1 e^{i\omega t}]; \quad w_1 = \mathbf{R}[w_1^1 e^{i\omega t}]$$

where $\mathbf{R}$ denotes the real part of the following argument. In the above expressions,

$$t_1^1 = \mu^* \frac{dw_1^1}{dy}$$

$$\mu^* = \int_0^\infty m(s, 0) \left(\frac{i - e^{-i\omega s}}{i\omega} \right) ds$$

$$m(s, 0) = \sum_{p=1}^{\infty} \frac{\eta_p}{(\lambda_{2p})^2} e^{-\frac{s}{\lambda_{2p}}}$$

and $w_1^1 = \dfrac{V_0 \sinh by}{\alpha \sinh bh}$

with $b^2 = i\omega\rho/\mu^*$.

(b)   When nonlinear effects are considered, oscillatory inputs can give rise to *steady* outputs. Show that, to $O(\alpha^2)$, normal stresses appear, and that one may expect them to contain both steady and oscillatory parts. Note that, for two complex variables z_1 and z_2,

$$\mathbf{R}(z_1)\,\mathbf{R}(z_2) = \tfrac{1}{2}\,\mathbf{R}[z_1 z_2 + \tilde{z}_1 z_2]$$

where $\sim$ denotes the complex conjugate.

12.4 Marshall and Metzner [85] have modeled flow through a porous medium by an assembly of spherical particles. For such a model, they recommend that the Deborah number can be estimated by

$$N_{\mathrm{Deb}} = 2.3\,\theta^{fl}\,\frac{V}{D^p}$$

where D_p is the particle diameter and V is the average pore velocity of the fluid.

In a polymer flooding operation, one might choose a displacement velocity of 10 ft per day through a rock that could be modeled by an assembly of spheres 40 μm in diameter. For a polyacrylamide flooding solution, the characteristic time θ_{fl} might be as high as 0.01 s. How close does one come to the critical Deborah number of Fig. 12.13?

12.5. Water ($\rho=1$ g/cm³, $\mu=1$ centipoise) is to be pumped through a Schedule 40 1-in pipe (actual inside diameter=1.049 in) at a flow rate of 25 US gallons per minute. Use Virk's correlation to estimate the percent reduction in power consumption if 30 weight parts per million of a polyacrylamide drag reducing agent are added. Relevant properties of the polyacrylamide are

$$R_g = 2500 \text{ Å}$$
$$M = 4.7 \times 10^6$$
$$\mathcal{N} = 1.3 \times 10^5$$

CHAPTER 13

SUSPENSION RHEOLOGY

13.1. INTRODUCTION

Ever since Einstein [1] developed an expression for the viscosity of a dilute suspension of rigid spheres, the rheology of multiphase systems has been an active topic for research. Two motivations can be identified. First, people have simply wished to know the rheological behavior of composite systems in terms of their component parts. Second, two-phase systems form an attractive model for rheologically complex materials. In this sense, in fact, the discussion of bead–spring and network models of Chapter 10 was "suspension rheology".

Following the first motivation the Einstein equation has been used and modified in a variety of ways. Unfortunately, the equation can be rigorously applied only to systems far more dilute than those likely to be of technological interest. Consequently, a great effort has been expended toward finding expressions for viscosity of suspensions under conditions of moderate to high concentration [2]. In spite of its great importance that subject has remained largely empirical. Our discussion here centers on the second motivation: a means of modeling from which one can infer something about fundamental processes which affect rheological behavior. The recent spate of activity in this subject [3] has resulted in a foundation from which one can proceed with some confidence to connect bulk stress to behavior of individual particles and thereby obtain at least a crude notion of the effect of molecular shape and deformability on bulk rheological properties. A thorough knowledge of the behavior of dilute suspensions also provides an appreciation of the problems associated with an extension to nondilute systems.

By "suspension rheology" (in the present context a "suspension" can have a solid, liquid, or gaseous discontinuous phase) we mean more than a study of the flow behavior of a multiphase system. The term refers to a modeling wherein explicit account is taken of the shape and mechanical properties of each phase to determine flow properties of the bulk. This distinguishes suspension rheology from the phenomenological continuum approach to rheology on the one hand, and from the macromolecular description of bead–spring and network models on the other. In the former one usually includes little or no provision for structure while in the latter the details of a coupled boundary-value problem between continuous and discontinuous phases are ignored. Shortcomings of each of these approaches, and by inference a statement of the purpose of suspension rheology, have been stated very well by Roscoe [4]:

"A phenomenological approach to the study of isotropic elastoviscous liquids is provided by the theory of 'simple fluids with fading memory' of Coleman and Noll, but this theory only gives limited information on account of its extreme generality. In order to obtain further results, it is necessary to construct theories which take some account of the structure of these liquids. Several such theories have been developed for liquid polymers and polymer solutions which take explicit account of the chain structure of the molecules. Difficulties arise, however, and these have only been resolved by making mechanical or hydrodynamic assumptions which are difficult to verify. When such theories fail to agree with observation, it is never clear whether the fault arises from the physical model or from these assumptions.

"A more satisfactory approach can be made by choosing a simplified physical model for which the mathematical calculations can be carried out exactly."

In this chapter we shall see that the models for which "calculations can be carried out [more or less] exactly" will often lead to constitutive relations of a form familiar from continuum or from macromolecular theories. The distinction of course is that in the present case the origin of each term is readily traced to the results of a mechanical analysis of each component phase, including conditions at boundaries between the two. We shall see, however, that there are two facets of suspension rheology which demand departure from a strictly deterministic analysis. One is the need for statistical averaging in order to infer bulk properties of the suspension from a detailed analysis of a small sample of the material. The other is the necessity, in many cases, of including the effect of Brownian motion. In Chapter 10 we went to some length to express the Brownian effect as a smooth function of position and time. We shall also employ this time-smoothed description in our treatment of suspensions.

We begin with the rationale for describing bulk properties in terms of microscopic behavior. This aspect of the subject has only recently been set out in an unambiguous fashion. Then we describe some of the classical models for suspensions of rigid particles. These models provide a base for discussion of the interesting behavior of deformable particles, including "elastic" effects such as stress relaxation and normal stress differences. We shall gain some insight into mechanisms for the great difference in response of suspensions to elongational and to shearing flows, and, finally, we shall consider some aspects of suspension rheology of nondilute systems (i.e., a suspension with interacting particles).

13.2. BULK PROPERTIES IN THE ABSENCE OF BROWNIAN MOTION: THE METHOD OF AVERAGING

In a fundamental paper Batchelor [5] has pointed out the conceptual advantages of using an ensemble average to relate microscopic to macroscopic properties. The ensemble average of some quantity, say $\mathbf{v}$, at position $\mathbf{r}$ and time t, is formed by performing a large number of experiments ("realizations") with the same macroscopic initial and boundary conditions, and measuring $\mathbf{v}$ at $\mathbf{r}$ at the same time relative to the initiation of each experiment. The average of these realizations is then the ensemble average. Use of the ensemble average implies of course that the same macroscopic boundary and initial conditions allow different microscopic conditions. This is the first sense, mentioned in the previous section, in which suspension rheology is indeterminate and must be treated statistically.

Although the ensemble average is conceptually convenient, it must be related to calculable averages if a connection with experiment is to be made, and it is at this point that the inevitable discussion, familiar from kinetic theory, of length scales arises. To proceed we must assume that the ensemble average is equal to an integral average over some region in the system characterized by length scale L, and that L is large compared to the average distance between particle centers but small with respect to a distance over which the average of the property in question, say $\mathbf{v}$, varies appreciably. In fact then we shall be computing volume (or surface) averages in the manner adopted by most workers in the field, though recognizing that, just as in kinetic theory, one can formulate the problem in terms of ensemble averages without a need to stipulate proper ordering of length scales.

Since we wish ultimately to develop a constitutive equation which relates bulk stress to bulk motion, definitions of some bulk properties are required. Over a volume V fulfilling the length–scale requirements just cited we define the average velocity gradient $\overline{\nabla \mathbf{v}}$ by

$$\overline{\nabla \mathbf{v}} = \frac{1}{V} \int_V \nabla \mathbf{v} \, dV \tag{13.1}$$

To relate this to measured quantities in a viscometer experiment one must be more specific. Suppose, for example, that we consider the motion induced in a suspension between two parallel planes in steady relative motion in the x-direction. If the planes are located at $y=0$ and $y=d$, the average (13.1) can be formed over some appropriate volume $(\Delta x)(\Delta z)d$. However, the suspension cannot be homogeneous in a statistical sense arbitrarily near to the walls. This means that (13.1), taken over a volume bounded by the planes $y=0$ and $y=d$, is not an ensemble average characteristic of the interior region because the sample volume is not statistically homogeneous. However, if we assume that the wall effect extends over a region small in extent compared to d, it is then easily shown that the average (13.1) over any volume between $y=0$ and $y=d$ should approach the ensemble average. Furthermore, it is readily shown that this average is just the shear rate $\kappa = [v_x(y=d) - v_x(y=0)]/d$. To do this, apply the divergence theorem to (13.1).

$$\overline{\nabla \mathbf{v}} = \frac{1}{(\Delta x)(\Delta z)d} \int \mathbf{v} \otimes \mathbf{n} \, dA$$

The only nonzero component is

$$\frac{\overline{\partial v_x}}{\partial y} = [v_x(y=d) - v_x(y=0)]/d = \kappa \tag{13.2}$$

This implies that the linear variation in average velocity between the two planes is the same as one would find for a single-phase fluid. Happily then, we may continue to use the same relation between shear rate and wall motion for conventional viscometers as has been familiar from single-phase continuum viscometry.

The development of an expression for bulk stress is more subtle. In fact, there is no unique definition for the quantity. We choose a definition consistent with the interpretation from kinetic theory of stress as a manifestation of momentum transport on a microscopic scale [6], and a definition which also leads to a familiar form for energy dissipation expressed in terms of bulk quantities. Thus in defining the average (ensemble or areal) stress across a surface element we take the sum of the contact forces exerted across the element plus any momentum flux transported by fluctuations $\mathbf{v}' = \mathbf{v} - \bar{\mathbf{v}}$, the overbar indicating an ensemble or an integral average, since we shall henceforth assume equivalence between the two. Then, following Batchelor [5], a bulk stress $\boldsymbol{\Sigma}$ is defined by

$$\boldsymbol{\Sigma} = \bar{\mathbf{S}} - \overline{\rho\mathbf{v}' \otimes \mathbf{v}'} \tag{13.3}$$

due regard being taken of the sign convention for stress. Let us suppose that the suspending fluid is Newtonian and incompressible. Then taking into account the meaning of the overbar in (13.3), one may write

$$\boldsymbol{\Sigma} = \frac{1}{V} \int_{V - \Sigma V_p} (-p\mathbf{I} + 2\mu\mathbf{D})dV + \frac{1}{V}\Sigma \int_{V_p} \mathbf{S}dV - \frac{1}{V}\int_V \rho\mathbf{v}' \otimes \mathbf{v}'dV \tag{13.4}$$

where V is a suitable suspension volume over which the average is being taken, V_p denotes the volume of any particle, and ΣV_p the sum over all particle volumes. The suspension is also taken to have constant density and to be free from external forces (or, equivalently, to be subject to a constant external force), but the possibility of an external torque is left open.

It is convenient to rewrite the second integral on the right-hand side of (13.4), using

$$\int_{V_0} \nabla \cdot (\mathbf{S}^T \otimes \mathbf{r})dV = \int_{A_0} \mathbf{n} \cdot (\mathbf{S}^T \otimes \mathbf{r})dA = \int_{V_0} (\nabla \cdot \mathbf{S}^T) \otimes \mathbf{r}dV + \int_{V_0} \mathbf{S}dV \tag{13.5}$$

where V_0 is the volume of a typical particle, and the other notation is obvious. Using the equation of motion, and recalling that the particles are force-free, one can write for any point in the suspension

$$\nabla \cdot \mathbf{S}^T = \rho\mathbf{a}' \tag{13.6}$$

where $\mathbf{a}'$ is the local acceleration. Now (13.4) may be rewritten as

$$\boldsymbol{\Sigma} = -\frac{1}{V}\int_{V - \Sigma V_p} p\mathbf{I}dV + 2\mu\bar{\mathbf{D}} + \boldsymbol{\Sigma}^{(p)} \tag{13.7}$$

Here

$$\boldsymbol{\Sigma}^{(p)} = \frac{1}{V}\Sigma \int_{A_p} [\mathbf{n} \cdot (\mathbf{S}^T \otimes \mathbf{r}) - \mu(\mathbf{v} \otimes \mathbf{n} + \mathbf{n} \otimes \mathbf{v})]dA$$

$$\tag{13.8}$$

$$-\frac{1}{V}\Sigma \int_{V_p} \rho\mathbf{a}' \otimes \mathbf{r}dV - \frac{1}{V}\int_V \rho\mathbf{v}' \otimes \mathbf{v}'dV$$

The average value of the deformation gradient $\bar{\mathbf{D}}$ signifies an average over both fluid and particles (which of course can also be a second fluid). We have used

$$\overline{\nabla\mathbf{v}} = \frac{1}{V}\int_{V - \Sigma V_p} \nabla\mathbf{v}dV + \frac{1}{V}\Sigma \int_{A_p} \mathbf{v} \otimes \mathbf{n}dA \tag{13.9}$$

The quantity of chief interest to us here is the so-called particle stress $\boldsymbol{\Sigma}^{(p)}$. Aside from an isotropic term, $\boldsymbol{\Sigma}^{(p)}$ represents departure of the suspension from behavior as a Newtonian fluid with the viscosity μ of the suspending fluid. In order to allow for the effect of an external couple on the bulk stress we have not assumed, initially, that $\mathbf{\Sigma}^{(p)}$ is symmetric. If $\mathbf{L}_{(0)}$ is a couple externally applied to one particle, conservation of angular momentum requires

$$\mathbf{L}_{(0)} = \int_{A_0} (\mathbf{S} \cdot \mathbf{n}) \times \mathbf{r} dA - \int_{V_0} \rho \mathbf{a}' \times \mathbf{r} dV \qquad (13.10)$$

But from (13.5) and (13.6) one finds that the stress averaged over a particle is asymmetric if there is an external torque on the particle. In component form

$$L_{i(0)} = e_i{}^{jk} \int_{V_0} S_{jk} dV \qquad (13.11)$$

Furthermore, from (13.7) and (13.8) one obtains

$$e_i{}^{jk} \sum\nolimits_{jk}^{(p)} = \frac{1}{V} \sum_p L_{i(p)} \qquad (13.12)$$

showing that any asymmetry of bulk stress is related to an externally imposed couple on the particles.

In applications to be discussed here we shall assume that inertial effects are negligible (but see [7]) and that no external couple is applied. Then the expression for particle stress reduces to

$$\mathbf{\Sigma}^{(p)} = \frac{1}{V} \sum \int_{A_p} [\mathbf{n} \cdot (\mathbf{S} \otimes \mathbf{r}) - \mu(\mathbf{v} \otimes \mathbf{n} + \mathbf{n} \otimes \mathbf{v})] dA \qquad (13.13)$$

In the case of identical particles and sufficient dilution so that particle interactions can be ignored, the integral in (13.13) need only be applied to one particle and multiplied by the number of particles per unit volume. An alternate scheme for expression of bulk stress of a dilute suspension of identical rigid particles has been developed by Brenner [8]. He expresses his results in terms of a group of tensors intrinsic to the particle.

13.3. AN EXAMPLE: THE VISCOSITY OF A DILUTE SUSPENSION OF RIGID SPHERES

In the present example the particles are dilute, meaning that there is negligible interaction between them, and consist of identical neutrally buoyant rigid spheres with radius a. The spheres are dispersed in an incompressible Newtonian fluid with viscosity μ. One wishes to compute, in terms of properties of each component, the viscosity of the two-phase mixture. This is the problem, already referred to, which Einstein solved. We adopt here a treatment which is illustrative of the approach taken in recent years with more complicated systems.

One first considers the fluid dynamic problem of a single sphere, upon which there is no net force or torque, placed in a Newtonian fluid. The origin of a coordinate system is placed at the center of the sphere and motion is imparted to the fluid such that

$$\left. \begin{array}{l} \text{as} \quad |\mathbf{r}| \to \infty : \mathbf{v}^{\infty}(\mathbf{r}) = \mathbf{Gr} \\ \qquad\qquad\qquad p = 0 \end{array} \right\} \qquad (13.14)$$

where $\mathbf{G}$ is a constant second-order tensor with $\mathrm{tr}(\mathbf{G}) = 0$. Thus the continuity equation is automatically satisfied as $|\mathbf{r}| \to \infty$. Near the sphere, velocity and pressure fields are distorted from (13.14), and one expects this alteration to lead to an increase in viscous dissipation relative to that of a flow described at all values of $\mathbf{r}$ by (13.14). The increased rate of viscous dissipation is a consequence of the effective viscosity of the suspension being greater than that of the Newtonian continuous phase. To obtain a quantitative value for the composite viscosity we solve the creeping-motion (i.e., inertia-free) equations of motion, along with the continuity equation,

$$\left.\begin{array}{l} 0 = -\nabla p + \mu \nabla^2 \mathbf{v} \\ \nabla \cdot \mathbf{v} = 0 \end{array}\right\} \tag{13.15}$$

to obtain a solution satisfying the boundary conditions (13.14), and the condition that there be no net torque or force on the sphere. Also, the no-slip condition is invoked on the surface of the sphere

$$\mathbf{v}(|\mathbf{r}| = a) = \mathbf{v}'(|\mathbf{r}| = a) \tag{13.16}$$

the prime referring to the discontinuous particle phase. The solution for the velocity field can then be used to evaluate the bulk stress according to (13.7) and (13.13). From that result one can assign a value of effective viscosity to the suspension.

Several procedures are possible for solution of (13.15) consistent with the boundary conditions. One can, for example, begin with a general solution to the creeping-motion equations obtained by separation of variables in spherical coordinates. The result, known as Lamb's general solution, is in terms of an infinite series of solid spherical harmonics. Happel and Brenner [9; 10, pp. 62 *et seq.*] have recorded a straightforward procedure for isolating those harmonics which can be nonzero in a properly posed boundary-value problem.

For the problem under consideration here we adopt an alternate approach [11]. Because of the linearity of the problem and the absence of imposed force or torque on the sphere, the velocity and pressure fields must be linear vector- and scalar-valued functions, respectively, of $\mathbf{G}$. The pressure field is governed by

$$\nabla^2 p = 0 \tag{13.17}$$

which is formed by taking the divergence of (13.15a) and applying (13.15b). A solution is postulated which has the form

$$p = \mu \sum_n \frac{A_n}{r^n} (\mathbf{Gr}) \cdot \mathbf{r}$$

where the A_n are to be determined. Substitution into (13.17) and application of (13.14b) immediately lead to the requirement that $A_n = 0$ for all values of n except $n = 5$. Then

$$\nabla p = \mu A \left\{ -\frac{5}{r^7} [(\mathbf{Gr}) \cdot \mathbf{r}]\mathbf{r} + \frac{1}{r^5} [\mathbf{Gr} + \mathbf{G}^T \mathbf{r}] \right\}$$

which may be substituted into (13.15a) to give a differential equation for $\mathbf{v}$. The equation is solved by finding particular and homogeneous solutions. From the arguments stated earlier it is clear that linearity in $\mathbf{G}$ requires both solutions to be composed of linear combinations of $[(\mathbf{Gr}) \cdot \mathbf{r}]\mathbf{r}$, $\mathbf{Gr}$, and $\mathbf{G}^T \mathbf{r}$. One can readily verify that a particular solution is

$$\mathbf{v}_p = \frac{A}{2r^5} \left[(\mathbf{Gr}) \cdot \mathbf{r} \right] \mathbf{r}$$

A solution to the homogeneous equation

$$\nabla^2 \mathbf{v}_h = 0 \tag{13.18}$$

must have the form

$$\mathbf{v}_h = \sum_n \left\{ \frac{B_n}{r^n} \left[(\mathbf{Gr}) \cdot \mathbf{r} \right] \mathbf{r} + \frac{C_n}{r^n} (\mathbf{Gr}) + \frac{D_n}{r^n} (\mathbf{G}^T \mathbf{r}) \right\}$$

Substitution into (13.18) reveals that the only nonzero coefficients consistent with (13.14a) are B_7, C_0, C_3, C_5, D_3, and D_5. Next, one forms

$$\mathbf{v} = \mathbf{v}_p + \mathbf{v}_h$$

and imposes the requirement (13.15b) and the condition of no torque on the sphere. Expressions for the velocity and pressure fields result.

$$\left. \begin{aligned} \mathbf{v} &= \frac{5}{2} a^3 \left((\mathbf{Gr}) \cdot \mathbf{r} \right) \mathbf{r} \left[\frac{a^2}{r^7} - \frac{1}{r^5} \right] + \left[1 - \frac{1}{2} \left(\frac{a}{r} \right)^5 \right] (\mathbf{Gr}) - \frac{1}{2} \left(\frac{a}{r} \right)^5 (\mathbf{G}^T \mathbf{r}) \\ p &= - \frac{5\mu a^3}{r^5} (\mathbf{Gr}) \cdot \mathbf{r} \end{aligned} \right\} \tag{13.19}$$

Furthermore, the no-torque condition leads to an expression for the angular velocity Ω of the sphere

$$\Omega = \frac{1}{2} \nabla \mathbf{x} (\mathbf{Gr}) \tag{13.20}$$

This full solution to the flow problem provides more information than is required for derivation of the effective viscosity of the suspension. We proceed by using properties of the solution at the sphere surface to evaluate (13.13). The stress $\mathbf{S}$ on the sphere can be computed from the constitutive equation for a Newtonian fluid

$$\mathbf{S} = -p\mathbf{I} + \mu(\nabla \mathbf{v} + (\nabla \mathbf{v})^T)$$

and substituted into (13.13).

It is often convenient to work with the deviatoric part of the bulk stress. From (13.7) and (13.13) one can write

$$\tau = \Sigma - \frac{1}{3} (\operatorname{tr} \Sigma)\mathbf{I} = 2\mu \overline{\mathbf{D}} + \Sigma^{(pd)} \tag{13.21}$$

where

$$\Sigma^{(pd)} = \Sigma^{(p)} - \frac{1}{3} \mathbf{I} \frac{1}{V} \int_{A_p} (\mathbf{Sr}) \cdot \mathbf{n} \, dA$$

Evaluation of the surface integrals can be facilitated by an identity given by Brenner [9] for integration over a sphere. One finds

$$\int \underbrace{\frac{\mathbf{r} \otimes \mathbf{r} \otimes \ldots \otimes \mathbf{r}}{r^{n+2}}}_{n\text{-times}} \, dA = \frac{4\pi}{(n+1)!} \underbrace{(\nabla\nabla \cdots \nabla)r^n}_{n\text{-times}} \tag{13.22}$$

for n even. If n is odd, the integral on the left is zero.

As a final result for the bulk stress one obtains the traceless stress

$$\tau = 2\mu\mathbf{D}\left(1 + \frac{5}{2}\phi\right) \tag{13.23}$$

where ϕ is the volume fraction of spheres in the suspension, and $\bar{\mathbf{D}}$ for the composite system has now been designated by $\mathbf{D}$. In the special case of plane Couette flow ($\mathbf{v}^\infty = (\kappa y, 0, 0)$) the only nonzero component of $\mathbf{G}$ is $G_{xy} = \kappa$. Then all components of τ vanish except

$$\tau_{xy} = \tau_{yx} = \left(1 + \frac{5}{2}\phi\right)\mu\kappa \tag{13.24}$$

One can express (13.23) very simply by writing a constitutive equation for the suspension, now viewed as a single phase,

$$\tau = 2\mu_{\text{sus}}\mathbf{D} \tag{13.25}$$

where

$$\mu_{\text{sus}} = \mu\left(1 + \frac{5}{2}\phi\right) \tag{13.26}$$

This is the celebrated Einstein result for the viscosity of a suspension of dilute rigid spheres Having looked at this simple result in some detail we can proceed to arrive at numerous extensions with a minimum of additional effort.

13.4. DEFORMABLE PARTICLES: THE SLIGHTLY DEFORMED DROP

A natural extension of Einstein's theory for a dilute suspension of rigid spheres is a comparable emulsion of Newtonian drops. A paper on this subject was published by Taylor in 1932 [12], the result being an analogue of Einstein's expression but applicable to the viscosity of a dilute emulsion (which in this chapter we also refer to as a suspension).

Taylor's results were extended in 1968 by Schowalter *et al.* [13]. To continue by analogy to the development of the previous section, let us first consider the conditions under which deformation of the drop from sphericity can be ignored. One can readily show from dimensional analysis that drop deformation D, suitably scaled, can be written

$$D = D\left(\frac{Ga\mu}{\sigma}, \lambda\right) \tag{13.27}$$

where σ is the interfacial tension; μ is the viscosity of continuous phase; G is the measure of magnitude of shearing, e.g. $[2\mathrm{tr}(\mathbf{GG})]^{1/2}$; and $\lambda = \mu^*/\mu$ is the viscosity ratio, where μ^* refers to the viscosity of the drop.

For a given value of λ, $Ga\mu/\sigma$ is a measure of the tendency of shearing to deform a drop relative to the tendency of interfacial tension to maintain a spherical drop shape. Thus one expects an analysis based on small departures from sphericity to apply when $Ga\mu/\sigma \to 0$.

Consider again the steady far-field flow (which we can also call the "undisturbed" or "basic" flow) of (13.14). The origin of the coordinate system is fixed in the center of a Newtonian spherical drop with radius a and viscosity μ^*. In contrast to the previous example, however, the creeping-motion equations now apply to *both* phases. Also, continuity of both velocity and stress must be considered in utilizing the drop/continuous-phase boundary conditions. The form of the solution for each fluid can again be found by requiring linearity in $\mathbf{G}$ and imposing boundary conditions at $r \to 0$ and $r \to \infty$ [14, p. 249], or one can begin with Lamb's general solution and find those terms which are needed to satisfy the boundary conditions [13]. In either case the interior and exterior solutions are joined by matching the velocity and tangential stress at $r=a$. For the flow exterior to the drop the velocity is given by

$$\mathbf{v}^0 = \nabla \Phi^0 + \frac{1}{2\mu} \mathbf{r} \, p^0 \tag{13.28}$$

where

$$\Phi^0 = -\frac{1}{2} B^0 \left(\frac{a}{r}\right)^5 [(\mathbf{Gr}) \cdot \mathbf{r}]; \quad A^0 = \frac{5\lambda + 2}{5(\lambda + 1)}$$

$$p^0 = -5 A^0 \mu \frac{a^3}{r^5} [(\mathbf{Gr}) \cdot \mathbf{r}]; \quad B^0 = \frac{\lambda}{\lambda + 1}$$

The superscript zero denotes that the results apply to a spherical drop. One can now obtain an expression, analogous to (13.26), for the viscosity of an emulsion [12]. Note, however, that up to this point there has been no requirement for a matching of normal stress at the drop surface. It is of course the normal stress condition that dictates the shape of the drop, and the next step is to determine, to first order, the departure from sphericity and the effect of deformation on bulk stress and hence on rheology.

We have seen that the deformation is governed by $Ga\mu/\sigma$ and $\lambda = \mu^*/\mu$, and that the deformation from sphericity should go to zero as $Ga\mu/\sigma \to 0$. Hence we define a deformation parameter, assumed small, by

$$D = \frac{Ga\mu}{\sigma} g(\lambda) \tag{13.29}$$

where $g(\lambda)$ is a function, assumed $O(1)$, of the viscosity ratio. The form of $g(\lambda)$ is found from the interfacial boundary conditions. An equation for the drop surface is given by

$$F(x_1, x_2, x_3) = r - 1 - D\tilde{f}(x_1/r, x_2/r, x_3/r) = 0 \tag{13.30}$$

Here the x_i refer to a rectangular Cartesian coordinate system of fixed orientation and origin in the drop center, and $r = (x_i x_i)^{1/2}$ is now the distance to the drop surface. D is assumed small and $\tilde{f} = O(1)$. Lengths have been scaled in terms of the equivalent drop radius a. To first order in the deformation it can be shown that the drop will deform into an ellipsoid.

$$r = 1 + D f_2(\theta, \phi) \tag{13.31}$$

where $f_2(\theta, \phi)$ is a surface spherical harmonic of degree two [15]. Next, one assumes that the velocity and stress can be expanded in terms of the deformation parameter. External to the drop we write

$$\left.\begin{array}{l} \mathbf{v} = \mathbf{v}^\infty + \mathbf{v}^0 + D\mathbf{v}^1 + \mathbf{O}(D^2) \\ \mathbf{S} = \mathbf{S}^\infty + \mathbf{S}^0 + D\mathbf{S}^1 + \mathbf{O}(D^2) \\ p = p^0 + Dp^1 + O(D^2) \end{array}\right\} \tag{13.32}$$

where $\mathbf{S} = \mathbf{T} - p\mathbf{I}$ and $\mathbf{T} = 2\mu\mathbf{D}$. Similar expansions can be written for variables $\mathbf{v}^*$, $\mathbf{S}^*$, and p^* in the drop interior.

To determine, to first order, the degree of drop deformation and its effect on bulk stress we shall assume the simplest possible statement for an expression of discontinuity of normal stress across the interface; namely, Laplace's equation relating interfacial curvature and interfacial tension. On the drop surface

$$\mathbf{n} \cdot [(\mathbf{S} - \mathbf{S}')\mathbf{n}] = \sigma \left(\frac{1}{r_1} + \frac{1}{r_2} \right) \tag{13.33}$$

where r_1 and r_2 are the principal radii of curvature at the drop surface and $\mathbf{n}$ is the outer unit normal to the drop surface. The unit normal is related to the equation of the drop surface (13.30) by

$$\mathbf{n} = \frac{\nabla F}{(\nabla F \cdot \nabla F)^{1/2}} \tag{13.34}$$

A solution for $\mathbf{v}^1$ can be found from Lamb's general solution [13]. The exterior solution has the form (13.28), where superscripts zero are replaced by one, and

$$\left.\begin{array}{l} \Phi^1 = 0; \qquad p^1 = p^{(1)} + p^{(2)} \\[2mm] p^{(1)} = \frac{20}{7} A^{(1)}_{-3} G\mu \left(\frac{a}{r} \right)^3 \left[\mathrm{tr}[(\mathbf{G})^2] - \frac{3}{r^2} (\mathbf{G}r)^2 \right] \\[4mm] p^{(2)} = 12 A^{(2)}_{-3} G\mu \frac{a^3}{r^5} (\boldsymbol{\xi} \times \mathbf{r}) \cdot (\mathbf{G}r) \end{array}\right\} \tag{13.35}$$

Here $\boldsymbol{\xi}$ is the dimensionless vorticity vector, $G\xi^i = -\frac{1}{4} e^{ijk}(G_{jk} - G_{kj})$. From the boundary conditions, including this time (13.33), one can determine that

$$\left.\begin{array}{l} f_2 = \frac{2[(\mathbf{G}r) \cdot \mathbf{r}]}{r^2}; \quad D = \frac{Ga\mu}{16\sigma} \left[\frac{19\lambda + 16}{\lambda + 1} \right] \\[5mm] A^{(1)}_{-3} = \frac{25\lambda^2 + 41\lambda + 4}{25(\lambda + 1)^2}; \quad A^{(2)}_{-3} = \frac{19\lambda + 16}{15(\lambda + 1)}; \quad g(\lambda) = \frac{19\lambda + 16}{16(\lambda + 1)} \end{array}\right\} \tag{13.36}$$

Now one can apply (13.13) and obtain an expression for bulk stress [13]. We note here only the special case of plane Couette flow ($\mathbf{v}^\infty = (\kappa y, 0, 0)$). One ultimately finds that the constitutive equation predicts both shear and normal stress effects. In matrix form

$$\begin{bmatrix} \tau_{11} & \tau_{12} & 0 \\ \tau_{21} & \tau_{22} & 0 \\ 0 & 0 & \tau_{33} \end{bmatrix} = \left[1 + \phi \left(\frac{5\lambda+2}{2(\lambda+1)} \right) \right] \mu \begin{bmatrix} 0 & \kappa & 0 \\ \kappa & 0 & 0 \\ 0 & 0 & 0 \end{bmatrix}$$

$$+ \frac{\mu^2 a \phi}{\sigma} \left[\frac{19\lambda+16}{20(\lambda+1)^2} \right] \left[\frac{25\lambda^2+41\lambda+4}{28(\lambda+1)} \right] \left\{ \begin{bmatrix} \kappa^2 & 0 & 0 \\ 0 & \kappa^2 & 0 \\ 0 & 0 & -2\kappa^2 \end{bmatrix} \right.$$

$$\left. - \frac{(19\lambda+16)}{4} \begin{bmatrix} -\kappa^2 & 0 & 0 \\ 0 & \kappa^2 & 0 \\ 0 & 0 & 0 \end{bmatrix} \right\} \tag{13.37}$$

The expression for shear stress $\tau_{12} = \tau_{21}$ is that of Taylor. However, by including the deformation of the drop from sphere to ellipsoid one obtains effects of interfacial tension which provide an elastic-like response, manifested in this case by appearance of nonzero normal stress differences. Consider, finally, the two asymptotic cases $\lambda \to 0$ (a gas in a liquid) and $\lambda \gg 1$ (a highly viscous drop). Reduction of (13.37) yields

(i) $\lambda \to 0$

$$\mathcal{N}_1 = \tau_{11} - \tau_{22} = \frac{32}{5} \frac{\mu^2 a \kappa^2 \phi}{\sigma}; \qquad \mathcal{N}_2 = \tau_{22} - \tau_{33} = - \frac{20}{7} \frac{\mu^2 a \kappa^2 \phi}{\sigma} \tag{13.38}$$

(ii) $\lambda \gg 1$

$$\mathcal{N}_1 = \frac{361}{40} \frac{\mu^2 a \kappa^2 \phi}{\sigma}; \qquad \mathcal{N}_2 = - \frac{551}{280} \frac{\mu^2 a \kappa^2 \phi}{\sigma} \tag{13.39}$$

Two features of this result are noteworthy. In spite of the vast physical difference between the model and a polymer solution, we find, in accord with experimental results for polymers, that $\mathcal{N}_1$ exceeds $\mathcal{N}_2$ in magnitude, and $\mathcal{N}_1 > 0$ while $\mathcal{N}_2 < 0$. Thus one may entertain the hope that suspension models have at least some qualitative similarity to rheological behavior of polymer melts and solutions.

The example of a deformable drop clearly illustrates that rheological properties of suspensions cannot, in general, be characterized by a single scalar parameter, such as a viscosity [4, 8]. In the case of rigid spheres there is sufficient particle symmetry so that the same result is obtained by deriving the full bulk stress tensor (Section 13.3) or by following Einstein's example and associating an effective viscosity with the (scalar) rate of dissipation of energy of a dilute suspension. Nonlinear effects, such as the normal stress differences just described, however, can only be identified from the full bulk stress tensor.

The development leading to (13.38) and (13.39) contains some shortcomings which were eliminated with the treatment of deformation of a drop in a general time-dependent flow by Cox [16] and the application of Cox's analysis by Frankel and Acrivos [17] to problems of suspension rheology. To appreciate fully the intricacies of their contribution would require more of a diversion into the subject of fluid dynamics at low Reynolds numbers than is warranted here. Nevertheless, several features of their work are pertinent.

Cox recognized that the expansion parameter D (13.29) contains an implication that the dimensionless group $k^{-1} = Ga\mu/\sigma$ predominates over λ as the relevant parameter for describing small deformations from sphericity, and thus it may be inconsistent to consider, for example, the limit of (13.39). He chose to describe the drop surface by

$$r = 1 + \varepsilon f (x_1/r, x_2/r, x_3/r) \tag{13.40}$$

where $\varepsilon \ll 1$ is a parameter characterizing the drop deformation. In this way no *a priori* ordering of k or λ is assumed. A complete statement of the problem, with coordinates relative to the drop center, is given below. Lengths have been scaled by a, velocities by Ga, and stresses by $G\mu^*$ or $G\mu$ inside or outside of the drop, respectively.

$$\nabla^2 \mathbf{v} = \nabla p; \qquad \nabla \cdot \mathbf{v} = 0 \qquad \text{for } r > 1 + \varepsilon f \qquad (13.41)$$

$$\nabla^2 \mathbf{v}^* = \nabla p^*; \qquad \nabla \cdot \mathbf{v}^* = 0 \qquad \text{for } r < 1 + \varepsilon f \qquad (13.42)$$

$$\mathbf{v} \to \mathbf{Gr} \quad \text{as} \quad |\mathbf{r}| \to \infty \qquad (13.43)$$

$$\mathbf{v} = \mathbf{v}^*; \qquad \mathbf{v} \cdot \mathbf{n} = K\varepsilon \frac{\partial f}{\partial t} \qquad \text{at } r = 1 + \varepsilon f \qquad (13.44)$$

$$(\mathbf{S} - \lambda \mathbf{S}^*) \cdot \mathbf{n} = \mathbf{n}k \left[\frac{1}{R_1} + \frac{1}{R_2} \right] \qquad \text{at } r = 1 + \varepsilon f \qquad (13.45)$$

where $K = |\nabla (r - \varepsilon f)|^{-1}$. In this more general formulation $\mathbf{G}$ is permitted to be time dependent. Thus the two additional features of Cox's analysis are the expansion without regard to the relative importance of k and λ, and the explicit allowance for time-dependent flows. Frankel and Acrivos [17] have shown how the resulting constitutive equations for a suspension are affected by the ordering of k and λ. Perhaps of more importance, however, is the effect of time dependence on the equations for bulk constitutive behavior, even if the macroscopic flow is steady and the time derivative enters only through the microscopic boundary conditions (13.44). It is found that the pertinent time derivative is the corotational or Jaumann derivative, $\mathscr{D}/\mathscr{D}t$, discussed at some length in Chapter 11. Here is an illustration of successful marriage of continuum and microscopic approaches to rheology. In Chapter 11 we saw the need for a properly invariant time derivative in constitutive equations but were unable to make a clear choice among several alternatives. Now by looking at a distinct suspension model, the proper choice of time derivative arises quite naturally. The appearance of a Jaumann derivative was inferred by Schowalter *et al.* [13] in their analysis of the steady-state problem, but the true origin of the derivative was made clear by the work of Cox and of Frankel and Acrivos. These last authors also show the consistency of an emulsion model with the constitutive equation posed by Oldroyd (eqn. (11.44)).

This is an appropriate place to remark on the stringency of the far-field boundary condition (13.14) for this and other examples. Strictly speaking, we have restricted the analysis to flows which at large distances from a particle or in the absence of a particle are homogeneous in the deformation rate. It is uniformly assumed, however, that the constitutive equations derived from the boundary condition (13.14) are applicable to non-homogeneous shear fields if the length scale over which the rate of deformation changes appreciably is large with respect to the characteristic particle length.

13.5. OTHER ANALYSES FOR DEFORMED PARTICLES

Suspension rheology has by no means been limited to a discontinuous phase which is either a rigid solid or a Newtonian liquid. In connection with Oldroyd's work on time differentiation we have already mentioned (Section 11.6) Fröhlich and Sack's analysis of a suspension of elastic spheres in a Newtonian matrix [18]. Oldroyd himself considered an extension of Taylor's problem for a dilute emulsion to conditions where time dependence may be important [19]. An extensive treatment of the rheology and the stress–optical behavior of a suspension of elastic spheres was published in 1951 by Cerf [20]. The fluid mechanical basis for several of his assumptions is not clear, but most of his results have been verified by later workers. In particular Roscoe [4] and Goddard and Miller [21] published, virtually simultaneously, papers dealing with the suspension rheology of deformable viscoelastic spheres, and they clarified some of Cerf's work. The basic models employed by Roscoe and by Goddard and Miller are essentially identical. The spheres are assumed to be solid-like, and, contrary to the analysis of the last section, interfacial tension is not a variable. Roscoe has shown how Jeffery's solution for flow in the vicinity of a *rigid* ellipsoid can be easily altered to apply to nonrigid ellipsoidal surfaces. An adaptation of Jeffery's results was also made by Goddard and Miller. The work of Goddard and Miller includes the possibility of a time-dependent basic shearing field, and their results include the Jaumann derivative in the constitutive equation. Recall that the Jaumann derivative was also present in the work of Frankel and Acrivos, discussed in the last section. Those authors in fact have remarked on the close similarity between one of the results obtained by them for a particular ordering of k and λ, and the constitutive equations of Goddard and Miller governing Hookean elastic particles. Furthermore, the results of both can be cast in forms familiar from continuum mechanics. The distinction here, then, is not in the rheological response of the viscous drop or elastic particle models, but only in the physical origin of elastic behavior. In the former elasticity is introduced through a time constant in which interfacial tension appears. In the latter the corresponding time constant contains the elastic modulus for the particles. Once more we see that the qualitative utility of these constitutive results goes beyond the specific model systems from which the results were developed.

13.6. A RIGID ELLIPSOID IN A HOMOGENEOUS SHEARING FIELD

Let us now consider the motion of a torque-free force-free rigid ellipsoid placed in a flow which has the same far-field boundary condition as before, namely (13.14). This problem was first considered by Jeffery in 1922 [22]. He restricted his analysis to the special case of a far-field flow

$$\mathbf{v}^\infty = (\kappa y, 0, 0) \tag{13.46}$$

that is, all $G_{ij} = 0$ except $G_{xy} = \kappa$.

The computations are far more tedious because of the appearance of elliptic integrals, and one might be tempted to view the ellipsoid problem as an inevitable but unimportant addition to Einstein's solution for rigid spheres. However, Jeffery's results have formed the basis for numerous analyses of both rigid and deformable particle suspensions, and more than a few researchers have been spared tedious analysis because of the availability of Jeffery's work.

A modern treatment of the ellipsoid problem is available from Brenner [8]. Although the solution for velocity and stress fields is cumbersome, the procedure is straightforward until one comes to relating the single-particle solution to the rheological properties of the suspension as a whole. At this point an important difference from the sphere problem becomes evident. Clearly, the stress field produced in the flow defined by, say, (13.46) depends on the orientation of the ellipsoid with respect to $\mathbf{v}^\infty$. Furthermore, that orientation will, in general, change with time and will be dependent on the specification of an ellipsoid orientation at some stated time, i.e., on an initial condition. The consequences for bulk stress of a dilute suspension of ellipsoids in, for example, a Couette viscometer are profound. One concludes that the bulk rheological properties are sensitive to the distribution function of ellipsoid orientations, and the distribution function is affected by the initial conditions of ellipsoid orientations at the time the flow was begun, no matter how far in the past that may have been. This highly unphysical violation of the principle of fading memory is due to the fact that the laws governing ellipsoid motion are fully deterministic and do not contain any randomizing influence, such as Brownian motion. We shall return to this interesting problem later, only mentioning now that, in lieu of Brownian motion, Eisenschitz proposed that the rheological properties of a suspension of ellipsoids of revolution, i.e., of spheroids, be computed on the basis of an initial condition for which all orientations of the spheroid are equally probable [23, 24]. Note that at steady state this does not imply a distribution function for which all orientations are equally probable. The major axis of the spheroid describes an orbit (sometimes called a Jeffery orbit) on a unit sphere centered at the center of the spheroid. In general, the spheroid spends more time in some parts of its orbit than in others because of a nonuniform orbit speed over the circuit. Orbit calculations for spheroids have been worked out in great detail. (See, for example, the review of Goldsmith and Mason [25].

13.7. ROTARY BROWNIAN MOTION

The reader will recall from Chapter 10 that Brownian motion contributed in an important way to the force balance on a bead in the Rouse–Zimm bead–spring model. For the substructure models being considered in this chapter we shall see that Brownian motion can influence suspension rheology in a rather different way. We are considering throughout a suspension which remains effectively spatially homogeneous in a statistical sense with respect to particle concentration over the region of interest (cf. Section 13.2). This restriction precludes the interesting clustering effects that can arise from interaction between particles. The interaction can be composed of both hydrodynamic and Brownian components [3, 26]. Thus there is no need to include a distribution function describing the *position* distribution of particles. Furthermore, the particles under consideration here are not subject to internal statistical fluctuations analogous to the Gaussian distribution of bead–bead separation distance of the Rouse–Zimm model. However, a distribution function is needed to describe the *orientation* of nonspherical particles, and the orientational motion of a particle is sensitive to Brownian effects. Just as in Chapter 10 we had to account for the noncontinuum effect of Brownian motion of translation, and hence developed an effective Brownian force, so now we shall find a noncontinuum correction to the angular motion of particles which can be equated to a Brownian torque. Rotary Brownian motion affects the suspension rheology problem in two ways. First, it enters into the torque balance in which the bulk stress figures. Second, it affects the differential equation governing the orientation distribution function.

A complete theory of Brownian motion would of course allow for simultaneous translational and rotational Brownian motion along with the possibility for coupling between the two. These theories are available [27], but become rather more involved than the exposition for translational Brownian motion given in the appendix to Chapter 10.

Rotational Brownian motion enters, as one would anticipate, into the conservation equation for the rotational distribution function. Let us describe orientation of a particle in terms of a single vector (a vector parallel to the major axis of a spheroid, for example). Then a distribution function $f(\theta, \phi)d\Omega$ is the probability that the orientation vector will intersect the unit sphere in some solid angle $d\Omega = \sin\theta d\theta d\phi$ about polar angle θ and azimuthal angle ϕ. The quantities are sketched in Fig. 13.1. The distribution function is normalized so that

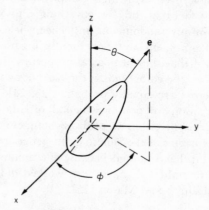

Fig. 13.1. Parameters for orientation distribution function.

$$\int_0^{2\pi} \int_0^{\pi} f \sin\theta d\theta d\phi = 1$$

It is also subject to a conservation equation analogous to (10A5.6). This can be written

$$\frac{\partial f(\theta, \phi)}{\partial t} + \nabla \cdot \mathbf{j} = 0 \tag{13.47}$$

where ∇ is a two-dimensional gradient over the unit sphere and $\mathbf{j}$ is a normalized rotational flux of particles with orientation vector $\mathbf{e}$ (Fig. 13.1) between θ and $\theta + d\theta$, and ϕ and $\phi + d\phi$. The flux $\mathbf{j}$ can be considered as a sum of convective, or "hydrodynamic", and diffusive contributions, a result entirely similar to the corresponding translation flux. Thus one may write

$$\mathbf{j} = \mathbf{j}_h + \mathbf{j}_d \tag{13.48}$$

where

$$\mathbf{j}_h = f \dot{\mathbf{e}}_h$$

The hydrodynamic flux is associated with $\dot{\mathbf{e}}_h$, the analogue of the mass average velocity employed in descriptions of conventional diffusion [28, ch. 16]. Continuing the analogy, if one assumes that the diffusive flux $\mathbf{j}_d$ is adequately described by a Fickian diffusion law [3], then

$$\mathbf{j}_d = f \dot{\mathbf{e}}_d = -D_r \nabla f \tag{13.49}$$

D_r is a "rotational" diffusion coefficient, assumed here to be a scalar constant. Combining (13.47), (13.48), and (13.49) one obtains the following conservation equation

$$\frac{\partial f}{\partial t} + \nabla \cdot (f\dot{\mathbf{e}}_h) = D_r \nabla^2 f \tag{13.50}$$

Because Brownian motion contributes to the rotational motion of the molecules, and because this rotational motion figures linearly in the calculation of bulk stress it is convenient to consider the hydrodynamic and Brownian effects separately. A hydrodynamic angular velocity $\mathbf{\Omega}_h$ is defined by

$$\dot{\mathbf{e}}_h = \mathbf{\Omega}_h \times \mathbf{e} \tag{13.51}$$

Analogously, a "Brownian" angular velocity $\mathbf{\Omega}_B$ may be defined, from (13.49), by

$$\dot{\mathbf{e}}_d = \mathbf{\Omega}_B \times \mathbf{e} = -D_r \nabla \ln f$$

Using the vector triple product and the fact that $\nabla \ln f$ must be in the surface of the unit sphere and hence orthogonal to $\mathbf{e}$, it is easy to verify that [8]

$$\mathbf{\Omega}_B = -D_r \mathbf{e} \times \nabla \ln f \tag{13.52}$$

In the appendix to Chapter 10 we saw that translational Brownian motion can be interpreted as an effective force if one writes a bead force balance for the Rouse–Zimm model. The analogue carries over to rotational motion. It can be shown that there is an effective Brownian torque $\mathbf{L}_B$ that must be considered in writing the particle torque balance [3, 29]

$$\mathbf{L}_B = \frac{kT}{D_r} \mathbf{\Omega}_B \tag{13.53}$$

It is now possible to identify the two separate contributions of D_r to the bulk stress of a suspension. Because of the effect of particle orientation on stress, an influence of D_r through (13.50) is expected. Less obvious, however, is a contribution through (13.53). Recall that one of the boundary conditions which determines the velocity field around a particle, and hence the stress field as well, is the requirement of no net torque on the particle (see, for example, (13.20)). Thus in the presence of rotational Brownian motion there must be a balance between the hydrodynamic torque and the Brownian torque (13.53).

Brenner [8] has presented explicit formulas for computing the rotational Brownian diffusion coefficients of spheroids and rigid spherical dumbbells. The latter result is particularly simple if the spheres are joined by a rigid rod with negligible hydrodynamic resistance and the center-to-center distance between spheres, $2l$, is such that $l/b \gg 1$, where b is the sphere radius. Then hydrodynamic interaction between spheres can be neglected. Brenner reports that $D_r = kT/(12\pi\mu b l^2)$, where k is the Boltzmann constant and T the absolute temperature.

A formalism has been developed for computation of specific constitutive equations [8, 30]. Solution of the distribution function equation (13.50) has, however, proved extremely difficult for arbitrary ratios of diffusion coefficient to shear rate. Hinch and Leal have exposed the full nature of the problem and solved it for several specific cases [31, 32]. The complete problem requires numerical solution and the results are difficult to illustrate. Asymptotic cases, however, can be discussed with more ease, and we consider some of these results.

It is worth noting briefly the state of the subject at the time of the Hinch and Leal papers. To be specific we consider the "undisturbed" simple shearing flow

$$\mathbf{v}^{\infty} = [\kappa y, 0, 0]$$

The rotational diffusion equation (13.50) can be written, for steady state,

$$\nabla \cdot (\kappa \mathbf{w} f) = D_r \nabla^2 f \tag{13.54}$$

where $\kappa \mathbf{w} = (0, \dot{\theta}, \dot{\phi} \sin \theta)$ is the hydrodynamically induced particle velocity. As early as 1938 Burgers [24] had given an approximate solution to (13.54) for large values of D_r/κ. He expressed f as a power series expansion in κ/D_r, the first approximation being the result of pure diffusion $\nabla^2 f = 0$. However, the other extreme, that of weak Brownian motion, posed a greater problem. We have already seen that if a meaningful rheological result is to be obtained, *some* influence of Brownian motion (lacking a nonphysical artifact) is needed in order to randomize the distribution function for particle orientation away from its dependence upon initial conditions. Recall that for the case of ellipsoids the rheology, in the absence of Brownian motion, is destined to depend upon the initial orientation of the ellipsoids because of the determinate nature of the problem. Leal and Hinch were able to allow for a small amount of rotational diffusion by integrating (13.54) for a spheroid around a Jeffery orbit and by following a procedure developed for an analogous fluid mechanics problem. A second contribution was their computation of the full bulk stress tensor. Except for Giesekus [33], prior workers had concentrated their attention solely on the shear stress.

By way of illustration we consider the results of Hinch and Leal for limiting cases of strong Brownian motion $(D_r/\kappa \gg 1)$ (done also by Giesekus) and weak Brownian motion $((r^3 + r^{-3}) D_r/\kappa \ll 1$, where r is a particle aspect ratio). Both hydrodynamic and Brownian contributions to the bulk stress were considered, and the results can be presented in the following final form:

(i) Strong Brownian motion

$$\left. \begin{aligned} t_{12} &= \tau(\kappa) = \phi \mu \kappa \left\{ f_0^S(r) + O(\kappa/D_r)^2 \right\} \\[1em] \mathcal{N}_1 &= \phi \mu \kappa \left\{ \frac{\kappa}{D_r} [f_1^S(r) - f_2^S(r)] + O(\kappa/D_r)^3 \right\} \\[1em] \mathcal{N}_2 &= \phi \mu \kappa \left\{ \frac{\kappa}{D_r} f_2^S(r) + O(\kappa/D_r)^3 \right\} \end{aligned} \right\} \tag{13.55}$$

where ϕ refers to volume fraction of particles. The functions $f_i^S(r)$ have been plotted in Figs. 13.2 and 13.3.

(ii) Weak Brownian motion

$$\left. \begin{aligned} t_{12} &= \tau(\kappa) = \phi \mu \kappa \left\{ f_0^W(r) + O(D_r/\kappa)^2 \right\} \\[1em] \mathcal{N}_1 &= \phi \mu \kappa \left\{ \frac{D_r}{\kappa} [f_1^W(r) - f_2^W(r)] + O(D_r/\kappa)^3 \right\} \\[1em] \mathcal{N}_2 &= \phi \mu \kappa \left\{ \frac{D_r}{\kappa} f_2^W(r) + O(D_r/\kappa)^3 \right\} \end{aligned} \right\} \tag{13.56}$$

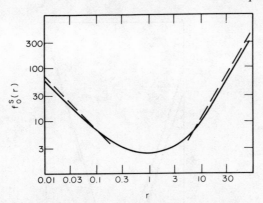

Fig. 13.2. The dependence of $f_0^s(r)$ on aspect ratio r [32]. Asymptotic values are indicated by dashed lines.

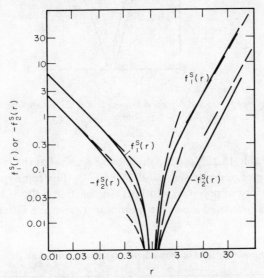

Fig. 13.3. The dependence of $f_1^s(r)$ and $f_2^s(r)$ on aspect ratio r [32]. Asymptotic values are indicated by dashed lines.

with $f_i^W(r)$ shown in Figs. 13.4 and 13.5. Asymptotic values for the functions $f_i^S(r)$ and

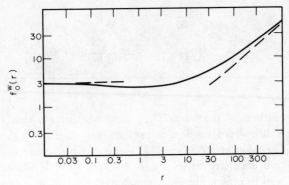

Fig. 13.4. The dependence of $f_0^W(r)$ on aspect ratio r [32]. Asymptotic values are shown by dashed lines.

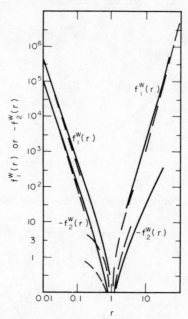

Fig. 13.5. The dependence of $f_1^W(r)$ and $f_2^W(r)$ on aspect ratio r [32]. Asymptotic values are shown by dashed lines.

$f_i^W(r)$ are given in Table 13.1. Note the interesting result that the normal stress differences (7.33) $\mathcal{N}_1$ and $\mathcal{N}_2$ are of opposite sign, and $|\mathcal{N}_1| > |\mathcal{N}_2|$. The same qualitative behavior has been found for a variety of polymer solutions (see Section 8.9). Subsequent work by Leal and Hinch [34] has indicated how the Jaumann derivative enters calculations for the

TABLE 13.1. Asymptotic values of shape factors for strong and weak Brownian motions [32]

	$f_0^S(r)$	$f_1^S(r)$	$f_2^S(r)$	$f_0^W(r)$	$f_1^W(r)$	$f_2^W(r)$
$r \to 0$	$\dfrac{32}{15\pi r}$	$\dfrac{4}{21\pi r}$	$-\dfrac{8}{105\pi r}$	3.13	$\dfrac{4}{3\pi r^3}$	$-\dfrac{1}{3\pi r^3}$
$r = 1 + \epsilon$ ($|\epsilon| \ll 1$)		$\dfrac{6}{35}\epsilon^2$	$-\dfrac{1}{35}\epsilon^2$		$\dfrac{216}{35}\epsilon^2$	$-\dfrac{36}{35}\epsilon^2$
$r \to \infty$	$\dfrac{4}{15}\dfrac{r^2}{\ln r}$	$\dfrac{2}{35}\dfrac{r^2}{\ln r}$	$-\dfrac{1}{105}\dfrac{r^2}{\ln r}$	$0.315\dfrac{r}{\ln r}$	$\dfrac{r^4}{4\ln r}$	

rheology of nearly spherical particles. They have also shown that their results for *rigid* near-spheres, when Brownian motion is present, are similar to those of Goddard and Miller [21] and of Frankel and Acrivos [17], both of which were discussed earlier in this chapter. An effective elasticity arises in these three models from rotational diffusion, particle elasticity, and interfacial tension, respectively.

13.8. ELONGATIONAL FLOW.
THE IMPORTANCE OF PARTICLE INTERACTIONS

In a paper cited earlier [8] Brenner calculated the stress, in the absence of Brownian motion, of a dilute suspension of rigid elongated particles in both uniaxial extensional and compressive flows. The extensional viscosity becomes infinite as prolate spheroids approach rigid rods of infinite aspect ratio, or as oblate spheroids approach flattened disks of negligible thickness relative to the diameter. Although we shall see that the assumption of a dilute suspension imposes severe restrictions, such a result seems intuitively reasonable. The rods and disks become aligned with the flow in such a way that the drag over the particle surface becomes infinite even for a single particle.

Batchelor [35] has approached a similar problem from a somewhat different point of view, and found that when considering this class of problems the notion of "diluteness" must be considered very carefully. His model is a suspension of rigid elongated particles in a Newtonian fluid under conditions such that there is no external force or torque on the particles. Inertial effects are assumed to be insignificant and Brownian motion is neglected. Then the expression (13.8) for bulk stress reduces to

$$\mathbf{\Sigma}^{(p)} = \frac{1}{V} \sum \int_{A_p} \mathbf{n} \cdot \mathbf{S}^T \otimes \mathbf{r} dA \tag{13.57}$$

Although there is no requirement at this stage to do so, it is convenient to imagine the particles to be circular cylinders ("rods") of length $2l$ and radius b. In considering the force transmitted between the particle and its surroundings we limit the discussion to the case $l \gg b$, and hence omit consideration of the end faces of the particle. If there is no net force or couple on the particle, only the tangential component of the stress vector can contribute to (13.57). Then one can write an approximate relation for the bulk stress in terms of the force $\mathbf{F}$ per unit length exerted by the particle on the surrounding fluid at position sl from the rod midpoint,

$$\mathbf{\Sigma}^{(p)} = -\frac{1}{V} \mathbf{e} \otimes \mathbf{e} \sum l^2 \int_{-1}^{1} \mathbf{F} \cdot \mathbf{e} s ds \tag{13.58}$$

where $\mathbf{e}$ is a unit vector aligned along the particle axis. We shall also assume that each rod is aligned in the same direction so that $\mathbf{e}$ need not figure in the summation of (13.58).

Of special interest here is the behavior of this suspension model in a uniform steady uniaxial extension of the type considered in Sections 9.4 and 12.6. We shall consider the basic bulk flow

$$\mathbf{v} = \left(dx^1, -\frac{d}{2}x^2, -\frac{d}{2}x^3 \right) \tag{13.59}$$

in a Cartesian coordinate system, and we picture the rods to be aligned in the direction x^1 of extension.

At this point a result derivable from a branch of fluid mechanics known as "slender-body theory" is accepted [36]. It can be shown that $(\mathbf{F} \cdot \mathbf{e})$ of (13.58) can be written

$$\mathbf{F} \cdot \mathbf{e} = -\mu \mathbf{e} \cdot \mathbf{De} \, l \, G(s)$$

where $G(s)$ is a known function and $\mathbf{D}$ is the rate of deformation tensor of the basic flow. In the coordinate system under consideration the nonzero components of $\mathbf{D}$ are $d_{11} = d$, $d_{22} = d_{33} = -\frac{1}{2}d$. For $l/b \gg 1$ one can write $G(s)$ approximately as [35]

$$G(s) \cong 2\pi s \varepsilon \tag{13.60}$$

where $\varepsilon = [ln(2l/b)]^{-1}$.

It is of interest to compute the contribution of the particles to the stress in the direction of extension and to compare this to the familiar Trouton viscosity $3\mu d$ of the suspending medium. One way to do this is to compute the quantity

$$T \stackrel{d}{=} \frac{\sum_{11}^{(p)} - \frac{1}{2}(\sum_{22}^{(p)} + \sum_{33}^{(p)})}{3\mu d} = \frac{1}{3V} \sum l^3 \int_{-1}^{1} G(s) s \, ds \tag{13.61}$$

For the limiting case of (13.60) one obtains

$$T = \frac{1}{3V} \left[\frac{4}{3} \pi \sum \varepsilon l^3 \right] \tag{13.62}$$

Equation (13.62) provides the suggestion that the influence of the particles on extensional viscosity is not properly characterized by the volume fraction of particles present. In fact one can infer from (13.62) that the influence on T is proportional to a fictitious volume fraction of spheres of radius $\varepsilon^{1/3}l$. Batchelor [35] has shown that this notion can be made somewhat more general. If one defines

$$Q(\varepsilon) = \frac{3}{4\pi \varepsilon} \int_{-1}^{1} G(s) s \, ds$$

then it follows from (13.61) that

$$T = \frac{1}{3} \alpha \tag{13.63}$$

where $\alpha = (1/V) \sum \frac{4}{3} \pi l^3 \varepsilon Q(\varepsilon)$ is the volume fraction of "effective spheres" with radius $[\varepsilon Q(\varepsilon)]^{1/3}l$. It is apparent that one can have extremely large values of α, and hence of T, even though the actual volume fraction of suspended particles $\sum (1/V) 2\pi b^2 l$ remains very small. However, this result must be interpreted with some caution.

One can show [35] that in the extensional flow under consideration the disturbance to the basic velocity field (13.59) at large distances r from a particle is of magnitude $\lambda/\mu r^2$, where

$$\lambda = l^2 \int_{-1}^{1} F_1(s) s \, ds = -\frac{4}{3} \pi \mu l^3 d\varepsilon Q(\varepsilon)$$

(The quantity λ is equivalent to a force doublet located at the origin.) Then the disturbance to $\mathbf{D}$ is of order $\lambda/\mu r^3$. Now let us inquire about the conditions such that each particle does not affect the force field around an adjacent particle, i.e., the suspension is "dilute" in the sense given earlier. If the suspension is to be dilute, then the distance r between particles must be such that

$$|\lambda|/\mu r^3 \ll d$$

If n is the number density of dilute particles separated, on average, by distance r between centers, $n \sim r^{-3}$ and the above condition becomes

$$n|\lambda| \ll \mu d$$

But we have $n|\lambda| = \alpha \mu d$, so the requirement becomes

$$\alpha \ll 1 \quad \text{or} \quad nl^3 \varepsilon \ll 1 \tag{13.64}$$

Thus we see that although (13.63) predicts large values of the Trouton ratio T for $\alpha \gg 1$, the assumption of noninteracting particles is only valid when $\alpha \ll 1$. A supplementary conclusion is that, for the model and flow under consideration, volume fraction is not a good criterion of applicability because each particle commands a sphere of influence comparable to an "effective" volume fraction α. Here is an excellent example of how constitutive model (in this case a noncontinuum model) and flow interact to produce unusual effects and surprising regimes of applicability.

Although the model showed a startling increase of extensional viscosity, indicated here by the relative measure T, under conditions for which the assumptions employed are inapplicable, one may nevertheless hope that the qualitative features of a sharply rising Trouton ratio with increasing l/b would also hold if a more rigorous analysis were applied. Batchelor has shown this to be the case. It is possible to obtain explicit results for the extensional viscosity under conditions such that $b \ll h \ll l$, where h is the average lateral spacing between the needle-like rods, all aligned in the direction of extension. One finds

$$T = \frac{4}{9} \phi \, \frac{(l/b)^2}{\ln(\pi/\phi)} \tag{13.65}$$

with the requirements $\phi^{1/2} \ll 1$ and $\phi^{1/2} l/b \gg 1$. Results obtained from (13.65) have been sketched, for selected values of l/b, in Fig. 13.6. The dramatic rise of T possible for small values of ϕ is not readily apparent because of the factor nl^3 in the denominator of the ordinate. For example, if $\phi = 0.01$ and $l/b = 10^3$, (13.65) predicts $T = 770$. Good agreement between experimental results and the predictions of (13.65) were found by Mewis and

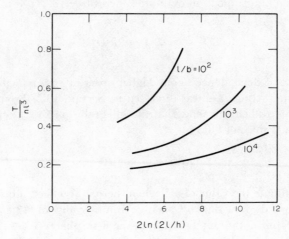

Fig. 13.6. Relative extensional stress for a suspension of rods in the concentration and geometric range $b \ll h \ll l$ [35].

Metzner [37] for asbestos fibers with $l/b = 282$ and 586. Agreement to within about 30% was reported for $l/b = 1259$.

13.9. NONDILUTE SYSTEMS

A special case illustrating the importance of particle interactions for slender particles in elongational flow was considered in the last section. Our concern here with nondilute suspensions, i.e., suspensions in which interparticle effects are significant, is more general in scope. Although empirical descriptions of suspension viscosity are available for moderate and high concentrations, rigorous treatments of suspensions, as we have already seen in this chapter, are almost always performed under the assumption that interparticle effects are negligible. Then of course any derived description of constitutive behavior will be linear in the concentration of suspended particles. Some progress has been made at the opposite end of the concentration scale, i.e., suspensions that amount to a packed structure with the surrounding continuum acting almost as a lubricating layer [38]. However, when particles are very close to each other nonhydrodynamic effects can become very important. An example is interaction from electrical charge effects on particle surfaces. These effects can be quite large over distances comparable to the spacing between particles [39].

A major effort over many years has been the extension of a formulation for dilute systems to one which accounts, to a first order of approximation, for interparticle effects. One would hope that there could be developed, at least in principle, an interaction scheme with which successive approximations to ever-increasing concentrations could be made. For the most part, however, these attempts have been distinguished by lack of success. An unambiguous development of constitutive behavior correct to order ϕ^2, ϕ being the volume fraction, is available [40], and some features of that work will be recounted here. The subject is highly technical. As we shall see, it is an example of a branch of science where those parts of the subject which an expositor would wish to relegate to "detail" become in fact the essence of the problem. Although the development need not be constrained to rigid spheres the notation is simpler if one imposes that restriction. Hence our attention will be fixed on the following problem: given the Einstein expression for the constitutive equation of a suspension of dilute spheres (see 13.21)

$$\boldsymbol{\tau} = \boldsymbol{\Sigma} - \frac{1}{3}\,(\mathrm{tr}\,\boldsymbol{\Sigma})\mathbf{I} = 2\mu\,\bar{\mathbf{D}}\left(1 + \frac{5}{2}\,\phi\right)$$

what is the nature of ϕ dependence to one higher order of approximation in ϕ? The bulk stress for the noninteracting case was discussed in Section 13.3. From that analysis and from (13.22) it is not difficult to show that the integral of (13.13) over a single sphere of radius a in an infinite mass of fluid is

$$\boldsymbol{\sigma}^{(p)} = \frac{20}{3}\,\pi a^3 \mu \mathbf{D} \tag{13.66}$$

We now wish to estimate the value of $\boldsymbol{\sigma}^{(p)}$ when there is an effect, albeit small, from other particles. In other words, how does $\boldsymbol{\sigma}^{(p)}$ change from its value in (13.66) when there exists a second sphere far, but not infinitely far, from the first sphere? It is convenient to express the particle stress (13.13) relative to the stress due to an isolated sphere. Thus we write, for a suspension of identical spheres,

$$\Sigma^{(p)} = n\bar{\sigma}^{(p)} = 5\phi\mu\bar{\mathbf{D}} + 5\phi\mu \left[\frac{\bar{\sigma}^{(p)}}{\frac{20}{3}\pi a^3 \mu} - \bar{\mathbf{D}} \right] \tag{13.67}$$

where n is the number density of particles and $\bar{\sigma}^{(p)}$ is an ensemble average of the stress over a sphere located at, say, position $\mathbf{x}_0$ in the (nondilute) suspension. $\bar{\mathbf{D}}$ is the ensemble average of $\mathbf{D}$ at $\mathbf{x}_0$, irrespective of whether or not there is a particle at $\mathbf{x}_0$. Since we are interested in the first-order approximation to particle interaction one would hope that it would suffice to consider only the effect of *one* particle on another, i.e., we are limiting our scope to two-body interactions, thus avoiding the need to write a probability distribution function for the simultaneous location of many particles in space. Let $P(\mathbf{x}_0+\mathbf{r}|\mathbf{x}_0)d\mathbf{r}$ be the probability that a particle is located with center in an element of volume $d\mathbf{r}$ about position $\mathbf{x}_0+\mathbf{r}$ when there is a particle (the reference particle) at $\mathbf{x}_0$. It would then seem appropriate to evaluate the first term in the square bracket of (13.67) by integrating

$$5\phi\mu \left[\frac{\sigma^{(p)}}{\frac{20}{3}\pi a^3 \mu} \right] P(\mathbf{x}_0 + \mathbf{r}|\mathbf{x}_0)d\mathbf{r} \tag{13.68}$$

over all values of $\mathbf{r}$, the bracketed expression being determined at $\mathbf{x}_0$. When the locations of the two particles are sufficiently far apart, i.e., $r/a \gg 1$, one would expect statistical independence of particle locations and

$$P(\mathbf{x}_0 + \mathbf{r}|\mathbf{x}_0) \cong P(\mathbf{x}_0 + \mathbf{r}) \to n$$

$P(\mathbf{x}_0+\mathbf{r})d\mathbf{r}$ is the probability that a particle center is located in an element of volume $d\mathbf{r}$ about position $\mathbf{x}_0+\mathbf{r}$ without regard to the presence or absence of a particle at $\mathbf{x}_0$. At this point a difficulty arises, however, because of the nature of $\sigma^{(p)}$. We can estimate its magnitude to be equal to the "dilute" value (13.66) corrected by a perturbation due to the presence of a second sphere a large distance r from $\mathbf{x}_0$. The form of this perturbation is readily found by computing the shear rate induced at $\mathbf{x}_0$ when a single sphere is located at $\mathbf{x}_0+\mathbf{r}$. Simple computation from (13.19) shows that the perturbation is $O(r^{-3})$. Since the perturbation does not decrease faster than the volume over which it is being integrated increases, one concludes that the integral formed from (13.68) is nonconvergent and therefore meaningless.

By drawing upon some earlier work dealing with sedimentation of interacting particles [41] Batchelor and Green were led to a reformulation of the probability-weighted volume average in terms of a convergent integral. The key is recognition that the integral expression for the average value of $\mathbf{D}(\mathbf{x}_0, \mathbf{x}_0+\mathbf{r})$ suffers from the same convergence deficiency as that for $\sigma^{(p)}(\mathbf{x}_0, \mathbf{x}_0+\mathbf{r})$. This fact can be used to advantage by rewriting (13.67) as

$$\Sigma^{(p)} - 5\phi\mu\bar{\mathbf{D}} = 5\phi\mu \int \left\{ \left[\frac{\sigma^{(p)}(\mathbf{x}_0, \mathbf{x}_0+\mathbf{r})}{\frac{20}{3}\pi a^3 \mu} - \bar{\mathbf{D}} \right] P(\mathbf{x}_0 + \mathbf{r}|\mathbf{x}_0) \right.$$

$$\left. - 5\phi\mu[\mathbf{D}(\mathbf{x}_0, \mathbf{x}_0+\mathbf{r}) - \bar{\mathbf{D}}]P(\mathbf{x}_0 + \mathbf{r}) \right\} d\mathbf{r} \tag{13.69}$$

Here $\mathbf{D}(\mathbf{x}_0, \mathbf{x}_0 + \mathbf{r})$ is the local deformation rate at $\mathbf{x}_0$ when there is a particle at $\mathbf{x}_0 + \mathbf{r}$. Because of similar asymptotic behavior at $r/a \gg 1$ of the first terms in each square bracket of (13.69), their ill behavior cancels in the composite form and the full expression is convergent. Note also that the first square bracket contributes nothing to the integral for $r < 2a$ since we must have $P(\mathbf{x}_0 + \mathbf{r}|\mathbf{x}_0) = 0$. This provides a residual in the second term of order ϕ^2 and is precisely the concentration effect that is being sought.

Equation (13.69) is then the basic expression with which one can find the ϕ^2 contribution to bulk stress. Computation is, however, quite another matter. One must of course have an expression for $P(\mathbf{x}_0 + \mathbf{r}|\mathbf{x}_0)$ and this is naturally dependent upon the details of the flow—still another concrete example of the interdependence of flow and constitutive equation. Calculations are difficult and at the time of Batchelor and Green's paper results were only available (approximately) for the case where one particle, relative to a second particle, comes from an infinite distance away. This excludes the important case of laminar shear flow because it is well known that there are regions of closed streamlines in which one particle orbits around another. On the other hand, extensional flow does qualify and an expression for extensional viscosity has been computed. Batchelor and Green found an approximate expression for the effective viscosity μ_{sus} in this special case, where $\tau = 2\mu_{\mathrm{sus}}\bar{\mathbf{D}}$ and

$$\mu_{\mathrm{sus}} = \mu \left[1 + \frac{5}{2}\phi + 7.6 \, \phi^2 \right] \tag{13.70}$$

The error in the coefficient of ϕ^2 was estimated to be about 10%.

Note that Brownian motion has been neglected. Its inclusion will in some ways be a complicating factor, but on the other hand it may remove the present dependence of the analysis on initial conditions of particle position.

13.10. CONCLUSION

It is hoped that this sketchy description of suspension rheology will, nevertheless, be sufficient to convince the reader that the subject contains many of the separate advantages inherent in the continuum and molecular bases of rheology. We have seen that a remarkable variety of suspension models exhibit similar qualitative phenomena, e.g., the presence of an elastic response manifested in a characteristic relaxation time, and normal stress differences. This similarity to some features of the flow of polymer melts and solutions indicates that suspensions can be used to model the bulk flow behavior of some polymer systems. Beyond the qualitative significance for polymer behavior, one must not lose sight of the value of a suspension model in its own right, precisely because it is a specific model. The words of Roscoe [4], quoted in Section 13.1 are again pertinent. When a model either describes or fails to describe a sought-for response, the mechanical bases for the results are always apparent.

Finally, we have been given a glimpse of the ϕ^2 problem and the difficulties inherent in it. In this writer's opinion the current flush of activity—and progress—in suspension rheology will either accelerate and come to new and far-reaching results, or will again soon settle into a dormant state. The former course can only be followed if we find means to understand and to predict the behavior of suspensions under nondilute conditions.

REFERENCES

1. Einstein, A., *Ann. Physik* **19**, 289 (1906); **34**, 591 (1911).
2. Rutgers, R., *Rheol. Acta* **2**, 202, 305 (1962).
3. Brenner, H., *Progress in Heat and Mass Transfer*, vol. V, p. 89 (W. R. Schowalter *et al.*, eds.), (Oxford: Pergamon Press, 1972).
4. Roscoe, R., *J. Fluid Mech.* **28**, 273 (1967).
5. Batchelor, G. K., *J. Fluid Mech.* **41**, 545 (1970).
6. Hirschfelder, J. O., Curtiss, C. F., and Bird, R. B., *The Molecular Theory of Gases and Liquids*, p. 457 (New York: Wiley, 1954).
7. Lin, C. J., Peery, J. H., and Schowalter, W. R., *J. Fluid Mech.* **44**, 1 (1970).
8. Brenner, H., *Chem. Eng. Sci.* **27**, 1069 (1972).
9. Brenner, H., *Chem. Eng. Sci.* **19**, 519 (1964).
10. Happel, J., and Brenner, H., *Low Reynolds Number Hydrodynamics* (Englewood Cliffs, New Jersey: Prentice-Hall, 1965).
11. Peery, J. H., PhD thesis, Princeton University, 1966.
12. Taylor, G. I., *Proc. Roy. Soc.* A, **138**, 41 (1932).
13. Schowalter, W. R., Chaffey, C. E., and Brenner, H., *J. Coll. Interface Sci.* **26**, 152 (1968).
14. Batchelor, G. K., *An Introduction to Fluid Dynamics* (Cambridge: Cambridge University Press, 1967).
15. Chaffey, C. E., and Brenner, H., *J. Coll. Interface Sci.* **24**, 258 (1967).
16. Cox, R. G., *J. Fluid Mech.* **37**, 601 (1969).
17. Frankel, N. A., and Acrivos, A., *J. Fluid Mech.* **44**, 65 (1970).
18. Fröhlich, H., and Sack, R., *Proc. Roy. Soc.* A, **185**, 415 (1946).
19. Oldroyd, J. G., *Proc. Roy. Soc.* A, **218**, 122 (1953).
20. Cerf, R., *J. Chim. Physique* **48**, 59 (1951).
21. Goddard, J. D., and Miller, C., *J. Fluid Mech.* **28**, 657 (1967).
22. Jeffery, G. B., *Proc. Roy. Soc.* A, **102**, 161 (1922).
23. Eisenschitz, R., *Z. physik. Chemie* A, **158**, 78 (1932).
24. Burgers, J. M., in *Second Report on Viscosity and Plasticity*, Verhandelingen der Koninklijke Nederlandsche Akademie van Wetenschappen, Eerste Sectie, Deel XVI, No. 4, pp. 113–184 (1938).
25. Goldsmith, H. L., and Mason, S. G., *Rheology*, vol. 4, p. 85 (F. R. Eirich, ed.), (New York: Academic Press, 1967).
26. Krieger, I. M., *Trans. Soc. Rheol.* **7**, 101 (1963).
27. Brenner, H., *J. Coll. Interface Sci.* **23**, 407 (1967).
28. Bird, R. B., Stewart, W. E., and Lightfoot, E. N., *Transport Phenomena* (New York: Wiley, 1960).
29. Condiff, D. W., and Brenner, H., *Phys. Fluids* **12**, 539 (1969).
30. Brenner, H., *Int. J. Multiphase Flow* **1**, 195 (1974).
31. Leal, L. G., and Hinch, E. J., *J. Fluid Mech.* **46**, 685 (1971).
32. Hinch, E. J., and Leal, L. G., *J. Fluid Mech.* **52**, 683 (1972).
33. Giesekus, H., *Rheol. Acta* **2**, 50 (1962).
34. Leal, L. G., and Hinch, E. J., *J. Fluid Mech.* **55**, 745 (1972).
35. Batchelor, G. K., *J. Fluid Mech.* **46**, 813 (1971).
36. Batchelor, G. K., *J. Fluid Mech.* **44**, 419 (1970).
37. Mewis, J. and Metzner, A. B., *J. Fluid Mech.* **62**, 593 (1974).
38. Frankel, N. A., and Acrivos, A., *Chem. Eng. Sci.* **22**, 847 (1967).
39. Hoffman, R. L., *J. Coll. Interface Sci.* **46**, 491 (1974).
40. Batchelor, G. K., and Green, J. T., *J. Fluid Mech.* **56**, 401 (1972).
41. Batchelor, G. K., *J. Fluid Mech.* **52**, 245 (1972).

AUTHOR INDEX

Abbott, T. N. G.   253, 255
Abernathy, F. H.   245
Ackerberg, R. C.   218, 220
Acrivos, A.   188, 205, 207, 208, 209, 261, 274, 275, 282, 286
Adams, N.   108
Allen, S. J.   195
Amundson, N. R.   131, 171
Aris, R.   18, 40, 41, 43, 164
Ashare, E.   231
Astarita, G.   70, 214, 221, 224, 250

Bagley, E. B.   106, 107
Balakrishnan, C.   246
Ballman, R. L.   139, 221, 222
Bargeron, K. G.   210
Barnes, H. A.   178, 228, 229, 230
Batchelor, G. K.   155, 265, 267, 272, 283, 284, 285, 286, 287
Beard, D. W.   209, 213, 214, 240
Bernstein, B.   195
Bhatnagar, R. K.   240
Bird, R. B.   38, 45, 71, 100, 138, 139, 152, 166, 168, 169, 199, 201, 202, 217, 225, 245, 247, 253, 266, 278
Block, H. D.   6
Blyler, L. L., Jr.   252
Boger, D. V.   210, 217
Bogue, D. C.   107, 195, 202, 210, 211, 212
Brennan, C.   245
Brenner, H.   217, 264, 268, 269, 270, 271, 272, 273, 274, 275, 276, 277, 278, 279, 283
Broadbent, J. M.   109
Brown, D. R.   108
Brumm, E. A.   227
Burgers, J. M.   277, 280

Carreau, P. J.   199, 200, 202, 227, 231
Carter, T. R.   210
Caswell, B.   218
Cerf, R.   276

Chaffey, C. E.   271, 272, 273, 275
Chan Man Fong, C. F.   223
Chandrasekhar, S.   160, 161, 162, 236, 238, 239, 241
Chartoff, R. P.   252
Chen, I-J.   195
Christiansen, E. B.   210, 250
Christopher, R. H.   225
Churchill, R. V.   150
Cole, J. D.   105
Coleman, B. D.   48, 63, 67, 68, 71, 76, 78, 80, 87, 94, 95, 105, 118, 120, 125, 142, 181
Collins, M.   210, 217
Colwell, R. E.   85, 86, 94, 95, 106
Condiff, D. W.   279
Cowsley, C. W.   109, 112
Cox, R. G.   274
Cox, W. P.   151
Crawley, R. L.   105
Curtiss, C. F.   38, 168, 266
Cygan, D. A.   218

Dauben, D. L.   226
Davies, D. M.   241, 243
Davies, M. H.   209, 240
Deardorff, J. W.   239
Denn, M. M.   112, 211, 212, 214, 216, 222, 241, 242, 243, 244, 250
Denson, C. D.   222
DeSilva, C. N.   195
DeWitt, T. W.   94, 96
Dierckes, A. C., Jr.   84, 94, 139
Dodge, J. S.   231
Duda, J. L.   210

Einstein, A.   183, 264
Eisenschitz, R.   277
Elliott, J. H.   249
Ellis, H. D.   245, 249
Elrod, H.   95
Ericksen, J. L.   18, 42, 43, 49, 141, 195, 202, 231

Eringen, A. C.   6, 42, 195
Etter, I.   142, 227, 228
Evans, D. C.   152, 166, 169, 201

Fabula, A. G.   221, 244
Feinberg, M. R.   213, 244
Ferry, J. D.   143, 144, 148, 151, 157
Fikhman, V. D.   222
Fosdick, R. L.   126
Frankel, N. A.   188, 274, 275, 282, 286
Fredrickson, A. G.   6, 71, 86, 100, 103, 106, 136,
   141, 151, 217, 227, 228, 231, 261
Fröhlich, H.   147, 183, 275
Fulks, W.   95

Gadd, G. E.   244, 245, 249
Gallegher, A. P.   239
Gatlin, C.   140
Gavis, J.   3, 108
Giesekus, H.   152, 218, 219, 220, 231, 232, 240
   241, 280
Ginn, R. F.   94, 241, 243
Glasscock, S. D.   105
Goddard, J. D.   130, 241, 276, 282
Goldin, M.   Fig. 1.1
Goldsmith, H. L.   277
Goldstein, C.   231, 253, 255
Goldstein, S.   210
Gordon, R. J.   246, 253
Graessley, W. W.   105
Green, A. E.   65, 179, 231
Green, J. T.   286
Green, M. S.   196
Greensmith, H. W.   87, 91
Gregory, D. R.   226
Griskey, R. G.   226, 250
Gross, B.   142, 143
Gruntfest, I. J.   108

Halmos, P. R.   10, 15
Happel, J.   217, 269
Harris, E. K., Jr.   253
Hayes, J. W.   97, 99, 241
Hermes, R. A.   217, 261
Herzog, R. O.   87
Higashitani, K.   109, 112
Hill, C. T.   232
Hinch, E. J.   192, 279, 281, 282
Hirschfelder, J. O.   38, 168, 266
Hoffman, K.   49, 129, 130
Hoffman, R. L.   97, 99, 100, 140, 286
Hopke, S. W.   218
Hoppmann, W. H., II   231
Houghton, W. T.   102, 139
Hoyt, J. W.   244
Huilgol, R. R.   80, 120, 123, 125, 253
Huppler, J. D.   97, 99
Hutton, J. F.   241

Jackson, R.   109, 112
James, D. F.   227

Jeffery, G. B.   276
Jensen, G. E.   250
Jerrard, H. G.   152
Johns, L. E., Jr.   80
Johnson, M. W., Jr.   217, 232
Jones, J. R.   228
Jones, W. M.   241, 243
Joseph, D. D.   108, 126
Joye, D. D.   222

Kalb, J. W.   227, 228
Kaloni, P. N.   218
Kapoor, N. N.   227
Kaye, A.   109, 112
Kearsley, E. A.   109, 112, 195
Kedzierska, K.   224
Kelsey, S. J.   210
Kennard, E. H.   160
Kim, K. Y.   85, 86, 94, 95, 106
Kirkwood, J. G.   160, 166
Kiser, K. M.   250
Klein, I.   250
Kotaka, T.   87
Kramer, J. M.   232
Krieger, I. M.   95, 96, 231, 277
Kundu, P. K.   245
Kunze, R.   49, 129, 130
Kuo, Y.   258
Kurata, M.   87
Kurtz, S. J.   252

Langlois, W. E.   218, 220
LaNieve, H. L.   107
Laurence, R. L.   108
Leaderman, H.   147
Leal, L. G.   192, 279, 281, 282
Leider, P. J.   209
Leigh, D. C.   6, 31, 34, 39, 48, 130
Lichnerowicz, A.   6, 10, 16, 18
Lightfoot, E. N.   45, 225, 245, 247, 278
Lilleleht, L. U.   209
Lin, C. C.   236
Lin, C. H.   244
Lin, C. J.   268
Lockett, F. J.   239
Lodge, A. S.   63, 108, 109, 125, 153, 156, 157, 158,
   166, 169, 173, 175, 188, 191, 195, 196, 197, 198,
   200
Lumley, J. L.   244, 245, 246
Lyons, J. W.   85, 86, 94, 95, 106

McConnell, A. J.   18, 21, 22, 27
Macdonald, I. F.   231
McIntire, L. V.   240, 244
McLaren, D. R.   227
Maerker, J. M.   222, 223
Maiti, M. K.   209
Markovitz, H.   48, 68, 71, 80, 87, 94, 95, 105, 108
Marrucci, G.   222
Marsh, B. D.   71, 109, 112, 231
Marshall, R. J.   226

Martin, B.   108, 251, 252
Mason, S. G.   277
Matsuo, T.   224
Maxwell, B.   252
Meissner, J.   222
Meister, B. J.   3
Menzie, D. E.   226
Mercer, A. McD.   239
Merrill, E. W.   246
Merz, E. H.   151
Metzner, A. B.   94, 102, 138, 139, 211, 212, 223, 224, 226, 286
Metzner, A. P.   224
Mewis, J.   286
Mickley, H. S.   246
Middleman, S.   6, 105, 108, 139, 141, 151, 225
Mihaljan, J. M.   238
Miller, C.   130, 241, 276, 282
Miller, C. E.   231
Modan, M.   3
Mollo-Christensen, E. L.   246

Nagano, Y.   224
Nakano, Y.   218
Noll, W.   39, 48, 60, 61, 62, 63, 65, 66, 67, 68, 70, 71, 76, 80, 87, 94, 95, 105, 118, 120, 124, 127, 135, 136, 141, 142, 179, 181, 184

Ohzawa, Y.   224
Oka, S.   218
Olabisi, O.   231
Oldroyd, J. G.   61, 65, 141, 183, 192, 205, 234, 276
Osaki, K.   157
Owen, J. R.   126, 130, 232

Padden, F. J.   94, 96
Paterson, R. W.   245
Pawlowski, J.   95, 217
Pearson, J. R. A.   109, 112, 138, 225, 250, 251, 252
Peery, J. H.   268, 269
Peterlin, A.   152
Petersen, E. E.   205, 207, 208, 209
Pfeffer, R.   Fig. 1.1
Pipkin, A. C.   80, 109, 112, 126, 130, 228, 232, 255
Poehlein, G. W.   222
Porteous, K. C.   243, 244
Powell, R. L.   105
Pritchard, W. G.   109, 112

Rabinowitsch, B.   87
Radushkevich, B. V.   222
Rajeswari, G. K.   209
Rama Murthy, A. B.   210, 217
Rathna, S. L.   209
Rea, D. R.   84, 94
Redheffer, R. M.   87, 100
Reed, X. B.   184
Reiner, M.   136, 147
Reischmann, M. M.   249

Richardson, J. C.   225
Richardson, S.   105
Riseman, J.   166
Rivlin, R. S.   65, 87, 91, 97, 136, 141, 179, 195, 211, 218, 220, 231
Roisman, J. J.   112, 242
Roscoe, R.   264, 274, 276, 288
Rouse, P. E., Jr.   155, 156, 200
Rutgers, R.   264

Sack, R.   147, 183, 275
Sailor, R. A.   102, 139
Sawyers, K. N.   231
Schechter, R. S.   217
Scheele, G. F.   3
Schlichting, H.   1, 205, 206, 207, 210, 214
Schowalter, W. R.   84, 94, 102, 139, 142, 205, 206, 210, 217, 222, 223, 227, 228, 231, 240, 244, 253, 255, 268, 271, 272, 273, 275
Schrag, J. L.   157
Schultz-Grünow, F.   109
Schümmer, P.   218
Schwarz, W. H.   218
Scott Blair, G. W.   138
Serrin, J.   43, 45, 137
Serth, R. W.   216, 250
Shah, M. J.   205, 207, 208, 209
Shah, Y. T.   225
Shaw, R.   109
Shinnar, R.   Fig. 1.1
Sidahmed, G. A.   250
Simmons, J. M.   201, 202
Siskovic, N.   226
Skelland, A. H. P.   210, 249
Slattery, J. C.   102, 206, 218
Smith, K. A.   246
Smith, L. S.   209
Sokolnikoff, I. S.   18, 27, 53, 87, 100, 180, 185
Sokolov, M.   240
Spencer, A. J. M.   65, 179
Spriggs, T. W.   200
Stark, J. H.   195
Stewart, W. E.   45, 225, 245, 247, 278
Stowe, F. S., Jr.   249
Sun, Z. S.   243
Sutterby, J. L.   217

Tadmor, Z.   250
Takami, A.   218
Tamura, M.   87
Tanner, R. I.   97, 99, 101, 109, 112, 140, 195, 201, 202, 221, 222, 240, 255, 256, 257, 258
Tao, F. S.   250
Taylor, G. I.   241, 271, 272
Thomas, M. C.   241, 243
Tiederman, W. G.   249
Tien, C.   218
Tobolsky, A. V.   160, 196
Toms, B. A.   244
Toupin, R.   47, 54, 57, 60, 63
Townsend, P.   178, 228, 229, 230
Treloar, L. R. G.   157, 160, 195, 200, 222

Truesdell, C.   29, 39, 47, 54, 57, 60, 61, 62, 63, 66, 70, 76, 80, 118, 124, 127, 131, 135, 136, 141, 179, 184, 211
Turian, R. M.   108

Uebler, E. A.   223
Ultman, J. S.   250

Vale, D. G.   109, 112
Van Dyke, M.   205
Van Wazer, J. R.   85, 86, 94, 95, 106
Vela, S.   228
Vinogradov, G. V.   222
Virk, P. S.   246, 247, 249
Volterra, V.   63, 127, 179
Vrentas, J. S.   210

Walters, K.   178, 182, 194, 209, 213, 214, 228, 229, 230, 240, 253, 255

Walters, T. S.   228
Wankat, P. C.   80, 222
Warner, H. R.   152, 166, 169, 201
Weissenberg, K.   87, 108
White, J. L.   102, 139, 195, 202, 211, 212, 213, 223
Williams, M. C.   231
Wineman, A. S.   255
Wissler, E. H.   226
Wu, Y.-J.   153, 156, 157, 158, 166, 169, 173, 175

Yamamoto, M.   196
Yates, B.   251
Yerushalmi, J.   Fig. 1.1
Yin, W.-L.   80

Zapas, L. J.   195
Ziabicki, A.   224
Zimm, B. H.   153, 155, 156, 157

SUBJECT INDEX

Adjoint of a tensor
  expressed as a matrix 130
Affine deformation 197
Aspect ratio 252
Average
  ensemble, definition of 265
  volume, definition of 266
Axial-annular flow 79, 97-101

Bagley end correction 106–8
Base vectors 11–12
  time differentiation of 184–7
Basis
  body-fixed 188, 189
  embedded 188, 189
  natural 15–16, 17
  reciprocal 14–15, 16
  rectangular Cartesian 12, 14
  transformation 12
  of a vector space 11–12
Blake-Kozeny equation 225
Body force 41
Boundary conditions
  in cone-and-plate viscometers 92–3
  in a free jet 103
  in parallel-plate viscometers 90
Boundary layers 205–11, 213–17
  for elastic fluids 213–17
  for a flat plate 207–8, 216–17, 261–2
  for flow past a wedge 207, 215
  and heat transfer from cylinders 208–9
  for inelastic fluids 205–11, 261–2
  integral method for 209
  similarity transformations for 207
  turbulent 210–11
Boussinesq approximation 238
Brinkman number 252
Brownian motion 153–4, 160–3, 169
  rotary, of particles 277–82
Buffer layer 246–7
Bulk stress *see* Stress, bulk

Cauchy-Green strain tensor
  in curvilineal flows 77
  in steady laminar shear flows 70
Cauchy's first law of motion 44
Cauchy's stress principle 42
Cayley-Hamilton theorem
  application to steady extension 123
  derivation of 130–1
Characteristic time
  for a flow 211
  for a fluid *see* Relaxation times
Christoffel symbols 25–7
Cofactor
  definition of 9
  of the elements of a matrix 131
Compliance, definition of 143
Cones, flow in 217–18, 226
Conservation
  of angular momentum 45
  of energy 45
  of linear momentum 41–4
  of mass 40–1
Constant stretch history
  definition of 117–20
  flows with 80, 117–25
  relation to viscometric flow 120, 133
Constitutive equations
  bead-spring model 152–7
  Bingham plastic 140–1, 205
  Bird-Carreau model 199–200, 222, 231, 240, 262
  BKZ model 195
  Coleman-Noll 194
  definition of 60
  differential models 181–2
  dumbbell model 152–7, 201
  Ellis model 139–40
  generalized Newtonian fluid 138, 205, 228–9
  Green-Rivlin 194
  incompressible Newtonian fluid 137
  inelastic fluid 134–41
  integral models 149–51, 193–4
  inviscid fluid 137
  linear viscoelastic fluid 142–51

295

Constitutive equations (*cont.*)
  Maxwell model   144–6, 150–1, 177, 212
    convected   220, 222–3, 243, 262
  molecular models   196–201
  for an nth order fluid   182
  network models   196–9, 200–1, 222
  network rupture model   222
  Oldroyd   183–4, 187–95, 227, 229
    liquid A   193
    liquid B   193
  Ostwald-de Waele model   *see* Constitutive
    equations, power-law model
  power-law model   108, 113, 138–9, 205–6, 218,
    225, 250, 261
  principles governing   60–4
  Reiner-Rivlin fluid   136–41, 231
  Rivlin-Ericksen fluid   141–2, 182
  Rouse model   152–7
  of rubber elasticity   196
  rubberlike liquid   157–8
  second-order fluid   142, 182, 217, 240, 241, 243
  simple fluid   65–7
  Stokesian fluid   136
  for a suspension of elastic spheres   183
  for a transversely isotropic fluid   203
  viscous fluid   134–41
  Voigt model   146–7
  Walters liquid A′   194
    liquid B′   194, 213
  Zimm model   152–7, 169–76
Contact force   41
Continuity equation   *see* Conservation, of mass
Continuum, definition of a   38
Coordinate invariance   60
Coordinate lines   15
Coordinate systems, inertial   60
Coordinates
  body-fixed   186, 189
  convected   186, 189
  distinguished from components   185–6
  embedded   186, 189
  space-fixed   186, 189
Couette flow
  cylindrical   75–6, 79
  plane   72
Covariant derivative   25–6
Creep   142
Cross product of vectors   28
Curvilineal flows   76–80
Cylinders, heat transfer from, to a non-Newtonian
    fluid   208–9
Cylindrical Couette flow   *see* Viscometers,
    cylindrical Couette

Darcy's law   225
Deborah number   212, 213, 214, 221, 222, 226,
    262–3
Deformation
  differentiability of a   48
  homogeneous   48
Deformation gradient
  properties of   47–9
  relative   52–3
  time derivatives of the   54–8

Derivative
  convected   191
  co-rotational time   192
  Fréchet   127
  Jaumann   192, 275, 282
  material   39–40
  Oldroyd   188–91
Determinant   *see* Tensors, determinant of
Differentiability of a deformation   48
Differentiation, chain rule of   23
Diffusion coefficient, rotational, due to Brownian
    motion   278–9
Dimension of a vector space   11
Dimensional analysis   211–13
Disk and cylinder, flow in   232–5
Distribution function for particles subject to rotary
    Brownian motion   277–9
Drag coefficient   208, 262
  See also Friction factor
Drag reduction
  diameter effect in   249
  in flow of fluids   217, 244–9, 263
Drops
  deformation of   271–5
  deformation due to shear   271–6

Effective viscosity   139
  measurement of   84–96, 105–9, 113–14
  See also Constitutive equations; Viscosity
Elasticity in fluids   136, 142–51
Ellipsoids
  from deformation of drops   272–3
  rigid, behavior in a shear field   276–7
Elliptical tube, flow through   231–2
Ellis fluid, viscous heating in a   108
Elongational flow   *see* Extensional flow
End effects, in viscometers   95, 105–8
Energy, balance of   45
Entry length, for non-Newtonian fluids   210
  See also End effects
Equations of motion   44
  in common coordinate systems   80–2
Equivalent motions   62
Equivalent processes   *see* Equivalent motions
Ergun equation   226
Euler's law of motion   41
Exchange of stabilities, principle of   238
Expansion
  retarded motion   182
  for simple fluids, differential   181–2
  for simple fluids, integral   179–80, 182
Extensional flow
  biaxial   125, 133, 222–4
  of a circular cylinder   123–5, 133, 220–5, 246
  of a dilute suspension of rods   283–5
  rectilinear, definition of a   120
  of a Rivlin-Ericksen fluid   141
  sheared extensions   125
  of a sheet   125
  of simple fluids   120–5, 133
  uniaxial   *see* Extensional flow of a circular
    cylinder
  See also Viscosity, extensional

Extensional viscosity *see* Viscosity, extensional
Extra stress, definition of 67, 136*n*

Fading memory, principle of 63
Flow
  axial-annular 79
  with constant stretch history *see* Constant
    stretch history
  Couette, cylindrical 75–6, 79
  Couette, plane 72
  extensional *see* Extensional flow
  helical 78–80, 83
  isochoric *see* Isochoric motion
  laminar, through a tube 73–6
  nearly viscometric 126–30, 232–5
  oscillatory 142, 227–31
  between parallel planes 75
  periodic, in a tube 227–31
  secondary 218–20, 231–5, 252
  slow 181, 234
  unsteady 227–31
  See also Curvilineal flows; Laminar shear flows;
    Porous media; Tube flow; Viscometric
    flows
Fluid
  definition of a 66
  dilatant 138
  pseudoplastic 138
  rheopectic 138
  shear thickening 138
  shear thinning 138
  thixotropic 138
  See also Constitutive equations
Fluid-like behavior 211–12
Force
  body 41
  contact 41
  in Zimm model
    due to Brownian motion 154, 160–3, 169
    due to Gaussian spring 154, 159–60, 169
    due to solvent 153–4, 169
Frame *see* Reference frame
Frame Indifference *see* Material objectivity
Fréchet differentiability 127
Free energy, Helmholtz 160, 197–9
Friction factor 225, 245–9
  See also Drag coefficient
Functional, description of a 62, 63

Gaussian spring 154, 159–60, 169
Gradient
  of a scalar 28
  of the velocity 45, 55
Graetz number 252
Gravity
  effect on a free jet 102
  effect on tube flow 74
Griffith number 252

Heat transfer
  from cylinders 208–9

in a fluid heated from below 236–40
  general description of non-Newtonian fluids
    249–50
Helical flow 78–80, 83, 94
History, difference, definition of 127
  See also Constant stretch history
Hydrodynamic interaction 154, 163–6
Hydrostatic pressure 68

Identity tensor 30
Inclined channel, flow down 255–8
Incompressible fluid, definition of an 67
Inertia
  in flows between rotating cylinders 241
  in flows for which effects are small 217–20
  in relation to secondary flow 88, 114
  in viscometers 88, 91, 97, 102
Interfacial curvature, and interfacial tension 273
Interfacial forces in a free jet 102
Interfacial tension, relation to interfacial curvature
    by Laplace's equation 273
Invariants, principal, of a tensor 33–6
Isochoric motion 67
  implications of, for a simple fluid in nearly
    viscometric flow 126, 128–9
Isotropic
  fluid 67
  material 67
  tensor function 123, 135
Jacobian 40
Jet devices for measurement of viscometric func-
    tions 102–5
Jet thrust, relation to normal stress differences
    104–5

Kronecker delta 14
  generalized 27

Laminar shear flows 68–76
  to measure the viscometric functions 84–97
Langevin equation 160
Leibniz' rule for differentiation of an integral 87,
    89
Linear dependence and independence 11
Linear transformations
  definition 9
  as tensors 29
Linear viscoelasticity *see* Viscoelasticity, linear
Local action, principle of 62
Loss tangent, definition of 143

Mapping 39
Mass, conservation of 40–1
Material derivative 39–40, 53
Material functions
  for steady laminar shear flows 71
  for a viscometric flow 71, 76
    applied to a nonviscometric flow 126–30
Material objectivity 61–2
  for extensional flows, 120–1
  for scalars 61

Material objectivity (*cont.*)
  for steady laminar shear flows 70
  for tensors 61
  for vectors 61
  for viscometric flows 127
Material point 185–7
  definition of a 38
Material surface 191
Matrix
  for the adjoint of a tensor 130
  cofactor of the elements of 131
  definition, in two dimensions 7
  inverse 9
  multiplication 9
  with tensor elements 131
  transpose 12
  unit 9
Metric tensor 22–3
Model *see* Constitutive equations
Modulus
  complex, definition of 143
  loss, definition of 143
  storage, definition of 143
Momentum
  angular, conservation of 45, 268
  linear, conservation of 41–4

Nearly viscometric flow of simple fluids 126–30, 232–5
Nilpotent tensor 120
No-drift hypothesis 169
No-slip condition 73
Nonviscometric flows of simple fluids 117–33
Norm of the difference history 127
Normal stress differences 68, 71, 80
  for a dilute suspension of deformable drops 274-5
  for a dilute suspension in the presence of Brownian motion 280–2
  effect on hydrodynamic stability 240, 242–3
  measurement of 84–112, 255–8
    by the method of Jackson and Kaye 109–12
  relation to tube flow pressure drop 107
  relative magnitudes of 103
  sign of 112, 255–8
Nusselt number 208

Objectivity *see* Material objectivity
Orthogonal rheometer 252–5
Orthogonal transformation 31
Oscillatory flow 142
Oseen tensor 166

Particle stress *see* Stress, particle
Peclet number 208, 252
Permeability 225
Polar decomposition theorem 49–50
Polyethylene, end corrections for 107–8
Polyisobutylene, effective viscosity of 113–14
Polymer processing 250–2

Polymers, extrusion of 250–2
Porous media, flow through 225–7, 262
Power-law fluid *see* Constitutive equations, power-law model
Pressure 68, 136n
Pressure gradient, effective, for flow through a tube 74–5, 86
Pressure-hole error 97, 108–9, 232, 255
Principal directions of a tensor 34–6
Principal invariants of a tensor 33–6
Principle of material objectivity *see* Material objectivity
Profile
  ultimate 247
  universal velocity 246–7

Rabinowitsch equation 87
Rate of deformation tensor *see* Rate of stretching tensor
Rate of spin tensor 56–7, 187
Rate of stretching tensor 56–7, 180, 187
  See also Stretch rate in steady extension
Rayleigh number 238–9
Rectangular Cartesian coordinate system 7, 11
Reference configuration 40, 52, 66–7
Reference frame 184–7, 188, 189
Relative deformation gradient, properties of 52–3
Relaxation of a fluid in a free jet 102
Relaxation times
  characteristic, of a fluid 245
  spectrum of 147–8
  superposition of 147–51
Rest history 68
Reynolds number 86, 206–7, 214, 215, 217–20, 225, 226, 234, 245, 248
Rheology, definition of 1
Rheology of suspensions, definition of 264
Rheometers *see* Viscometers
Rheopexy 138
Rivlin-Ericksen tensor 57–8, 141, 181, 190, 213, 214, 233
  in a curvilinear flow 77
  in steady laminar shear flows 69–70

Second-order fluid 109
Shear flows, some steady laminar 68–76
Shear rate in steady laminar shear flows 69
Simple fluids
  correspondence to real fluids 84–5
  definition of 65–7
  differential expansion for 181–2
  in extensional flow 120–5
  integral expansion for 179–80, 182
  in nearly viscometric flows 126–30
  nonviscometric flows of 117–33
Simple material 66
Siphon, tubeless 221
Slender-body theory 283
Solid-like behavior 212, 224
Sphere, flow past a 218
Spherical harmonics 269, 272–3
Spin tensor 57

Spinning, of a fiber 224
Squires' theorem 239, 240
Stability, hydrodynamic 235–44
  in cylindrical Couette flow 240–3
  of a fluid heated from below 236–40
  nonlinear analysis of 244
  in Poiseuille flow 243–4
  principle of exchange of stabilities 238
Stagnation flow 209
Stagnation point 208, 209
Steady extensional flow 120, 122–5, 133
Strain rate *see* Rate of stretching tensor
Strain tensor 51–2, 53, 70, 77, 233
  infinitesimal 180
Straingth 201
Stream function 215
Stress, bulk
  in absence of Brownian motion and inertia 283
  asymmetric 268
  definition of 266–7
  of a dilute suspension of rigid ellipsoids 276–7
  of a dilute suspension of rigid spheres 271
Stress, determinism of the 61
Stress, particle, in absence of Brownian motion 267–8
  See also Stress, bulk
Stress relaxation 142
Stress tensor 43
  extra stress, definition of 67, $136n$
  sign convention 85
  viscometric form 79
Stress vector 42–3
Stretch rate in steady extension 122, 220–5, 262
Stretch tensor 50–1
Stretching of a fluid filament 220–5, 262
Sublayer
  elastic 247
  viscous 246–7
Summation convention 8
Superposition, in viscoelasticity 147–51, 180
Suspensions, nondilute 286–8

$\tau(\kappa)$, measurement of *see* Effective viscosity
Taylor number 241
Tensor product 18
Tensors
  absolute 21
  addition of 21–2
  adjoint of, expressed as a matrix 130
  antisymmetric 56
  associated 24
  characteristic equation for 33
  components of 18
    contravariant 19
    covariant 19
    mixed 19
    physical 28–9
    time derivatives referred to time-dependent bases 189–90
    transformation of 19–21
  contraction of 22
  covariant derivative of 25–6
  definition of 18–19

deformation gradient, properties of 47–9
  determinant of 8, $32n$
  double 49
    contraction of 52
  eigenfunctions of 34–6
  eigenvalue equation for 33
  eigenvalues of 34–6
    as exponents 118
  Hermitian 35
  identity 30
  inner multiplication of 22
  inverse of 30
  isotropic 33, 37
  isotropic tensor function 123, 135
  magnitude of 119
  material objectivity for 61
  metric 22–3
    invariance to covariant differentiation 27
  nilpotent 120
  orthogonal 31
  Oseen 166
  outer multiplication of 22
  positive $49n$
  positive definite $49n$
  principal coordinates for 36
  principal directions of 34–6
  principal invariants of 33–6
  principal values of 34–6
  quotient rule for 22
  rate of deformation *see* Rate of stretching tensor
  rate of rotation *see* Rate of spin tensor
  rate of spin *see* Rate of spin tensor
  rate of stretching *see* Rate of stretching tensor
  real 35
  relation to linear transformation 29–30
  relative 21
  relative deformation gradient, properties of 52–3
  Rivlin-Ericksen 57–8, 141, 181, 190
  rotation of 33
  singular 30
  spin 57
  strain 51–2, 53, 70, 77, 233
  strain rate *see* Rate of stretching tensor
  stress *see* Stress tensor
  stretch 50–1
  symmetric 23, 31
  tensor product of 18
  time differentiation of 53
  trace of 33
  transformation of 31
  transpose of 30
  velocity gradient 45, 55
  vorticity 57
Thixotropy 138
Time differentiation
  of the deformation gradient 54–8
  of tensors of arbitrary order 53
Toms phenomenon 244
Transformation
  of a basis 12
  improper 35
  invertible 15, 20
  linear *see* Linear transformation

Transformation (*cont.*)
  orthogonal   31
  proper   34
  of tensor components   19–21
  of tensors   31
  of vectors   8–9, 12–13
Trouton ratio   223
Trouton viscosity   125, 284–5
Tube flow
  enhancement by periodic flow   229–30
  laminar   73–6
  shutdown   227
  startup   227
  See also Viscometers, capillary
Turbulent core   246–7

Uebler effect   223
Unit vectors   28

Variational techniques   217
Vector product   28
Vector space
  basis   11–12
  definition of a   10
  dimension   11
  dual of   15
  product of   18, 19
Vectors
  addition   9
  angle between   13
  base   *see* Base vectors
  components of   11–12
    distinguished from coordinates   185–6
  contravariant components   14, 17
  covariant components   14, 17
  cross product   16
  definition of   7, 10
  divergence of   28
  dot multiplication   13–14
  inverse transformation   9
  linear dependence and independence   11
  magnitude   7, 14
  material objectivity for   61
  multiplication by a scalar   10
  orthogonal   14
  orthonormal   14
  physical components of   28–9
  reflection of   32
  rotation of   28–9
  scalar multiplication   13–14
  stress   42–3
  transformation of   8–9, 12–13
  transformation of components   7, 12–13
  unit   28
  See also Basis
Velocity gradient   45, 55

Viscoelasticity
  finite linear   180–1, 182–3
  linear   142–51, 180–1, 182–3, 253–4, 262
Viscometers   84–112
  based on nonviscometric flows   102–5
  capillary   86–7, 106–8
  centrifugal effects in   *see* Inertia, in viscometers
  coaxial cndeylir   94–7, 106, 116
  cone-and-plate   91–4, 109–12, 115, 232
  cylindrical Couette   94–7, 106, 116
  end effects in   95, 105–8
  experimental difficulties with   105–9
  jet devices   102–5
  orthogonal rheometer   252–5
  parallel plate   87–8, 114
  viscous heating in   108–9
  Weissenberg rheogoniometer   102–5
  See also Axial-annular flow
Viscometric flows
  definitions of   76
  measurement of the viscometric functions   84–112
  relation to flows with constant stretch history   120, 133
  of a Rivlin-Ericksen fluid   141
  See also Curvilineal flows; Helical flow; Laminar shear flows
Viscometric functions, measurement of the   84–112
  See also Viscometers
Viscosity   138
  complex   143, 254–5, 262
  of a dilute suspension of deformable drops   274
  of a dilute suspension of rigid spheres   268–71
  dynamic   143
  extensional   125, 133, 221, 223, 262
    of a dilute suspension of rods   285
  intrinsic   156
  of non-dilute suspensions   286–8
  of suspensions, general discussion   264–5
  Trouton   125, 284–5
  zero-shear   218
  See also Constitutive equations; Effective viscosity
Viscous dissipation   269
  See also Energy, balance of; Viscometers, viscous heating in
Viscous heating   *see* Viscometers, viscous heating in
Vortex inhibition   246
Vorticity tensor   57

Weissenberg hypothesis   103, 104
Weissenberg number   212, 214, 215
Weissenberg rheogoniometer   102–5

Yield stress   140